BARRON'S

NEW JERSEY

ASK 8 MATH TEST

THIRD EDITION

Judith T. Brendel, M.Ed.
District Supervisor of Mathematics and Art
Pascack Valley Regional High School District
Montvale, New Jersey

About the Author

Judith T. Brendel is District Supervisor of Mathematics and Art at the Pascack Valley Regional High School District in New Jersey, and she chairs their K–8 Math Articulation Committee. She has taught art and math in all K–12 grade levels, plus graduate math- and technology-education courses at Fairleigh Dickinson University in N.J., Mercy College in N.Y., and at the Rutgers New Math/Science Teacher Institute. Judith is also a member of the Board of Directors of NJPSA (Principal Supervisors Association) and former board member of AMTNJ (Association of Math Teachers of New Jersey.) She is a frequent presenter at national and state conferences and is an evaluator of math textbooks for middle and high school students. Formerly, Judith was a member of the New Jersey Coalition of Concerned Math Educators, recommending changes to the Pre-K to 12 Core Curriculum Content Math Standards and, more recently, to the new Common Core state standards.

All inquiries should be addressed to:

Barron's Educational Series, Inc.
250 Wireless Boulevard
Hauppauge, New York 11788
www.barronseduc.com

ISBN: 978-1-4380-0052-7

Library of Congress Control Number: 2011941771

Printed in the United States of America
9 8 7 6 5 4 3 2 1

10% POST-CONSUMER WASTE
Paper contains a minimum of 10% post-consumer waste (PCW). Paper used in this book was derived from certified, sustainable forestlands.

Contents

Important note: In Spring 2011, the New Jersey ASK 8 Math Test was changed to reflect revised performance indicators released by the New Jersey Department of Education. This book will provide a strong framework for students preparing for the test, but please consult **www.state.nj.us/education/assessment** for all the latest information on the exam.

Introduction

ABOUT THE NJ ASK 8 MATHEMATICS EXAM

This New Jersey 8th grade mathematics test is a state-required exam for all students (regular education, special education, and LES/ESL). It is designed to indicate the progress students are making in mastering the knowledge and skills specified in the Core-Curriculum Content Standards and needed to pass the High School Proficiency Assessment (HSPA). The New Jersey ASK 8 is a primary indicator for identifying 8th grade students who may need instructional intervention in three content areas: language arts literacy, mathematics, and science. This book deals with the mathematics component of those standards.

The ASK 8 assesses knowledge and skills in four content areas, called "clusters":

I—Number Sense, Concepts, and Applications

- Macro A: Make appropriate estimations and approximations.
- Macro B: Understand numbers, our numeration system, and their applications in real-world situations.
- Macro C: Use ratio, proportions, and percents in a variety of situations.

II—Spatial Sense and Geometry

- Macro A: Recognize, identify, and represent spatial relationships and geometric properties.
- Macro B: Apply principles of congruence, similarity, symmetry, geometric transformations, and coordinate geometry.
- Macro C: Apply the principles of measurement and geometry to solve problems involving direct and indirect measurement.

III—Data Analysis, Probability, Statistics, and Discrete Mathematics

- Macro A: Predict, determine, interpret, and use probabilities.
- Macro B: Collect, organize, represent, analyze, and evaluate data.
- Macro C: Apply the concepts and methods of discrete mathematics to model and explore a variety of practical situations.
- Macro D: Use iterative patterns and processes to describe real-world situations and solve problems.

IV—Patterns, Functions, and Algebra

- Macro A: Recognize, create, and extend a variety of patterns and use inductive reasoning to understand and represent mathematical and other real-world phenomena.
- Macro B: Use algebraic concepts and processes to concisely express, analyze, and model real-world situations.

ABOUT THIS BOOK

This book has been designed for use by students in school, at home, or for tutorial or remedial sessions. It is designed to help them review the math skills and applications they should master before their ASK 8 exam, which is generally taken in the spring of their 8th grade; it is also a resource for those 9th grade students who are in a high school remedial math program because they scored less than proficient on their ASK 8 or by Pre-Algebra, Basic Algebra, or Resource Room math students. It also is recommended for private and parochial school 7th and 8th graders who plan to attend a public high school in New Jersey.

This book is divided into four clusters that match the material covered on the New Jersey ASK 8:

I—Number Sense, Concepts, and Applications
II—Spatial Sense and Geometry
III—Data Analysis, Probability, Statistics, and Discrete Mathematics
IV—Patterns, Functions, and Algebra

Each chapter focuses on one cluster at a time and includes the three different types of questions used on the ASK 8: (MC) multiple choice, (SCR) short constructed response (noncalculator), and (ECR) extended constructed response questions. Each chapter introduces and reviews all material; gives examples with detailed diagrams, explanations, and solutions to review and study; and then follows with practice questions. At the end of each chapter there are new sections with additional short constructed response (SCR) noncalculator questions, extended constructed response (ECR) questions, and a one-period Cluster Test. Cluster tests are designed to be completed in 35 minutes.

TIP

Throughout this book you will find many sample Extended Constructed Response Questions. On these questions you are required to show your work.

At the end of the four chapters, there are two full-length sample ASK 8 tests with selected material from all four clusters. These tests are in official ASK format and are designed to simulate the actual test. Each test is divided into six parts. Students may *not* use a calculator on Parts I, II, and III.

BREAKDOWN OF TEST PARTS

Item Type Abbreviations			
MC – multiple choice, 1 raw score point SCR – short constructed response, 1 raw score point ECR – extended constructed response, 3 raw score points			
Part I	20 minutes	10 SCR questions (noncalculator)	10 pts
Part II	22 minutes	8 MC questions and 1 ECR question (noncalculator)	11 pts
Part III	22 minutes	8 MC questions and 1 ECR question (noncalculator)	11 pts
Part IV	25 minutes	10 MC questions and 1 ECR question	13 pts
Part V	25 minutes	10 MC questions and 1 ECR question	13 pts
Part VI	19 minutes	6 MC questions and 1 ECR question	9 pts
Total	133 minutes		67 points

MULTIPLE-CHOICE QUESTIONS (MC)

The multiple-choice questions are not the standard multiple-choice questions of the past; they require more thought, more work, and a true understanding of basic concepts. It is expected that each multiple-choice question will take a minute or two to answer. Students can write in their official test booklet but must bubble-in their answers on an answer sheet. Each question is worth one point. These multiple-choice answer sheets are scored by a computer.

SHORT CONSTRUCTED RESPONSE QUESTIONS (SCR)

Short constructed response questions are questions where the student must write the answer without having answer choices from which to choose. There is no partial credit given for SCR questions. Students may not use a calcualtor on these quesitons. Below is a sample short constructed response question that was copied from the NJ Department of Education's website.

> Point "P" has the coordinates (–2, 3). What are the coordinates of its image if it is translated 3 units to the left and then reflected over the x-axis?
>
> Solution:
>
> The correct answer is (–5, –3)

Note: Without having answer choices from which to choose, this item tests a higher level of understanding than a similar question with multiple choices given.

EXTENDED CONSTRUCTED RESPONSE (ECR)

Extended constructed response questions are different. They require students to write, chart, or graph their responses to questions and often require them to write an explanation as well. Each ECR question usually has two or three parts and is worth a total of three points. Because partial credit is given, a student may receive 0, 1, 2, or 3 points for each ECR question.

SCORING AND USE OF TEST RESULTS

School districts receive the results of the ASK 8 before September of the student's 9th grade. These records are forwarded to the high school district and are used by guidance counselors, case managers, and supervisors to help determine the best placement for students in their 9th grade courses. If a student scores "less than proficient" in mathematics, he or she will likely be placed in a remedial math program in grade 9 (in addition to the regular math class) or may be recommended for a summer program.

TIPS FOR SUCCESS WITH EXTENDED CONSTRUCTED RESPONSE QUESTIONS

- Your testing proctor may tell you to do the extended constructed response questions last. If you do this, it is a good idea to leave 10 minutes to answer the extended constructed response question at the end of each part of the test.
- Another method is to answer the ECR question first and then go back to the multiple-choice questions. If you do this, spend about 10 minutes on the ECR question and be careful where you bubble-in your answers for the remainder of the test. Make sure the question numbers match your answer-sheet numbers.
- Read each extended constructed response question twice: once for general understanding and the second time to get specific information and details. Each question will have at least two parts.
- Underline or circle important information.
- Even before you answer the question, write the beginning of each final answer at the bottom of the page so you remember to answer each part. For example:
  - The area of the circle is ______ sq. ft.
  - The perimeter of the room is ______ ft.
  - The reason they can afford the carpet is because ______ .
- Most extended constructed response NJ ASK 8 questions can be answered using a chart or table. (see page 5.)

Example

If a coat is priced at $85 and is reduced by 10% each day, when will Sarah be able to buy the coat if she has $60.00?

Each Day	Work Shown (computation is done with a calculator but "work" is still written here)	Price
Original price of coat		$85.00
Day #1 of sale Coat is 10% less	100% – 10% = 90%; She pays 90% (85) (0.90) =	$76.50
Day #2 of sale	(76.50) (0.90) =	$68.85
Day #3 of sale	(68.85) (0.90) =	$61.965 or $61.97
Day #4 of sale	(61.97) (0.90) =	$55.768 or $55.77
Sarah will be able to buy the coat on day #4 of the sale. On that day, it will cost $55.77 (or less than $60.00).		

REMINDERS

- Show all work, even if you use a calculator.
- Label all diagrams, graphs, and answers. Remember, *perimeter or circumference is in units* (feet or yards, centimeters or meters), *area is in square units* (sq. in., sq. ft), and *volume is in cubic units* (cubic cm, cubic yd).
- Circle or bullet final answers.
- Read your final answers to see if they make sense.
- These questions are scored by hand. The people scoring them work very hard all day reading hundreds of student answers. They spend very little time reading each one so it is important to be neat and to write clearly. Put a bullet or circle around your answers. In mathematics, using a table or a chart to outline information is often better than writing a long paragraph.
- Even if you cannot answer all parts of the question, answer as much as you can. Every point counts. You must get some points on the extended constructed response questions if you are to receive a proficient (passing) score on the ASK 8.

HELPFUL RESOURCES

NJ ASK 8 MATHEMATICS REFERENCE SHEET

Use the *NJ ASK 8 Mathematics Reference Sheet* on page 265 of this book and complete the *Scavenger Hunt* activity on page 263. You will be provided with the same *ASK 8 Mathematics Reference Sheet* when you take the real ASK 8 test. It is not necessary to memorize the formulas, but you should become familiar with the information here so you can use it during the test and not waste time looking for information.

STRAIGHTEDGE

You will also have a straightedge/ruler. This will be part of your *NJ ASK 8 Mathematics Reference Sheet*. Use it to measure and draw straight lines. It is especially helpful when drawing geometric shapes, setting up a table, or connecting points on a coordinate grid.

CALCULATORS

The New Jersey Department of Education permits any calculator on test Parts IV, V, and VI, except ones with a QWERTY keyboard (similar to a typewriter or computer keyboard). A graphing calculator is not needed nor is it recommended. Your calculator should have

- Built-in algebraic logic (follow the correct order of operations)
- Exponent and square root keys
- Change-in-sign key
- Percents
- Fraction keys
- Reset button (a way to clear all memory)

You are encouraged to use a calculator on the ASK 8, but it is not essential for many questions and may actually waste time on others. You will find your calculator helpful in the following situations:

1. Adding more than two or three numbers. Especially when you are finding an average:

   85, 88, 83, 73, 75, 90, 95, 78, 82, 55, 78, 80, 82, 92, 98, 65

2. Multiplying decimal numbers. For example, (.30) (.10) is 0.03, or 3% (not .3 as you might say if you did this quickly without a calculator).

3. Manipulating any expression that includes a combination of addition, multiplication, subtraction, and so on. Because your calculator follows the "algebraic order of operations," it knows whether to add or multiply first. Be sure to put in parentheses when necessary.

$$23 - 4 \times 8 + 5(56 - 19)^2$$

4. Converting between factions and decimals. Remember $\frac{2}{5}$ means 2 divided by 5 and 0.95 is the same as $\frac{95}{100}$ or 95%.

5. Finding the square root of a number. For example, $\sqrt{38}$ is ________ ?

6. Squaring or cubing a number. For example: What is the value of $18^3 - \sqrt{169}$?

- Remember to use a calculator you are comfortable and familiar with. A solar-powered scientific calculator is recommended. If you use a battery-operated calculator, be sure to put new batteries in your calculator before the test.
- Press in numbers carefully and slowly; do not rush when putting numbers or symbols into your calculator.
- Even when a calculator is permitted, you must decide if using it will be helpful or not. Have a calculator available as you work in this book and make notes to yourself about when it was helpful and when it was not. Practice and learn what works best for *you*!

OTHER TIPS AND STRATEGIES FOR SUCCESS

1. **Draw and label diagrams** to help you visualize the problem and solutions.

Example

The length of a rectangle is twice the width. If the length is increased by 4 and the width is diminished by 1, the new perimeter is 198. Find the dimensions of the original rectangle.

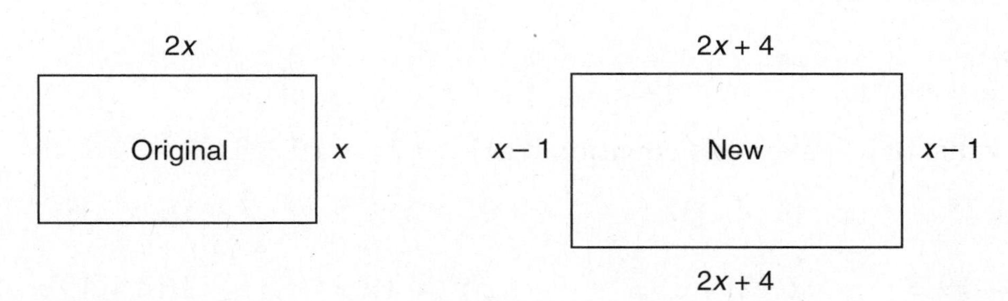

2. **Be aware of distractors.** Every multiple-choice question has distractors. These are incorrect choices the authors have put there because they know common mistakes students will make. Be careful; don't select one of the choices just because it is there.

Example

What is the value of the expression given? $2 + 3(5 - 5)(10)$

A. 2 B. 12 C. 17 D. 32

The correct answer is A. Notice that $(5 - 5 = 0)$ and anything times zero is zero. So, all that is left is the 2. Why would the author include B.12 as a choice? Why would 17 be a choice? Why 32?

- Usually the correct choice is not one of the numbers in the question itself.
- If you do not know the answer, do not leave it blank. Guess if you are not sure of an answer.

3. See if you can **eliminate one or two choices.** Often you can tell that the answer cannot be a negative number, or you know it must be smaller or larger than a particular number. Put a line through those choices then make an educated guess from the remaining choices.

Example

If there is a 25% chance of snow on Sunday and a 20% chance of snow on Monday, what is the probability that it will snow on *both* days?

A. 45% B. 70% C. 25% D. 5%

You should know that the possibility is *less* than 25% or 20% that it will snow on *both* days. So, looking at the choices, you can eliminate A and B. You can eliminate C, also, since it probably is not the same for both days as for one day. Yes, D is correct: $(0.25)(0.20) = 0.05 = 5\%$.

4. If you do not know how to solve for x, then **use the choices given to PLUG-IN to the problem.**

Example

What value of x makes this equation true? $2x + 19 = 31$

A. $x = 25$	B. $x = 12$	C. $x = 9$	D. $x = 6$
$2(25) + 19 =$ *too big*	$2(12) + 19=$ *too big*	$2(9) + 19 =$ *close*	$2(6) + 19 =$ *just right*

5. **Try a similar problem but use small numbers;** do not use zero or 1.

Example

What is the area of a triangle that has a base of 58 ft and is 29 ft high?

A. 2,842 sq. ft B. 174 sq. ft C. 841 sq. ft D. 420.5 sq. ft

If you have forgotten the formula for area of a triangle, you might draw a small triangle on a grid, count the squares, and then see how the base and height can give you that number or just notice that the triangle is half the size of the rectangle.

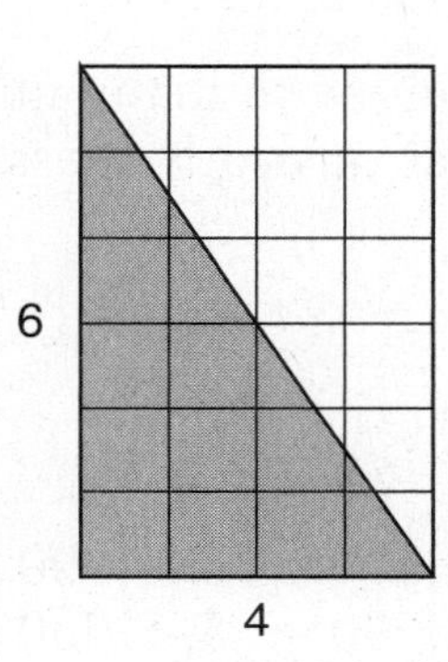

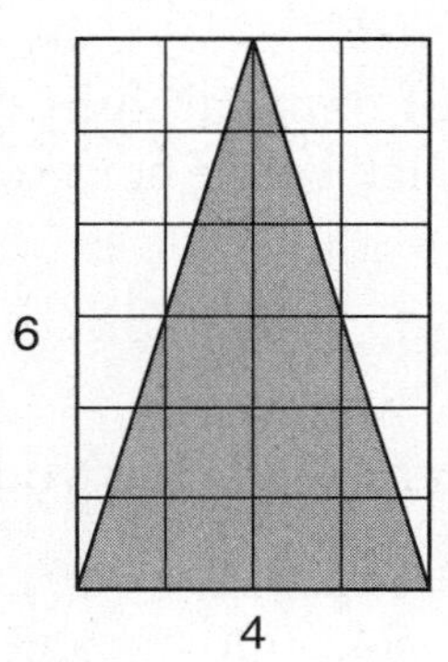

Here you see the area of the triangle is 12 square units, which seems be

$\frac{\text{(base)(height)}}{2}$ or $\frac{(6)(4)}{2} = 12$. Now you can do $\frac{(58)(29)}{2} = \frac{1682}{2} = 841$, which is

choice C. Correct!

6. **Use numbers instead of variables.**

Example

What are three consecutive odd integers whose sum is 579?

If you don't know how to do this, then try small consecutive odd integers like 3, 5, and 7. How would you represent them algebraically?

First number is	3
Second number is	$3 + 2 = 5$
Third number is	$3 + 4 = 7$
Therefore you can	let x be the smallest number
	let $x + 2$ be the next number
	let $x + 4$ be the biggest number

Therefore,

$$
\begin{aligned}
x + x + 2 + x + 4 &= 579 \\
3x + 6 &= 579 \\
3x &= 579 - 6 \\
3x &= 573 \\
x &= 191
\end{aligned}
$$

The three consecutive odd integers are 191, 193, and 195.
Always check; use your calculator: 191 + 193 + 195 = 579. Correct!

7. **Manage lots of data carefully.** When working with many numbers or items, be sure to put a tick mark next to each number or neatly cross out each number as you use it.

Example

Put these numbers in order so you can find the *median* salary.

~~\$38,000~~ \$95,000 \$72,000 \$50,000 ~~\$25,000~~ \$46,000
~~\$29,000~~ \$45,000 \$82,000 \$42,000 \$48,000 \$55,000

As you write these in order, put a line through the numbers you use so you don't forget any:

25,000 29,000 38,000 ...

It also is a good idea to *count* the numbers (12 numbers). Notice that I omitted the \$ signs for now to keep things easier to see.

8. **Learn from your mistakes.** After you complete a set of *Practice Examples* or a *Practice Test*, check your answers with the solutions provided and mark the examples where you had the wrong answer or where you guessed (either circle the number or star (*) the example). Study the details of the given solution, find out where you made your mistake, and keep a list of these examples on a separate sheet of paper in a binder or folder, or in a separate notebook. After you review the chapter again, do these examples over to be certain you now understand them.

Chapter 1

Cluster I: Number Sense, Concepts, and Applications

WHAT DO ASK 8 NUMBER SENSE, CONCEPTS, AND APPLICATIONS QUESTIONS LOOK LIKE?

MULTIPLE-CHOICE QUESTION (MC)

Example 1: The regular price of a pair of sneakers is $79.00. They are on sale for 15% off. If Jose buys them on sale, how much change will he get from $70?

A. $ 9.00
B. $ 2.85
C. $ 3.85
D. $ 67.15

Example 1: Strategies and Solutions
Read the *question part* of the example twice to be sure you understand what is being asked.

First, determine the sale price.
$79 × 0.15 (the % off) = $11.85
79.00 – 11.85 = $67.15 (sale price) or
79 × .85 (the % he would pay) = $67.15 (sale price)

Now, determine the change he would receive from $70.00.
70.00 – 67.15 = $2.85 (Watch your computation; a careless error might make you select $3.85.)

The correct answer is B $2.85.

There are two ways to solve this problem:

- Determine how much is 15% of the original price: [7] [9] [×] [.] [1] [5] = $11.85. Then subtract this discount from the original price: [7] [9] [–] [1] [1] [.] [8] [5] = 67.15. Jose's change would be $70.00 – 67.15 = $2.85.
- A quicker way would be to determine the percentage you would pay. If the sale is 15% off the regular price, that is the same as saying you would pay 100% – 15% or 85%. Therefore, 85% of the original price is what you would pay [7] [9] [×] [.] [8] [5] = $ 67.15. Jose's change would be $70.00 – 67.15 = $2.85.

SHORT CONSTRUCTED RESPONSE QUESTION (SCR)

Example 2: (No calculator permitted.)

Mr. Wieland gave his classes a homework assignment that included responding on the Internet to a blog about that night's homework. Of Mr. Wieland's 120 students, 90 responded by 8:00 P.M.

What percentage of his students responded by 8:00 P.M.?

Answer: ___________

Example 2: Solution

$$\frac{\text{Favorable outcome}}{\text{Total}} = \frac{\text{90 students responded}}{\text{120 students total}}$$

$$\frac{90}{120} \text{ reduces to } \frac{9\not{0}}{12\not{0}} = \frac{9}{12} =$$

$$\frac{3}{4} = 0.75 = 75\%$$

Answer: 75%

Example 3: (No calculator permitted.)

Last week, Kelsey worked 40 hours at $10.00 per hour. Then she worked 10 hours overtime on Saturday. She gets paid time-and-a-half for her overtime hours.

What was her total salary that week?

Answer: ___________

Example 3: Solution

40 hrs. × \$10 = (40)(10) = \$400

10 hrs. × (\$10+5) = (10)(15) = \$150

Total salary: \$550

Answer: \$550.00

You'll notice that in the SCR (short constructed response) questions you are NOT permitted to use a calculator, so the numbers used are not complicated.

EXTENDED CONSTRUCTED RESPONSE QUESTION (ECR)

Example 4: Nancy wants to buy a television. The regular price is \$325. The store is having a special sale and is reducing all television sets by 8% each Monday for the next five weeks.

- If Nancy can spend \$250, when will she be able to buy a television?
- How much will it cost her?
- How much more will she save if she waits one more week?

Example 4: Strategies and Solutions
Underline or circle important information in the question. Make a table to help you organize data. Remember, almost every ECR NJ ASK 8 question can be organized with a chart or table. Show your work even if you use a calculator.
If the TV is reduced by 8% each week, the cost will be 92% of the previous week's price.
(100% – 8% = 92%)

Original Price	Show Work	\$325.00
1st Monday	325 × 0.92	\$299.00
2nd Monday	299 × 0.92	\$275.08
3rd Monday	275.08 × 0.92	\$253.07
4th Monday	253.07 × 0.92	\$232.83
5th Monday	232.83 × 0.92	\$214.20

- On the 4th Monday, Nancy will be able to buy a television.
- It will cost her \$232.83.
- If she waits one week, she will save \$18.63 more.

Week 4 price: \$232.83
Week 5 price: –214.20
More savings: \$ 18.63

Important Reminder: The people who score these extended constructed response questions must work very fast. They spend only a short time reading each one. Therefore, it is very important that your work be neat, clear, and easy to read. On the math portion of the test, it is better to use a chart or a table to show your work than to write a complicated paragraph. Always remember to answer ALL parts of the question. There usually are two or three parts to these extended constructed response questions.

OUR NUMBER SYSTEM (NUMBER SETS)

INTEGERS

Before we begin, let's review some vocabulary you should remember. The first set of numbers people used to count their possessions was the *natural numbers,* or *counting numbers.*

- Counting numbers = {1, 2, 3, 4, 5, . . . , 100, 101, 102 . . . }
- Whole numbers counting numbers and zero = {0, 1, 2, 3, . . . , 100, 101, 102, . . .}
- Positive numbers whole numbers like 24 or 531
- Negative numbers whole numbers like – 6 or –3,000
- Integers positive and negative whole numbers

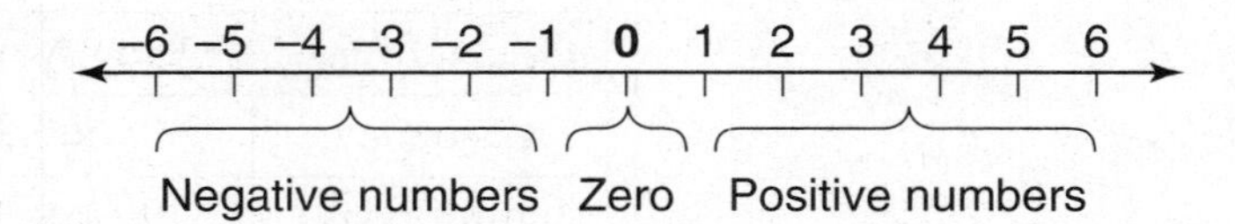

(Note: Zero is neither negative or positive.)

- Even numbers –2, . . ., 4, 6, . . ., 212, . . ., 254, . . . the last digit is always divisible by 2
- Odd numbers –5, –3, . . ., 7, 9, . . ., 273, . . ., 209, . . . the last digit is **not** divisible by 2

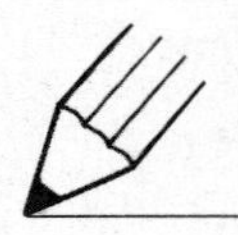

PRACTICE: Compare and Order Integers

(For answers, see page 57.)

Hint: Draw a number line to help you "see" the order.

1. Which of the following integers is in order from least to greatest?

A. –2, 2, 6 B. 5, –1, –8 C. –2, 3, – 4 D. 0, –2, –3

2. Which of the following integers is in order from greatest to least?

A. – 4, – 6, –8 B. –5, –2, 0, 1 C. –3, 4, –5 D. –2, –3, 0, 1

3. Which of the following integers is in order from least to greatest?

A. –10, 4, 6, –8 B. – 4, – 6, 0, 5 C. –3, 0, 2, 5 D. 4, 5, – 6, –8

Examples

Read the following real-life situations and think about the integer that matches each situation.

A. 50-yard gain	+50
B. 25 feet under water	–25
C. A typical July temperature in Italy is 37°C	+37
D. 14 degrees below zero	–14
E. 2,000 hits on a Web site	+2,000
F. 3,000 fewer viewers watched *Adventure Joe* this week	–3,000

PRACTICE: Positive and Negative Integers

(For answers, see page 57.)

Questions 1–5. *Select the integer that matches each situation.*

1. A loss of $ 2.00

A. –200 B. +20 C. –2 D. +2

2. You receive a gift of $10

A. –1 B. +1 C. –10 D. +10

3. 1,200 feet above sea level

A. –12 B. +12 C. –1,200 D. +1,200

4. The team is ahead 2 goals.

Write the number that represents this. __________

5. I have 45 minutes left on my cell phone.

What integer represents this? __________

6. Dan and his brother Jay each started with $100. Dan spent $16 for a t-shirt and then mowed a lawn and received $20; he spent $8 on gas for the car and bought a calculator for $12.50 and a CD for $14; next a friend paid him back $6. Jay bought books for $28 and then worked at school unpacking boxes and received $30; he bought a video for $17 and flowers for his mom's garden for $12.00. Who has the most money now?

A. Dan B. Jay
C. they both have the same D. not enough information given

INEQUALITIES

Often in mathematics symbols are used instead of words. Instead of saying *less than* or *greater than,* the following symbols are used.

$<$ symbol means *is less than*	$10 < 500$	$-6 < 10$
$>$ symbol means *is greater than*	$8 > 2$	$4 > -2$

Examples

Replace * with the correct symbol $>$ or $<$ or $=$ to make each statement true.

A. $5 * 8$ *Answer:* $5 < 8$
B. $10 + 2 * 8 + 4$ *Answer:* $10 + 2 = 8 + 4$
C. $-6 * 10$ *Answer:* $-6 < 10$
D. $-20 * 0$ *Answer:* $-20 < 0$
E. $50 * 0$ *Answer:* $50 > 0$

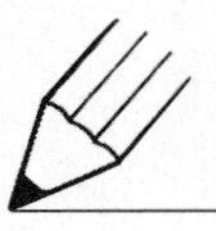

PRACTICE: Inequalities and Equals

(For answers, see page 57.)

Replace * with the correct symbol $>$ or $<$ or $=$ to make each statement true.

1. $-4 * 0$
2. $10 * 3$
3. $0 * 3$
4. $2 * -2$
5. $(5)(3) * (3)(5)$
6. $16 - 3 * 13 + 6$
7. $0 * -7$
8. $10 - 2 * 2^3$

ABSOLUTE VALUES

Another term you should understand is *absolute value.* The absolute value of a number is the distance of that number from 0 on the number line. The absolute value of a number is written as $|n|$. $|5|$ is read as "the absolute value of five"; $|-6|$ is read as "the absolute value of negative six." Because the absolute value of a number represents distance, it is always positive. $|-6| = 6$.

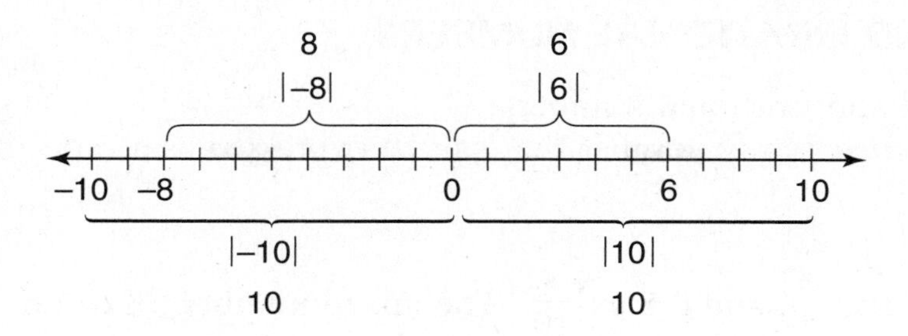

- The distance from 0 to −10 is the same length as the distance from 0 to +10.
- The distance from −8 to zero is greater than the distance from zero to +6; |−8| > |6|.

Examples

A. |− 6| The absolute value of − 6 could represent the number of yards the quarterback ran back before throwing a pass. He ran 6 yards.

B. |−15| The absolute value of −15 could represent the number of feet Pete swam under water. He swam 15 feet under water.

C. |−12| = |12| Whether I drive 12 miles south, or 12 miles north, I still drive 12 miles.

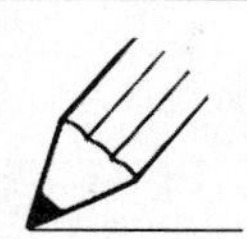

PRACTICE: Absolute Values

(For answers, see page 57.)

1. Which is the longer distance, |−16| or |5|?

A. |−16| B. |5|

2. What value is greater |−20| or |− 6|?

A. |− 6| B. |−20|

3. Which absolute value expression would best describe the distance the shark dove underwater?

A. |30| B. |−20| C. |20| D. |10|

RATIONAL AND IRRATIONAL NUMBERS

What are rational and irrational numbers?

A *rational number* is any number that can be expressed as a ratio of two integers.

Examples

4 can be expressed as $\frac{4}{1}$, and 0.8 as $\frac{8}{10}$. The mixed number 3½ can be written as $\frac{7}{2}$.

Rational Numbers	Decimal Form	Decimal Type
$\frac{1}{2}$	0.5	**Terminating**
$\frac{1}{6}$	0.1 6666 (written as $0.1\dot{6}$ or rounded up to 0.2)	Nonterminating, **repeating**
$\frac{2}{11}$	0.18181818 (written as $0.\overline{18}$)	Nonterminating, **repeating**

Any number that *cannot* be expressed as a ratio of two integers or as a repeating or terminating decimal is called an *irrational number.*

Irrational Numbers	Decimal Form	Decimal Type
$\frac{1}{17}$	0.0588235 ...	**Nonterminating, nonrepeating**
$\sqrt{3}$	1.7320508 ...	**Nonterminating, nonrepeating**
π	3.1415926 ...	**Nonterminating, nonrepeating**

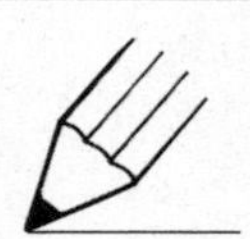

PRACTICE: Rational and Irrational Numbers

(For answers, see page 57.)

Investigate the following numbers and decide if they are rational or irrational. Ask yourself, does the the pattern repeat?

1. $\sqrt{6}$ Use your calculator to find $\sqrt{6}$. A. rational B. irrational

2. $\sqrt{8}$ Use your calculator to find $\sqrt{8}$. A. rational B. irrational

NUMERICAL OPERATIONS

ADDING AND SUBTRACTING INTEGERS

Remember these rules!

- Positive + Positive = Positive $\quad 6 + 3 = 9$

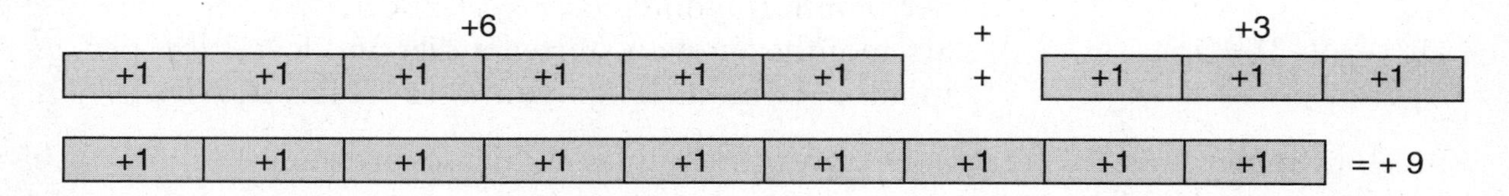

- Negative + Negative = Negative $\quad -4 + -6 = -10$

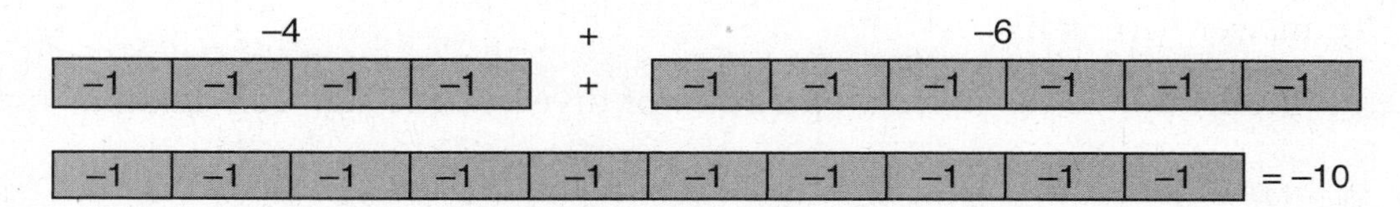

- Positive + Negative **or** $\quad 6 - 8 = -2$ or
  Negative + Positive $\quad -8 + 6 = -2$
  (Take the difference of the two numbers and use the sign of the larger number.)

Remember that +1 and −1 cancel each other out.
+1 plus −1 = zero $\quad$ 1+ −1 = 0

−1	−1	−1	−1	−1	−1	−1	−1	=	−8
+1	+1	+1	+1	+1	+1			=	+6
0	0	0	0	0	0	−1	−1	=	−2

Note: $-(-6)$ really means $(-1)(-6)$; therefore, $-(-6) = +6$.
Also, $-16 - (-3) = -16 + (-1)(-3) = -16 + 3 = -13$.

MULTIPLYING AND DIVIDING INTEGERS

- Positive × Positive = Positive $\quad$ (Positive)(Positive) = Positive $\quad (8)(2) = 16$
- Negative × Negative = Positive $\quad$ (Negative)(Negative) = Positive $\quad (-8)(-2) = 16$
- Positive × Negative is the same as
  Negative × Positive = Negative $\quad$ (Positive)(Negative) = Negative $\quad (2)(-8) = -16$
  and
  $(-2)(8) = -16$

Examples of Multiplying with Integers

A. $3 \times 12 = 36$ (I have 3 boxes of donuts and each box has 12 donuts.)

B. $-4 \times 2 = -8$ (Mary owes $4 to two different friends. She owes a total of $8.)

C. $(10)(-120) = -1{,}200$ (If we rented a storage facility for 10 months at $120 per month it would cost us $1,200.)

D. $(-6)(-3) = 18$ Six months ago you started a diet and lost 3 lbs. per month. Six months ago you were 18 lbs. heavier.

E. A swimmer dove 10 feet under water three times. Which expression represents this?

A. $(-3)(-10)$ B. $3|-10|$ C. $(3)(10)$ D. $-3|10|$

The answer here is B.

Since multiplication is the inverse operation of division, the sign rule remains the same.

- $\frac{\text{Positive}}{\text{Positive}} = \text{Positive}$ Positive ÷ Positive = Positive $\frac{24}{2} = 12$ or $24 \div 2 = 12$
- $\frac{\text{Negative}}{\text{Negative}} = \text{Positive}$ Negative ÷ Negative = Positive $\frac{-15}{-3} = 5$ or $-15 \div -3 = 5$
- $\frac{\text{Positive}}{\text{Negative}} = \text{Negative}$ $\frac{18}{-2} = -9$ or $\frac{\text{Negative}}{\text{Positive}} = \text{Negative}$ $\frac{-18}{9} = -2$

Examples of Dividing with Integers

A. $16 \div 2 = 8$ If Ryan has 16 cans of oil and put them evenly into 2 boxes, he would have 8 cans in each box.

B. $10 \div -2 = 5$ If Diane had $10 and the books at the library sale cost $2 each, she could buy 5 books.

C. $-\$30 \div 5 = -\$6$ Dan is at the zoo for 5 hours. He spends $30 all together. Dan spent an average of $6 per hour.

D. $-45 \div -5 = 9$ If Kevin wants to lose 45 pounds and planned on losing 5 pounds each month, he would reach his goal in 9 months.

Study the following chart to remind yourself how to add, multiply, or divide when you are working with positive and negative integers.

Operations with Integers

Adding Integers		Subtracting Integers	Multiplying & Dividing Integers	
Same sign + + − −	Integers with the different signs	Add the opposites.	Same sign + + − −	Integers with the different signs
Add the numbers and take the original sign.	Take the difference of the two numbers and the sign of the larger number.	Follow rules for addition.	Positive	Negative
$6 + 4 + 5 = 15$	$-9 + 6 = -3$	$(+5) - (-16) =$ $5 + 16 = 21$	$(3)(4) = 12$ $(-6)(-2) = 12$ $-(-12) = 12$	$(3)(-4) = -12$ $(-6)(2) = -12$
$-4 - 3 - 2 = -9$	$-2 + 7 = 5$	$-30 - (-20) =$ $-30 + 20 = -10$	$21 \div 3 = 7$ $(-21) \div (-3) = 7$	$(21) \div (-3) = -7$ $(-21) \div (3) = -7$

PRACTICE: Adding, Subtracting, Multiplying, and Dividing Integers

(For answers, see page 58.)

Perform the operations shown, and simplify each expression. No calculator is needed here!

1. 52 + (+20) =

A. –72
B. +72
C. +32
D. –32

2. 61 – 41 =

A. –20
B. +20
C. –102
D. +102

3. –30 + 15 =

A. 45
B. –45
C. –15
D. 15

4. –20 + 30 =

A. 50
B. –50
C. –10
D. 10

5. –3 + (–7) =

A. 10
B. –10
C. –4
D. 4

6. –4 – (–9) =

A. –13
B. 13
C. 5
D. –5

7. 6 – (–3) =

A. 9
B. –9
C. –3
D. 3

8. (3)(4.5) =

A. –13.5
B. 13.5
C. –12.5
D. 12.5

9. (–3)(5.2) =

A. 2.2
B. –15.6
C. 15.6
D. 4.9

10. (–2)(–5)=

A. –7
B. 7
C. –10
D. 10

11. (3)(–6)= _______

12. 16 ÷ 2 = _______

13. 25 ÷ 5 = _______

14. –30 ÷ –6 = _______

15. 100 ÷ –10 = _______

EXPONENTS, SQUARE ROOTS, AND CUBE ROOTS

In mathematics, *exponents* are used as a short way of saying a number is multiplied by itself a certain number of times.

- Instead of writing 5×5, we can write 5^2, which means 5 times itself 2 times ($5^2 = 25$).
- Instead of writing 10×10, we can write 10^2, which means 10 times itself 2 times ($10^2 = 100$).

We read these numbers two different ways.

- We can say 5^2 is *five squared*,
- or we can say 5^2 is *five raised to the second power.*

Examples

A. Instead of writing $2 \times 2 \times 2$, we can write 2^3, which means 2 times itself 3 times ($2^3 = 8$).

We can say that 2^3 is *two cubed*, or we can say that it is *two raised to the third power.*

B. Instead of writing $1 \times 1 \times 1$, we can write 1^3, which means 1 times itself 3 times ($1^3 = 1$).

We can say that 1^3 is *one cubed*, or we can say that it is *one raised to the third power.*

A number written in *exponential form* has a base and an exponent. The following are samples of numbers written in exponential form:

$$5^2 \qquad 4^3 \qquad 6^2 \qquad 2^4$$

- In the term 5^2, the 5 is the base and the 2 is the exponent.
- In the term 4^3, the 4 is the base and the 3 is the exponent.

Note: Any number raised to the zero power equals 1.

$$59^0 = 1, \qquad 2{,}980{,}00^0 = 1, \qquad (1/5)^0 = 1$$

PRACTICE: Exponents

(For answers, see page 58.)

1. Write *six squared* in exponential form.

   A. 2^6
   B. 6^2
   C. 6×2
   D. $(6)(6)$

2. Which statement means the same as 5^3?

   A. five squared
   B. three cubed
   C. five cubed
   D. five times three

3. Which expression = 3^4?

   A. 3×3
   B. 4×3
   C. $4 \times 4 \times 4$
   D. $3 \times 3 \times 3 \times 3$

4. Which represents the largest number? (Show your work; do not guess.)

   A. 4^2
   B. $(4)(2)$
   C. 24^0
   D. $8^2 - 4^2$

5. Which expression is equal to 9^2?

   A. $3^2 + 3^2$
   B. $(3^3)(6)^0$
   C. $(9)(3)(3)$
   D. $(2)(2)(9)$

6. How would you write two cubed?

   ____

7. How would you write eight squared?

   ____

8. Write $8 \times 8 \times 8$ in exponential form.

   ____

Using a Calculator

Many calculators have an x^2 key so you can easily *square a number.* Try these on your calculator:

- Find the value of 4^2. Use your calculator. Even though this is an easy one (you probably know automatically $4^2 = 16$), try it on your calculator to see which buttons you need to press and in what order. Remember this so you can work with more difficult numbers easily.
- Find the value of 29^2. Use your calculator again and remember how your calculator works.

The correct solution is 841.

- Try a few more to practice pressing the correct buttons.

What is the value of 16^2?
$16^2 = 256$

What is the value of 42^3?
$42^3 = 74{,}088$

What is the value of $(19.85)^2$?
$(19.85)^2 = 394.0225$

If you ask other students in your class how their calculators work, you may find that other models work differently. This is one reason it is important for you to use the same calculator you practice with when you take the official ASK 8 test.

Now let's work in reverse!

The following are some *perfect square numbers:*

1, 4, 9, 16, 25, 36, 49, 64, 81, 100

Perfect square numbers can be represented as perfect square shapes. Remember that the formula for the *area of a square* is *length* × *width*. See the following examples and diagrams.

- A square with length 1 and width 1 has an area of 1 square unit; 1 is a perfect square number.
- A square with length 2 and width 2 has an area of 4 square units; 4 is a perfect square number.
- A square with length 3 and width 3 has an area of 9 square units; 9 is a perfect square number.
- A square with length 4 and width 4 has an area of 16 square units; 16 is a perfect square number.
- A square with length 5 and width 5 has an area of 25 square units; 25 is a perfect square number.
- A square with length 6 and width 6 has an area of 36 square units; 36 is a perfect square number.
- A square with length 7 and width 7 has an area of 49 square units; 49 is a perfect square number.

These represent *perfect square* numbers:

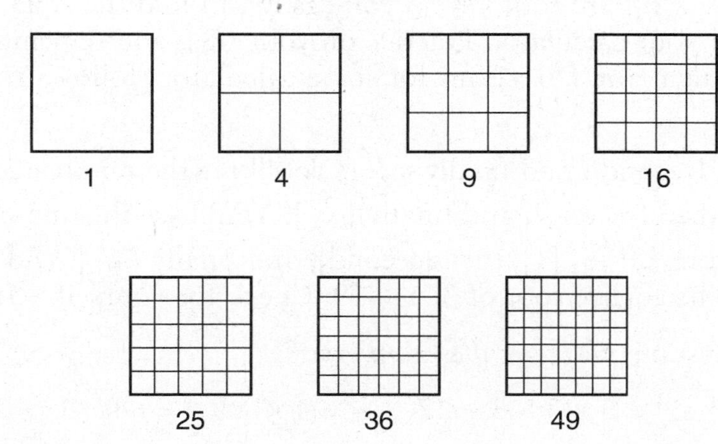

You should be familiar with these perfect square numbers and their *square roots*.

Perfect square number	The square root is	Because	
1	$\sqrt{1} = 1$	$1 \times 1 =$	$1^2 = 1$
4	$\sqrt{4} = 2$	$2 \times 2 =$	$2^2 = 4$
9	$\sqrt{9} = 3$	$3 \times 3 =$	$3^2 = 9$
16	$\sqrt{16} = 4$	$4 \times 4 =$	$4^2 = 16$
25	$\sqrt{25} = 5$	$5 \times 5 =$	$5^2 = 25$
36	$\sqrt{36} = 6$	$6 \times 6 =$	$6^2 = 36$
49	$\sqrt{49} = 7$	$7 \times 7 =$	$7^2 = 49$
64	$\sqrt{64} = 8$	$8 \times 8 =$	$8^2 = 64$
81	$\sqrt{81} = 9$	$9 \times 9 =$	$9^2 = 81$
100	$\sqrt{100} = 10$	$10 \times 10 =$	$10^2 = 100$
121	$\sqrt{121} = 11$	$11 \times 11 =$	$11^2 = 121$
144	$\sqrt{144} = 12$	$12 \times 12 =$	$12^2 = 144$
169	$\sqrt{169} = 13$	$13 \times 13 =$	$13^2 = 169$
196	$\sqrt{196} = 14$	$14 \times 14 =$	$14^2 = 196$
225	$\sqrt{225} = 15$	$15 \times 15 =$	$15^2 = 225$

Using a Calculator

Most calculators have a *square root key* so you easily can find the square root of any number. Try these on your calculator. Practice on your own and remember the steps that work for your calculator. Directions for some calculators follow. Yours may work differently.

- Press [4] then [second], and finally [√]. You'll see the number 2.
- Press [8] [1], then [second], and finally [√]. You'll see the number 9
- Press the buttons [3] [6] [1], then [second], and finally [√]. Did you get the number 19? The square root of 361 = 19. Check for yourself. Multiply 19 × 19.

The following are some *perfect cube numbers:*

$2^3 = 2 \times 2 \times 2 = 8$ $3^3 = 3 \times 3 \times 3 = 27$ $4^3 = 4 \times 4 \times 4 = 64$ $5^3 = 5 \times 5 \times 5 = 125$

If you think of cubed numbers as geometric figures, you will see that a perfect cubic number represents the *volume of a cube.*

$$\text{Volume} = \text{Length} \times \text{Width} \times \text{Depth}$$

- If a cube had length = 2, width = 2, depth = 2, its volume would = $2 \times 2 \times 2 = 8$ cubic units.

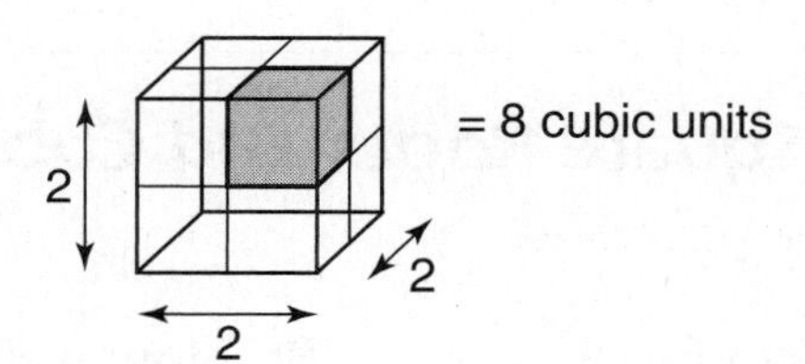

- If a cube had length = 4, width = 4, depth = 4, its volume would = $4 \times 4 \times 4 = 64$ cubic units.

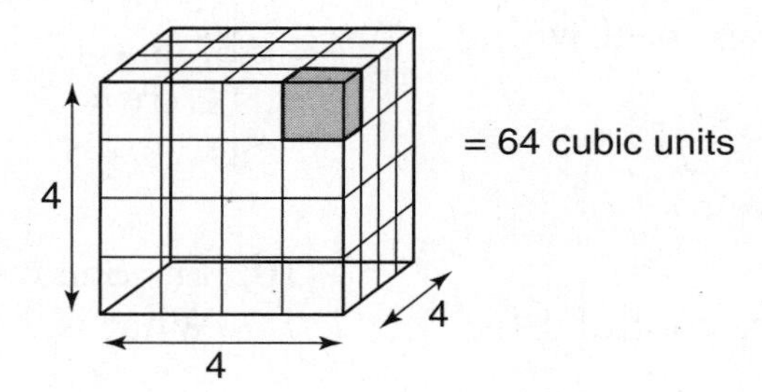

Examples

A. What number times itself will equal 16? *Answer*: 4
Because $4 \times 4 = 16$ or $4^2 = 16$

B. What number times itself will equal 100? *Answer*: 10
Because $10 \times 10 = 100$ or $10^2 = 100$

In mathematics we say this another way. We say, "What is the *square root* of 100?"

C. What is the *square root* of 16? *Answer*: 4
Because $4 \times 4 = 16$ or $4^2 = 16$

D. What is the *square root* of 100? *Answer*: 10
Because $10 \times 10 = 100$ or $10^2 = 100$

E. What is the *cube root* of 8? *Answer*: 2
Because $2 \times 2 \times 2 = 8$ or $2^3 = 8$

Still, in mathematics we show this another way. We use a symbol to represent the words *the square root of*. The symbol is $\sqrt{\ }$. Examples: $\sqrt{25} = 5$, $\sqrt{81} = 9$, $\sqrt{144} = 12$.

F. $\sqrt{16}$ is read as *the square root of 16.*

Answer: $\sqrt{16} = 4$

Because $4 \times 4 = 16$ or $4^2 = 16$

G. $\sqrt{100}$ is read as *the square root of 100.*

Answer: $\sqrt{100} = 10$
Because $10 \times 10 = 100$ or $10^2 = 100$

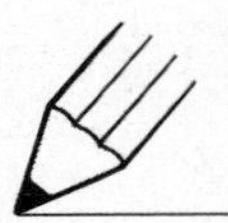

PRACTICE: Square Roots and Cube Roots

(For answers, see page 58.)

1. What number times itself will equal 25?

2. What number times itself will equal 81?

3. What is the square root of 36?

4. What is the square root of 144?

5. What is $\sqrt{49}$?

6. What is $\sqrt{100}$?

7. What do you think the $\sqrt{25}$ plus $\sqrt{100}$ equals?

A. $25 + 100$
B. $5 + 10$
C. $5^2 + 10^2$
D. $50 + 10$

8. What do you think $\sqrt{36} + \sqrt{25}$ equals?

A. $6 + 50$
B. $36 + 25$
C. $6^2 + 5^2$
D. 11

9. Now we'll combine both ideas. What is $4^2 + \sqrt{25}$?

A. $16 + 5$
B. $8 + 10$
C. $16 + 7$
D. $8 + 5$

10. Try one more combination. What is $2^3 + \sqrt{49}$?

A. $6 + 7$
B. $5 + 7$
C. $8 + 7$
D. $8 + 13$

11. Which of the following is NOT equal to 7?

A. $\sqrt{1} + \sqrt{36}$
B. $1^2 + 2^2 + 2$
C. $\sqrt{49}$
D. $\sqrt{100} - 1^2$

12. All of the following equal 16 except

A. $2^3 + (4)(2)$
B. $\sqrt{4}$
C. $5^2 - 3^2$
D. $6^2 - (10)(2)$

ESTIMATING SQUARE ROOTS

Some numbers are not perfect squares. For example the number 30 is not a perfect square, but we can still estimate its square root. Remembering those *perfect square numbers* will help you here.

Examples

A. $\sqrt{30}$ is between the perfect squares $\sqrt{25}$ and $\sqrt{36}$. Therefore, $\sqrt{30}$ is between 5 and 6.

B. $\sqrt{90}$ is between the perfect squares $\sqrt{81}$ and $\sqrt{100}$. Therefore, $\sqrt{90}$ is between 9 and 10.

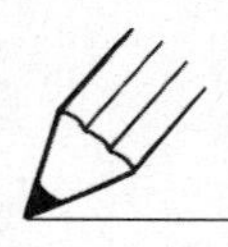

PRACTICE: Estimating Square Roots

(For answers, see page 59.)

1. $\sqrt{50}$ is between

A. 3 and 4
B. 2 and 3
C. 6 and 7
D. 7 and 8

2. $\sqrt{15}$ is closest to the number

A. 3
B. 4
C. 5
D. 6

3. Which of the following square roots is between 5 and 6

A. $\sqrt{10}$
B. $\sqrt{20}$
C. $\sqrt{30}$
D. $\sqrt{40}$

4. Which statement is true about $\sqrt{115}$? It is

A. less than 10
B. almost 8
C. between 10 and 12
D. greater than 20

5. $\sqrt{100}$ is equal to ALL of the following except

A. (2)(5)
B. 100 divided by 2
C. $\sqrt{4} + 2^3$
D. $4^2 - 6$

SCIENTIFIC NOTATION

Before reviewing scientific notation, let's review the number 10 written with different exponents. The following table shows some examples of this, which we call powers of ten.

Powers of 10	Meaning	Value	Power	Number of Zeros
10^1	10	10	Power of 1	One zero
10^2	10×10	100	Power of 2	Two zeros
10^3	$10 \times 10 \times 10$	1,000	Power of 3	Three zeros
10^4	$10 \times 10 \times 10 \times 10$	10,000	Power of 4	Four zeros

- Multiply by powers of 10

  $33 \times 10^5 = 33 \times 10 \times 10 \times 10 \times 10 \times 10 = 33 \times 100{,}000 = 3{,}300{,}000.$

- Multiply by powers of 10 mentally

25×10^9 *Think!* This is just 25 with 9 zeros added on to the end: 25,000,000,000.

3,300,000 and 25,000,000,000 are written in the *standard form* for large numbers. However, very large numbers become confusing and very difficult to write and to read. Scientific notation is a shorthand way of writing very large and very small numbers.

Standard Form	*Scientific Notation Form*
9,670,000,000,000	is written as 9.67×10^{12}

By the definition of *scientific notation*, there can be only one digit to the left of the decimal point. In this number, the first digit must be greater than or equal to 1 but less than 10. So we move the decimal point and then count how many places the decimal point was moved (count all of the zeros and the 6 and 7) 9,670,000,000,000 and you have 12 (the number of the exponent).

Examples

A. 540,000,000 (Remember there is really a decimal point at the end.) In scientific notation form, this number is written as 5.4×10^8. We moved the decimal point 8 places to the left.

B. 24,000,000,000 — In scientific notation form, this number is written as 2.4×10^{10}.

C. 2,000,000,000 — In scientific notation form this number is written as 2×10^9.

Here we'll work in reverse. We'll begin with a number in scientific notation form and change it to a number in standard form.

D. 6×10^7 — Means 6 times $10 \times 10 \times 10 \times 10 \times 10 \times 10 \times 10$ or 60,000,000 (7 zeros).

E. 3.8×10^5 — We must move the decimal point 5 places; but we don't have the necessary zeros. Think 3.8 is the same as 3.80000; so $3.8 \times 10^5 = 380{,}000$ so we can add the zeros we need.

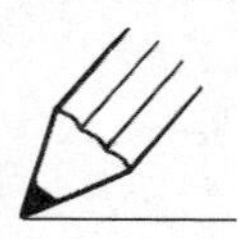

PRACTICE: Scientific Notation

(For answers, see page 59.)

1. Fill in the missing parts in the table.

Standard Form	How Many Places Will You Move the Decimal Point?	Scientific Notation Form
650,000,000,000,000	14	
2,000,000,000		2.0×10^9
		4.5×10^{11}
	9	3.25×10^9
12,000,000,000		

2. In space, light travels about 9,450,000,000,000 kilometers per year. This is called a light year. Write this number in scientific notation form.

A. 9.45×10^{12} B. 94.5×10^{11} C. 945×10^{10} D. 9×10^{12}

3. The number 650 billion is equal to 650,000,000,000. Write this number in scientific notation form.

A. 650×10^{10} B. 65×10^{11} C. 6.5×10^{11} D. 6.5×10^{10}

4. The air pressure at the bottom of the deepest ocean is said to be 1.1×10^8 pa (pascal). What number below is equivalent to this number?

A. 1.0000001 B. 110,000,000 C. 110.000,000 D. 1.10000000

5. In this example, Rachel says the answer is A; Barbara says it is B. Who is correct? Explain your answer.

$$280{,}000{,}000 = ?$$

A. 28×10^7 B. 2.8×10^8

ALGEBRAIC ORDER OF OPERATIONS

Please Excuse My Dear Aunt Sally

When you get dressed you put your socks on before your shoes. In life there are some things that are best done in a certain order. In math there also is a correct order for doing some things.

We'll be working with *expressions*. Expressions are made up of numbers and operations. The numbers might be 16, −4, 3.4, $\frac{1}{2}$; operations look like × ÷ + − or ( ).

If different people were asked to *evaluate* the expression $\mathbf{3(4-6)+12 \div 2-16(-1)^0+4^2}$, they must follow certain rules of order if they are to get the same answer. Should you divide by 2 first? Maybe you should add the four-squared first. This can get confusing. Is there a rule to follow? Yes! Just remember the phrase ***Please Excuse My Dear Aunt Sally (PEMDAS)*** and evaluate your expressions in this order: ***Parentheses, Exponents,*** then ***M**ultiplication* and ***D**ivision* (work from left to right), and finally ***A**dd and **S**ubtract* (work from left to right).

Examples

Think of the operation you would do first in each of the following expressions; then evaluate each expression. Remember PEMDAS.

A. $(9-2)\times 4-4$	First, work inside **P**arentheses	$7\times 4-4$
	Do **M**ultiplication next	$28-4$
	Finally, **S**ubtract	24
B. $46-4^2+3$	This expression has no parentheses, so	
	First, work with **E**xponents	$46-16+3$
	Next **A**dd and **S**ubtract	$30+3$
		33
C. $(28\div 4)+5^2(3-1)$	First, work inside both **P**arentheses	$7+5^2(2)$
	Work with **E**xponents next	$7+25(2)$
	Then do **M**ultiplication	$7+50$
	Finally, **A**dd	57
D. $50\times 2+12-2\times 4$	This expression has no parentheses or exponents, so	
	First, do both **M**ultiplications	$100+12-8$
	Next, **A**dd and **S**ubtract	$112-8$
		104
E. $5+6^2\times 10$	First, work with the **E**xponent	$5+36\times 10$
	Then **M**ultiply	$5+360$
	Finally, **A**dd	365

F. $[(6 \times 5) + 3](3)$

What do you do if you have parentheses inside parentheses or parentheses inside brackets? Keep the same order of operations. Do the inside parentheses first.

	$[(6 \times 5) + 3](3)$
First Multiply 6 and 5	$[(30) + 3](3)$
Next, continue working inside the brackets and Add inside parentheses	$[33]\ (3)$
Last, Multiply	99

G. $(5 + 3 \times 20) \div 13 + 3^2$

Work inside Parentheses first	$(5 + 3 \times 20) \div 13 + 3^2$
Multiply inside parentheses	$(5 + 60) \div 13 + 3^2$
Add inside parentheses next	$(65) \div 13 + 3^2$
Exponents are next	$65 \div 13 + 9$
Divide	$5 + 9$
Add	14

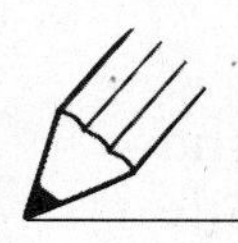

PRACTICE: Algebraic Order of Operations

(For answers, see page 59.)

Do NOT use a calculator to practice these examples!

1. $27 - 18 \div 6$

What do you do first?

A. subtract
B. divide

Evaluate the expression.

A. 9
B. 24
C. 1.5
D. 25

2. $36 + 18 \div 2 \times 3 - 8$

What do you do first?

A. add
B. subtract
C. divide
D. multiply

Evaluate the expression.

A. 96
B. 73
C. 32
D. 55

3. Evaluate the expression

$$(72 \div 9 - 2) + 2 \times 3$$

A. 12
B. 16
C. 24
D. 36

4. Evaluate the expression

$$(20 + 8 \div 4) + (3^2 \times 4)$$

A. 19
B. 43
C. 58
D. 71

5. Evaluate the expression

$$[2(5 - 1) + 3(2 \times 2)] - 4$$

A. 7
B. 16
C. 15
D. 20

6. Evaluate the expression

$$(6 - 4)^3 - 4$$

A. –5
B. 2
C. 4
D. 12

7. To evaluate this expression, what should you do first?

$$(9 - 2 \times 4) + 2^3(6)$$

A. $2 \times 2 \times 2$
B. $9 - 2$
C. $2(6)$
D. 2×4

8. Add parentheses to this expression $2 + 3 \times 2 \times 5 - 1$ so its evaluation is 49.

A. $(2 + 3) \times 2 \times 5 - 1$
B. $2 + (3 \times 2) \times 5 - 1$
C. $2 + 3 \times (2 \times 5) - 1$
D. $2 + 3 \times 2 \times (5 - 1)$

9. Add parentheses to this expression $4 + 3 \times 2 \times 5 - 1$ so its evaluation is 28.

A. $(4 + 3) \times 2 \times 5 - 1$
B. $4 + (3 \times 2) \times 5 - 1$
C. $4 + 3 \times (2 \times 5) - 1$
D. $4 + 3 \times 2 \times (5 - 1)$

10. Jaime says the evaluation of the following expression is 28; Billy says it is 24.

$$2(2 + 5 \times 2)$$

Who is correct? Explain your answer and show all work.

In the following examples we've combined positive and negative integers with + – × ÷ and ( ).

11. $(-3 - 4) + (-6 - 2) =$

A. –15
B. 15
C. –1
D. –24

12. $(-9 + 2) + (-30 + 5) =$

A. –18
B. 36
C. –32
D. 32

13. $(-8 + 2) + (-4 + 8) =$

A. −10
B. −2
C. − 48
D. 36

14. $(-16 + 2)(2) =$

A. −28
B. − 64
C. −36
D. 36

15. $(-2 - 4)(5 - 2) =$

A. 18
B. −18
C. 24
D. −3

16. $(-18) \div (10 - 4) =$

A. −3
B. −2
C. 3
D. −4

17. $(-16 + 4) \div (-2 - 2) =$ _______

18. $(10 - 2) \div (2^2) =$ _______

19. $2 + (-4 + 12 - 6) =$ _______

20. $-6 + 4 + (2)(5 - 3) =$ _______

FRACTIONS, DECIMALS, AND PERCENTS

The terms "fractions," "decimals," and "percents" are really different ways of describing the same things. Each of the following figures represents one whole, with a part shaded in. The shaded-in part can be described as a fraction, as a decimal, and as a percent or even as money if the whole shape = $1.00.

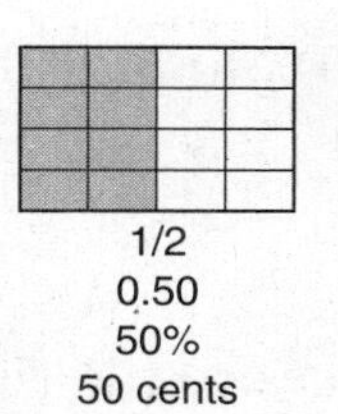
1/2
0.50
50%
50 cents

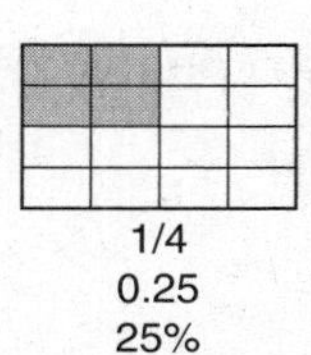
1/4
0.25
25%
25 cents

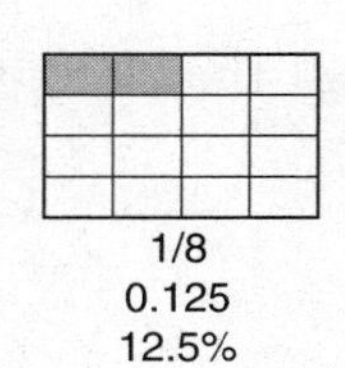
1/8
0.125
12.5%

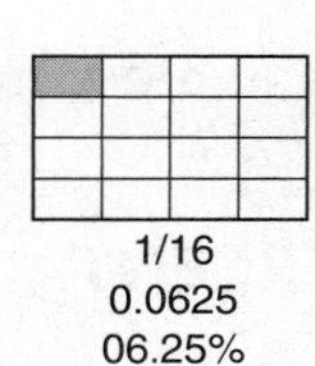
1/16
0.0625
06.25%

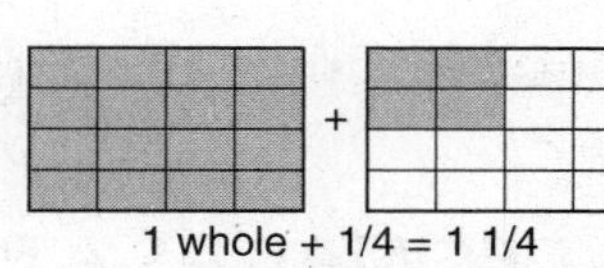

1 whole + 1/4 = 1 1/4
1 whole + 0.25 = 1.25
100% + 25% = 125%
one dollar and 25 cents

WRITE EACH PERCENT AS A DECIMAL

Read and study this chart.

USING SYMBOLS			USING WORDS	
Percent	Is the same as	Decimal	Percent	Decimal
75% =	$\frac{75}{100}$ =	0.75 =	Seventy-five percent =	Seventy-five hundre**dths**
150% =	100% + 50% =	1.50 =	One hundred fifty percent =	One and fifty hundre**dths**
325% =	300% + 25% =	3.25 =	Three hundred twenty-five percent =	Three and twenty-five hundre**dths**

WRITE EACH FRACTION AS A DECIMAL

Examples

$$\frac{1}{2} = 0.50; \quad \frac{2}{5} = 0.40; \quad 1\frac{3}{4} = 1.75$$

Remember that the *fraction bar* really means divide. Any fraction can be expressed as a decimal by dividing the numerator by the denominator. In $\frac{3}{4}$, 3 is the numerator; 4 is the denominator.

- $\frac{5}{8}$ means 5 ÷ 8 — Use your calculator and press [5][÷][8][=]. You will see [.625].

- $1\frac{2}{3}$ means $1 + \frac{2}{3}$ — Use your calculator to see how to represent the $\frac{2}{3}$ as a decimal.

  [2][÷][3] = .666 Therefore $1\frac{2}{3}$ = 1.666.

A decimal like .625 is called a *terminating decimal.* It is an exact answer. 5 ÷ 8 = .625 exactly. You can check this because (.625)(8) = 5 exactly. A decimal like 0.6666 is called a **repeating decimal.** It is not an exact answer. [2][÷][3] = 0.666666 . . . . If you check this you will see that (0.666)(3) = 1.998. It is very close to 2, but it does not equal 2; it is not an exact answer.

Note: We put a line over the parts of a decimal that keeps repeating: $0.\overline{6666}$ or $0.\overline{34}$.

WRITE A DECIMAL AS A PERCENT

To write a decimal as a percent, simply move the decimal point 2 places to the right. (We really are multiplying the decimal number by 100 to get the equivalent percent number.)

For example,

0.25 = 25%, 0.355 = 35.5%, 2.5 = 250%

Examples

A. Change the decimal number 0.66 to a percent. 0.66 = 66%

Whole = 1

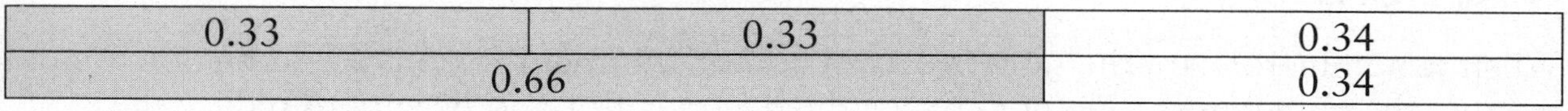

0.33	0.33	0.34
0.66		0.34

Whole = 100%

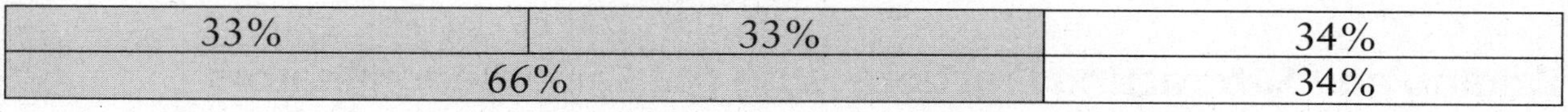

33%	33%	34%
66%		34%

B. Change the decimal number 1.25 to a percent. 1.25 = 125%

Whole = 1 or 100%

$\frac{1}{4}$ or 0.25 or 25%

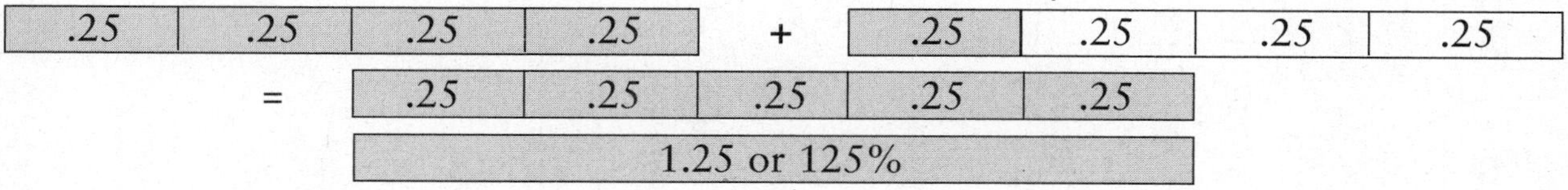

.25	.25	.25	.25	+	.25	.25	.25	.25

= | .25 | .25 | .25 | .25 | .25 |

1.25 or 125%

C. Change the decimal number 1.8 to a percent. 1.8 is the same as 1.80, so 1.80 = 180%

In this case we couldn't move the decimal point two places to the right until we added a zero. $0.8 = \frac{8}{10}$; $0.80 = \frac{80}{100}$ which reduces to $\frac{8}{10}$. Therefore we can add the zero to the right and not change the value of the decimal number.

WRITE A FRACTION AS A PERCENT

Examples

$$\frac{1}{2} = 50\%, \quad \frac{3}{4} = 75\%, \quad \frac{1}{5} = 20\%, \quad \frac{2}{5} = 40\%$$

Because you can use a calculator on some parts of the ASK 8 test, just remember that the fraction bar really means to divide. For example, $\frac{3}{8}$ means 3 ÷ 8. Use your calculator and press [3][÷][8][=]. You will see [.375]. Therefore, $\frac{3}{8}$ = 0.375 (the decimal number) = 37.5% (the percent number). Remember to move the decimal point two places to the right to change a decimal number to a percent (or remember that the decimal number × 100 = the percent number).

Judy said, "$\frac{5}{6}$ is more than 50%." Estimate to see if she is correct. Think, $\frac{5}{6}$ is more than half (since $\frac{3}{6} = \frac{1}{2}$), so $\frac{5}{6}$ is more than $\frac{1}{2}$, or more than 50%. Her answer makes sense.

FRACTIONS

When denominators are the same, you can add or subtract fractions by adding or subtracting the numerators. Do not change the denominator and be sure to reduce the solution to lowest terms when possible.

Adding and Subtracting Fractions with the Same Denominators

Examples

A. $\frac{3}{12} + \frac{5}{12} = \frac{8}{12}\left(\frac{\div 4}{\div 4}\right) = \frac{2}{3}$

B. $\frac{5}{36} + \frac{7}{36} = \frac{12}{36}\left(\frac{\div 12}{\div 12}\right) = \frac{1}{3}$

C.
$$\begin{array}{r} 6\frac{2}{5} \\ +\ 7\frac{4}{5} \\ \hline 13\frac{6}{5} \end{array}$$
which simplifies to $14\frac{1}{5}$

D.
$$\begin{array}{r} 35\frac{9}{10} \\ -\ 20\frac{3}{10} \\ \hline 15\frac{6}{10} \end{array}$$
which simplifies to $15\frac{3}{5}$

E.
$$\begin{array}{r} 8\frac{2}{9} \\ -\ 4\frac{5}{9} \\ \hline \end{array}$$
change to
$$\begin{array}{r} 7\frac{11}{9} \\ -\ 4\frac{5}{9} \\ \hline 3\frac{6}{9} \end{array}$$
which simplifies to $3\frac{2}{3}$

(You cannot subtract $\frac{5}{9}$ from $\frac{2}{9}$ so we renamed $8\frac{2}{9}$ as $7\frac{11}{9}$)

Adding and Subtracting Fractions with Different Denominators

Examples

When fractions have different denominators; select a Lowest Common Denominator (LCD) and rewrite each fraction using that LCD.

A. $\frac{1}{4} + \frac{5}{12} = \frac{3}{12} + \frac{5}{12} = \frac{8}{12}$ which simplifies to $\frac{2}{3}$ (*Note:* $\frac{1}{4}$ was rewritten as $\frac{3}{12}$.)

B. $\frac{9}{10} - \frac{2}{3} = \frac{27}{30} - \frac{20}{30} = \frac{7}{30}$ (*Note:* Here we had to change both fractions.)

C.

$$\begin{array}{r} 28\frac{7}{10} \\ -\ 13\frac{2}{10} \\ \hline 15\frac{5}{10} \end{array}$$

which simplifies to $15\frac{1}{2}$

D.

$$\begin{array}{r} 14\frac{3}{8} \\ -\ 9\frac{5}{8} \\ \hline \end{array} \quad \text{change to} \quad \begin{array}{r} 13\frac{11}{8} \\ -\ 9\frac{5}{8} \\ \hline 4\frac{6}{8} \end{array}$$

which simplifies to $4\frac{3}{4}$

(You cannot subtract $\frac{5}{8}$ from $\frac{3}{8}$ so we renamed $14\frac{3}{8}$ as $13\frac{11}{8}$)

MULTIPLYING AND DIVIDING FRACTIONS

Here you simply multiply all the numerators for a new numerator; then you multiply denominators for a new denominator. You do not need common denominators when multiplying or dividing.

Examples

A. $\frac{3}{4} \times \frac{2}{3} \times \frac{5}{9} = \frac{30}{108}$ simplifies to $\frac{15}{54} = \frac{5}{18}$

B. $\frac{1}{2} \times \frac{3}{5} \times \frac{4}{5} = \frac{12}{50}$ simplifies to $\frac{6}{25}$

When you have mixed numbers, change them to improper fractions and then multiply.

C. $5\frac{2}{3} \times 4\frac{3}{5}$ change to $\frac{17}{3} \times \frac{23}{5} = \frac{391}{15}$ (A calculator may save time here.) $= 26\frac{1}{15}$

Before we review division with fractions we need to review *reciprocals*. The reciprocal of $\frac{2}{3}$ is $\frac{3}{2}$. The reciprocal of $\frac{5}{12}$ is $\frac{12}{5}$. When you divide with fractions, you keep the first fraction the way it is given, change the fraction after the division sign to its reciprocal, and then multiply.

Examples

A. $\frac{3}{4} \div \frac{2}{3} = \frac{3}{4} \times \frac{3}{2} = \frac{9}{8} = 1\frac{1}{8}$

B. $\frac{3}{4} \div \frac{1}{2} = \frac{3}{4} \times \frac{2}{1} = \frac{6}{4} = 1\frac{2}{4} = 1\frac{1}{2}$

PRACTICE: Adding, Subtracting, Multiplying, and Dividing Fractions

(For answers, see page 60.)

1. $\frac{3}{8} + \frac{3}{8} =$

A. $\frac{3}{16}$ B. $\frac{6}{16}$ C. $\frac{3}{4}$ D. $\frac{9}{8}$

2. $\frac{5}{12} - \frac{1}{3} =$

A. $\frac{1}{12}$ B. $\frac{9}{12}$ C. $\frac{3}{4}$ D. $\frac{1}{3}$

3. $6\frac{1}{3} + 4\frac{2}{3} =$

A. $10\frac{1}{3}$ B. $10\frac{2}{3}$ C. 11 D. $11\frac{2}{3}$

4. $5\frac{1}{8} - 2\frac{3}{8} =$

A. $3\frac{3}{4}$ B. $2\frac{3}{4}$ C. $3\frac{2}{8}$ D. $2\frac{5}{8}$

5. $\frac{4}{5} \times \frac{1}{2} =$

A. $\frac{2}{5}$ B. $\frac{4}{7}$ C. $2\frac{1}{2}$ D. $\frac{13}{10}$

6. $\frac{5}{8} \div \frac{1}{2} =$

A. $\frac{1}{2}$ B. $1\frac{1}{4}$ C. $\frac{5}{16}$ D. $\frac{3}{5}$

7. $2\frac{1}{3} \times 4\frac{2}{3} =$

A. $\frac{98}{9}$ B. $2\frac{1}{3}$ C. $8\frac{1}{3}$ D. 9

8. $\frac{2}{3} \times \frac{1}{2} \times \frac{4}{5} =$

A. $\frac{8}{30}$ B. $\frac{12}{30}$ C. $\frac{3}{4}$ D. $\frac{7}{15}$

DECIMALS

Reading a decimal number aloud will help you understand the number. For example, read 0.3 as three tenths, 0.56 as fifty-six hundredths, 0.235 as two hundred thirty-five thousandths. Remember what each place value means in a decimal number.

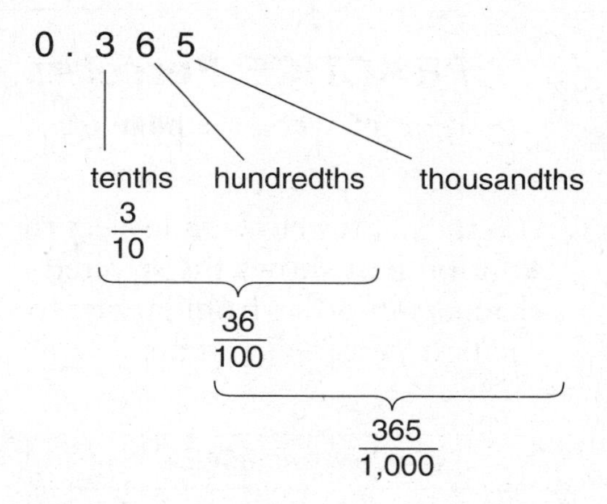

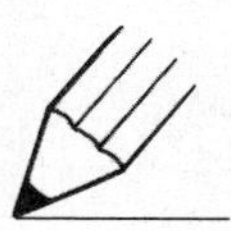

PRACTICE: Ordering Decimals by Size

(For answers, see page 60.)

1. Use the chart below and select the solution that shows the different sounds in order from loudest to quietest (the larger the number, the louder the sound).

Sound	Power (watts/sq. meter)
Door slamming in a room	0.3
Pain threshold; jet aircraft taking off 328 feet (100 meters) away	1.0
Falling leaves 3.28 feet (1 meter) away	0.00009
Rock concert	3.0
Normal conversation	0.03
Permanent ear damage, rocket taking off 328 feet (100 meters) away	300

A. 0.3, 3.0, 0.00009, 0.03
B. 3.0, 0.3, 0.03, 0.00009
C. 0.00009, 0.03, 0.3, 3.0
D. 3.0, 0.03, 0.3, 0.00009

2. The chart below describes the ability of different substances to conduct heat. The lower the conductivity number, the poorer that substance is as a conductor of heat. (Low number = bad conductor of heat.) What substance would be the worst conductor of heat?

Substance	Conductivity
Copper	385.0
Brick	0.6
Water	0.6
Wool	0.040
Gold	296.0
Air	0.025
Wood (oak)	0.15

A. wood
B. water
C. wool
D. air

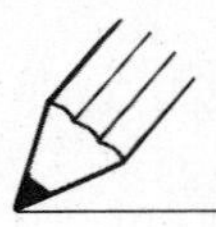

PRACTICE: Percents

(For answers, see page 61.)

1. Use the chart below and select the solution that shows the selected elements in order from largest to smallest percentage of air.

Composition of air	
Element	% of air
Argon (Ar)	0.93%
Carbon dioxide (CO_2)	0.03%
Helium (He)	0.0005%
Krypton (Kr)	0.00001%
Neon (Ne)	0.0018%
Nitrogen (N_2)	78%
Oxygen (O_2)	21%
Other gases	0.037695%

A. Krypton, Neon, Oxygen, Helium
B. Oxygen, Neon, Helium, Krypton
C. Oxygen, Neon, Krypton, Helium
D. Krypton, Helium, Neon, Oxygen

2. Use the circle graph below to determine the decimal number equal to the amount of oxygen plus carbon in our body.

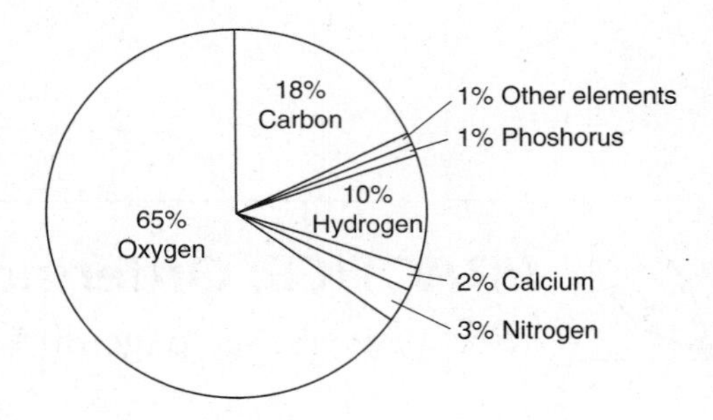

A. 8.30
B. 0.83
C. 0.083
D. 0.0083

3. The circle graph below shows the composition of the Earth's crust. What fraction is equivalent to the amount of *calcium* in the Earth's crust?

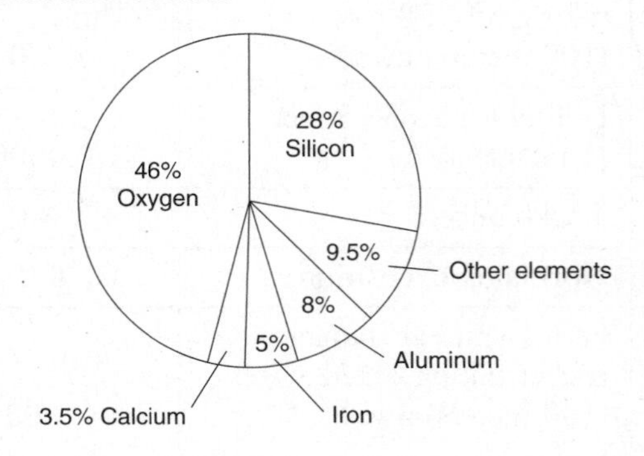

A. $\frac{35}{10}$

B. $\frac{35}{100}$

C. $\frac{35}{1{,}000}$

D. $\frac{35}{10{,}000}$

ESTIMATION

Most times in real life we don't have a pencil and paper or a calculator handy. Still, we need to know about how much money we'll need when we're selecting items to buy, or deciding if we have enough milk for tomorrow's breakfast. At these times, we use *estimation* to help us make decisions. You can become a very good estimator if you have a good number sense and if you learn a few strategies.

Here are some strategies:

- Plan, rearrange, group, and add 10s — $2 + 4 + 8 + 6 + 2 + 3 + 5$
  - Rearrange — $2 + 8 + 4 + 6 + 2 + 3 + 5$
  - Group — $2 + 8 \quad 4 + 6 \quad 2 + 3 + 5$
  - Add 10s — $10 + 10 + 10 = 30$
- Round numbers — $198 + 206 + 549$
  $200 + 200 + 550 =$ about 950
- Round decimals — $14.7 + 10.2 + 4.8 + 200.045$
  $15 + 10 + 5 + 200 =$ approximately 230
- Round money — You want to be sure you have enough money, so it is a good idea to round up when you are estimating money you will need.
  $\$16.59 + \$2.50 + 9.60$
  $17 + 3 + 10$
  $=$ approximately \$30
- Fractions — Sometimes you can compare fractions to $\frac{1}{2}$ to put them in order by size.

$\frac{4}{6}$	$\frac{3}{6}$ would $= \frac{1}{2}$	so $\frac{4}{6}$ is a **little more than** $\frac{1}{2}$
$\frac{7}{8}$	$\frac{4}{8}$ would $= \frac{1}{2}$	so $\frac{7}{8}$ is **much more than** $\frac{1}{2}$
$\frac{3}{10}$	$\frac{5}{10}$ would $= \frac{1}{2}$	so $\frac{3}{10}$ is **less than** $\frac{1}{2}$

From smallest to largest:		$\frac{1}{2}$			**1 whole**
Compare to $\frac{1}{2}$ or *Compare to 1 whole*		$\frac{3}{6}$	$\frac{4}{6}$		$\frac{6}{6}$
		$\frac{4}{8}$		$\frac{7}{8}$	$\frac{8}{8}$
	$\frac{3}{10}$	$\frac{5}{10}$			$\frac{10}{10}$

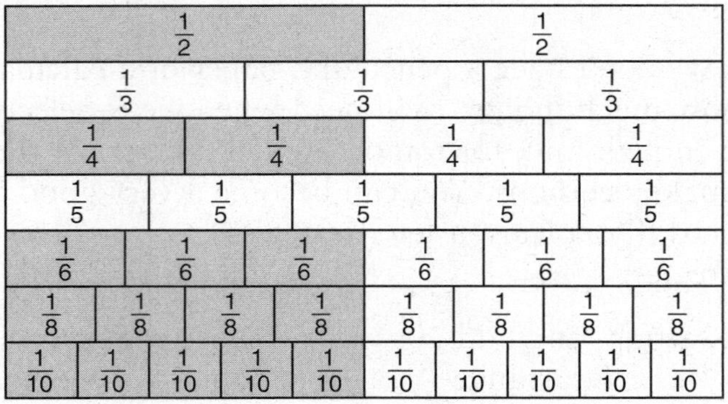

$\frac{1}{2}$ $\frac{1}{2}$

$\frac{1}{3}$ $\frac{1}{3}$ $\frac{1}{3}$

$\frac{1}{4}$ $\frac{1}{4}$ $\frac{1}{4}$ $\frac{1}{4}$

$\frac{1}{5}$ $\frac{1}{5}$ $\frac{1}{5}$ $\frac{1}{5}$ $\frac{1}{5}$

$\frac{1}{6}$ $\frac{1}{6}$ $\frac{1}{6}$ $\frac{1}{6}$ $\frac{1}{6}$ $\frac{1}{6}$

$\frac{1}{8}$ $\frac{1}{8}$ $\frac{1}{8}$ $\frac{1}{8}$ $\frac{1}{8}$ $\frac{1}{8}$ $\frac{1}{8}$ $\frac{1}{8}$

$\frac{1}{10}$ $\frac{1}{10}$ $\frac{1}{10}$ $\frac{1}{10}$ $\frac{1}{10}$ $\frac{1}{10}$ $\frac{1}{10}$ $\frac{1}{10}$ $\frac{1}{10}$ $\frac{1}{10}$

COMPARING AND ESTIMATING RATIONAL NUMBERS

Fractions, Decimals, and Percents: Always, always, always compare things that are the same type of number. If you need to compare fractions with fractions, first change all to the same denominator. If you need to compare fractions and decimals, change all to fractions **or** all to decimals. If you need to compare decimals and percents, change all to decimals **or** all to percents, then compare them.

Examples

A. What number is closest to $\frac{1}{2}$?

A. 0.48 B. 0.051 C. 0.620 D. 0.58

Remember to change all numbers to the same type of number. Change the fraction to a decimal number. ($\frac{1}{2}$ = .50 or .500) Then write all the decimal numbers to three places 0.480 0.051 0.620 0.580 so you can compare them easily. Notice that 0.480 is closest to 0.500, so the answer is A.

B. **Fractions:** Put these fractions in order from smallest to largest: $\frac{2}{3}$ $\frac{4}{5}$ $\frac{1}{4}$

Because you can use a calculator on some parts of the NJ ASK 8 test, it is easiest to change all to decimal numbers and then compare them. Remember the fraction bar means divide.

$\frac{2}{3}$ = 2 divided by 3 = 0.666 $\frac{4}{5}$ = 4 divided by 5 = 0.80 $\frac{1}{4}$ = 1 divided by 4 = 0.25

If you have a decimal number to three places (like .666), then write your other decimal numbers to three places, too. This way you can compare them easily.

The correct order is 0.250 0.666 0.800

or using the original fractions $\frac{1}{4}$ $\frac{2}{3}$ $\frac{4}{5}$

C. **Decimals:** Put these in order from smallest to largest:

40.009 40.30 40.12 40.05

Because all of these are 40 + the decimal component, just look at the numbers to the right of the decimal point. Write them all to three places, and then compare:

.009 .300 .120 .050

The correct order is 40.009, 40.050, 40.120, 40.300

D. **Fractions and Decimals:** Put these in order from lowest to highest:

$.53 \quad \frac{1}{4} \quad \frac{3}{5}$

- One strategy would be to *change all numbers to fractions with 100 as the denominator.*

$0.53 = \frac{53}{100}$

$\frac{1}{4}$ = 1 divided by 4 = 0.25 which = $\frac{25}{100}$

$\frac{3}{5}$ = 3 divided by 5 = 0.60 which = $\frac{60}{100}$

The correct order from smallest to largest is $\frac{25}{100}$ $\frac{53}{100}$ $\frac{60}{100}$

Using the original numbers the correct order is $\frac{1}{4}$ 0.53 $\frac{3}{5}$

- Another strategy would be to *change all numbers to decimals.*

0.53 $\frac{1}{4}$ = 1 divided by 4 = 0.25 $\frac{3}{5}$ = 3 divided by 5 = 0.60

The correct order is 0.25 0.53 0.60

Using the original numbers the correct order is $\frac{1}{4}$ 0.53 $\frac{3}{5}$

E. **Fractions and Percents:** Arrange these in order: 59% $\frac{2}{3}$ 25% $\frac{1}{5}$

When you are working with fractions and percents, it is usually easiest to change all to decimals and then compare them. Remember, the fraction bar means divide. Let your calculator help you here.

59% = 0.59 or 0.590 $\frac{2}{3}$ = 2 divided by 3 = 0.666

25% = 0.25 or 0.250 $\frac{1}{5}$ = 0.2 or 0.200

These are the decimals in order from smallest to largest: 0.200 0.250 0.590 0.670

These are the original numbers in order from smallest to largest: $\frac{1}{5}$ 25% 59% $\frac{2}{3}$

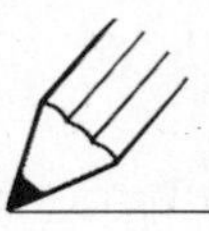

PRACTICE: Comparing and Estimating Rational Numbers

(For answers, see page 61.)

Show all work as you solve these problems. Do not use a calculator here.

1. Maria is inviting friends over after school and she wants to buy 3 bottles of soda, 2 bags of pretzels, and 1 bag of chips. Soda costs $1.05 per bottle, pretzels are $.89 a bag, and chips are $1.98 per bag. Use your estimation skills to decide the least amount of money she needs:

A. $8.00
B. $7.00
C. $6.00
D. $5.00

2. Willy has $5. Which of the combinations listed below would he NOT have enough money to buy?

Menu	
Hot dog	1.50
Burger	2.50
Extra cheese	.50
Soda	1.10
Milk	1.20
Apple	.70
Pretzel	1.10

A. one hot dog, one burger, and one soda
B. one burger, one extra cheese, and one soda
C. one burger, one milk, and one apple
D. one burger, one pretzel, and one soda

3. Jane and her mom just had dinner at a local restaurant. The bill is $16.40. If the tax is 8%, about how much money will they need? Use your estimation skills here.

A. $17.00
B. $18.00
C. $19.00
D. $20.00

4. What plan would be best to estimate the total number of miles for John's daily walk this week?

John's Daily Walks	
Monday	4.6 miles
Tuesday	5.3 miles
Wednesday	2.8 miles
Thursday	6.7 miles
Friday	2.5 miles

A. $(4 + 5) + (2 + 6 + 2) =$
$9 + 10 = 19$

B. $(4 + 5) + 2 + (7 + 3) =$
$9 + 2 + 10 = 21$

C. $(5 + 5) + 3 + (7 + 3) =$
$10 + 3 + 10 = 23$

D. $(5 + 5) + (3 + 6) + 2 =$
$10 + 9 + 2 = 21$

5. Use the chart to answer the following questions.

Item	Amount Recycled
Tires	$\frac{3}{4}$
Glass	$\frac{2}{5}$
Cans	$\frac{5}{8}$
Paper	$\frac{7}{10}$

a. Which item do we recycle the most?

A. tires
B. glass
C. cans
D. paper

b. Which has a recycle rate less than $\frac{1}{2}$?

A. tires
B. glass
C. cans
D. paper

c. Select the answer with the least and most recycled.

A. glass/cans
B. glass/paper
C. cans/paper
D. glass/tires

6. Jeff's dad has drill bits labeled in inches as follows: $\frac{7}{16}$, $\frac{3}{8}$, $\frac{5}{32}$, $\frac{9}{16}$, and $\frac{1}{4}$. His dad asked him to arrange them in order from least to greatest. Which arrangement is correct?

A. $\frac{1}{4}$ $\frac{5}{32}$ $\frac{3}{8}$ $\frac{7}{16}$ $\frac{9}{16}$

B. $\frac{5}{32}$ $\frac{1}{4}$ $\frac{3}{8}$ $\frac{7}{16}$ $\frac{9}{16}$

C. $\frac{1}{4}$ $\frac{3}{8}$ $\frac{7}{16}$ $\frac{9}{16}$ $\frac{5}{32}$

D. $\frac{9}{16}$ $\frac{7}{16}$ $\frac{3}{8}$ $\frac{1}{4}$ $\frac{5}{32}$

7. Jeff needs to find a drill bit just a little larger than $\frac{1}{2}$ inch. Which would be the best one to choose? (Remember, you can change all to decimals to compare them easily.)

A. $\frac{1}{4}$ inch

B. $\frac{3}{8}$ inch

C. $\frac{5}{32}$ inch

D. $\frac{9}{16}$ inch

8. Which fraction is larger than $\frac{1}{4}$ and smaller than $\frac{2}{5}$?

A. $\frac{1}{3}$
B. $\frac{5}{9}$
C. $\frac{2}{3}$
D. $\frac{1}{2}$

9. Carlos, Jim, Terrell, and Zak are in a special 18-mile marathon. Carlos plans to run $4\frac{1}{4}$ miles, Jim will run $5\frac{1}{3}$ miles, and Terrell will run $3\frac{1}{8}$ miles. How many miles will be left for Zak to run?

A. about 4 miles
B. just over 5 miles
C. more than 6 miles
D. about 13 miles

10. Mikaela works part time at the public library. She has kept a list of the books that were borrowed last week and she made a circle graph of the information.

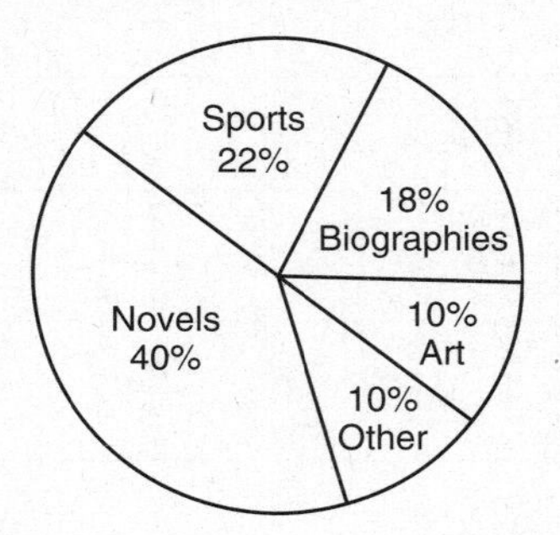

If 450 books were borrowed last week *about* how many more novels were taken out than biographies?

A. 80
B. 100
C. 260
D. 180

RATIO AND PROPORTION

A fraction is a ratio. Two ratios connected with an equals sign is called a *proportion*. You can understand the concept of proportion by looking at some drawings.

Examples

A. **Proportion:** In Figure A, the person is not in proportion to the house. When you compare the height of the person to the door, the person looks too big. In Figure B, the person seems to be in proportion to the door. In this figure, the person and the house seem to be in the correct proportion.

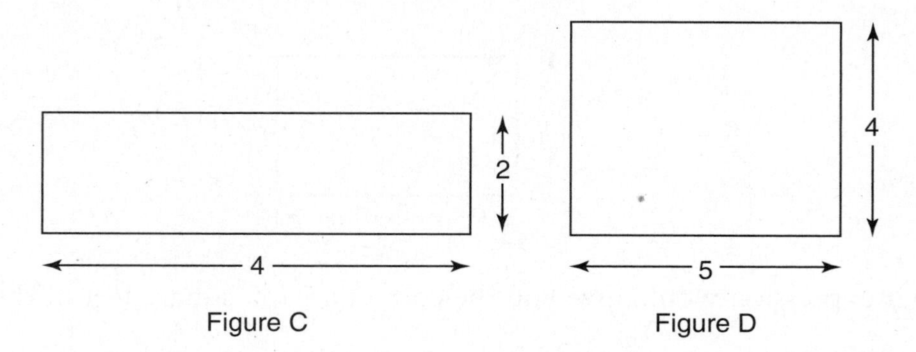

Figure C Figure D

B. Figure C is a long thin rectangle; Figure D is almost a square. These shapes are *not* in proportion. In Figure C, the rectangle has length = 4 and height = 2. In Figure D, the rectangle has length = 5 and height = 4.

Figure C: $\frac{\text{Length } 4}{\text{Height } 2}$ Figure D: $\frac{\text{Length } 5}{\text{Height } 4}$ $\frac{4}{2} \neq \frac{5}{4}$

These shapes are *not* in proportion.

C. Ratio and proportion are often used in drawings of maps or architectural drawings. For example, if the architect's plans are in the scale of $\frac{1}{3}$, it means that 1 inch on the plans equals 3 feet in real life. How long is a garage in reality if the plans show it to be 8 inches long?

$\frac{1}{3} = \frac{8}{x}$ In this proportion, both numerators represent the scale from the plans (1 inch and 8 inches). The denominators give you the actual size in real life (3 feet and x feet). We can solve this problem by cross multiplying

$\frac{1}{3} \times \frac{8}{x}$ $\frac{\text{inches}}{\text{feet}}$ $(1)(x) = (3)(8)$

$x = 24$ feet

So, if 1 inch represents 3 feet long, then 8 inches represents 24 feet. The garage is 24 feet long.

D. The builder is showing plans of a new housing development to a future buyer. The buyer notices that the plans show that the scale used is $\frac{1}{4}$ inch = 25 feet. Look at the scale drawing below; this represents the lot for one single-family house.

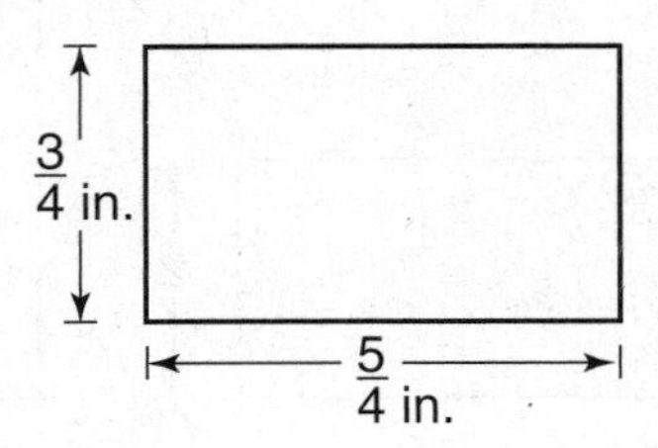

What expression would give you the correct area in square feet of this lot?

A. 125×75 B. $\frac{3}{4} \times \frac{5}{4}$ C. 3×25 D. 75×5

Since $\frac{1}{4}$ = 25 ft, the width = $\frac{3}{4}$ inch on paper and 75 feet in reality and the length is $1\frac{1}{4}$ inches or $\frac{5}{4}$ inches on paper and 125 feet in reality.

$$\text{Area of a rectangle} = \text{Length} \times \text{Width} = 125 \times 75$$

Solution A is correct.

E. You need 1 cup of pancake mix to make 6 pancakes. How many cups of mix will you need to make 18 pancakes? Which is the correct proportion to solve this problem?

A. $\frac{1}{6} = \frac{18}{x}$ B. $\frac{1}{6} = \frac{x}{18}$ C. $\frac{6}{1} = \frac{x}{18}$ D. $\frac{6}{1} = \frac{12}{x}$

$$\frac{1 \text{ cup mix}}{6 \text{ pancakes}} = \frac{x \text{ cups mix}}{18 \text{ pancakes}} \qquad (1)(18) = (x)(6)$$

$$18 = 6x$$

$$3 = x$$

(3 cups of mix will give you 18 pancakes)

The correct answer is B $\frac{1}{6} = \frac{x}{18}$; $6x = 18$, $x = 3$

PRACTICE: Ratio and Proportion

(For answers, see page 62.)

1. If there are 14 girls in the class and 11 boys, what is the ratio of girls to the total number of students in the class?

   A. $\frac{25}{14}$ B. $\frac{14}{25}$ C. $\frac{14}{11}$ D. $\frac{11}{14}$

2. If 4 out of 7 people in a New Jersey town use Brand X detergent, find the approximate number of people that use Brand X if there are 5,271 people in the town.

   A. 1,757
   B. 3,012
   C. 5,250
   D. 9,975

3. At college, Will gets discounted meals; he spends $2.00 for breakfast $3.00 for lunch, and $7.00 for dinner each weekday. If he makes $100 each week at his library job, write a ratio that shows the amount of money spent on food Monday through Friday as compared to the total amount of money he makes each week.

   A. $\frac{12}{100}$ B. $\frac{60}{100}$ C. $\frac{72}{100}$ D. $\frac{84}{100}$

4. Charleen is enlarging the figure shown so that the base of the new figure changes from 9 to 18 centimeters.

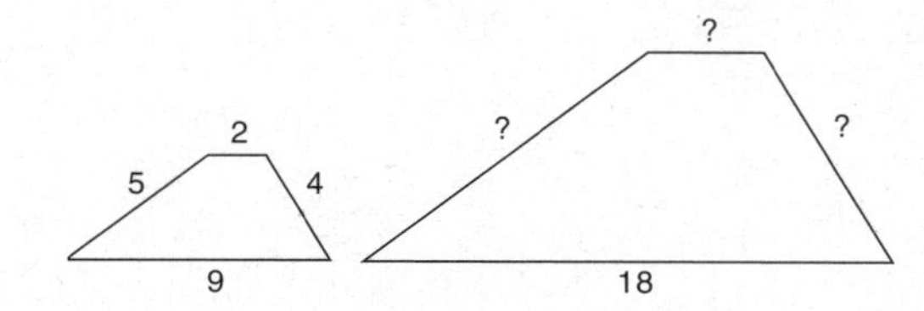

Note: Figures are not drawn to scale.

   What is the perimeter of the enlarged figure?

   A. 29 cm
   B. 38 cm
   C. 40 cm
   D. 42 cm

5. If triangles are similar, then their sides are in proportion. Look at these two *similar* triangles. The small triangle has a height of 3 feet. How tall is the larger triangle?

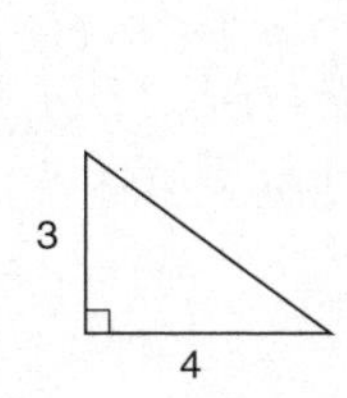

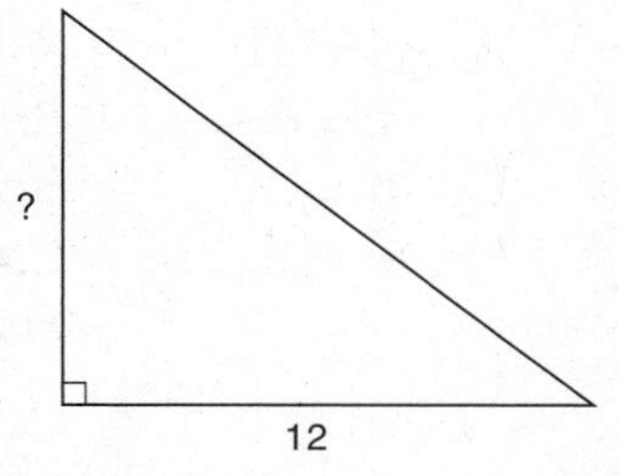

   A. 4 feet
   B. 6 feet
   C. 8 feet
   D. 9 feet

6. Trey wants to design a model car. The diameter of the real tire is 28″, and the diameter of the hubcap is 15″. If he makes a drawing where the tire is 7″ in diameter, which proportion would help him find the diameter of the hubcap in his drawing?

A. $\frac{28}{15} = \frac{x}{7}$

B. $\frac{28}{15} = \frac{7}{x}$

C. $\frac{15+x}{28+7}$

D. $\frac{\text{Real tire}}{\text{Real hubcap}} = \frac{\text{Drawn hubcap}}{\text{Drawn tire}}$

7. Jaime Escalante became a well-known math teacher from California because of his ability to inspire his students to succeed even though they thought they could not. Solve the following proportion example to find out the name of the movie made about this group.

$$\frac{8}{2} = \frac{m}{2.5}$$

A. $m = 10$ *Stand and Deliver*
B. $m = 6.4$ *Blackboard Jungle*
C. $m = 20$ *Beverly Hills High*
D. $m = 8.25$ *California Black Hawks*

8. Solve the following proportion example for c.

$$\frac{6}{c} = \frac{0.3}{9}$$

A. $c = 0.2$ B. $c = 2.0$
C. $c = 180$ D. $c = 18.0$

9. In real life, Mike's ladder is 24 feet tall. At the base, it is leaning 5.4 feet away from the wall of his house. In the smaller drawing, the ladder is only 6 inches tall. How far away from the house should it be drawn to keep the picture in proportion to the real thing?

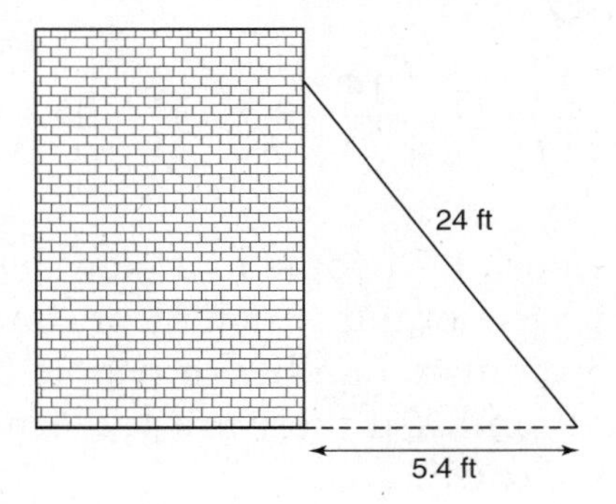

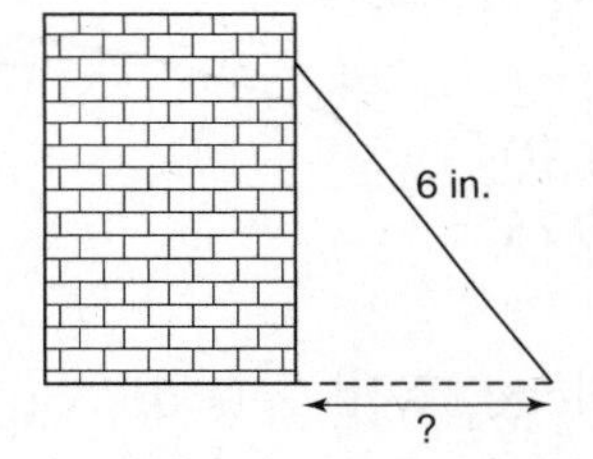

Note: Figures are not drawn to scale.

A. 0.135″ B. 1.35″
C. 0.266″ D. 2.66″

10. Solve this proportion problem to find out the name of the architect who designed the John F. Kennedy Library in Boston, Massachusetts.

$$\frac{5}{j} = \frac{12}{3} \qquad j = ?$$

A. 20 Jin H. Kinoshita
B. 1.25 Ieoh Ming Pei
C. 0.80 An Wang
D. 8.0 Maya Ying Lin

PRACTICE SCR NONCALCULATOR QUESTIONS

Each question is worth 1 point. No partial credit is given.
(For answers, see page 63.)

1. Using the correct algebraic order of operations, what would be the first step you should do to simplify this expression?

$$10 + 8 \div 2 + 3(5 - 2) - 4^2$$

Answer: __________

2. To add the fractions below, Jessica says you should use 12 as the lowest common denominator but Sarah says that 18 is the lowest common denominator. Who is correct?

$$\frac{1}{6} + \frac{2}{9}$$

Answer: __________

3. The center of the moon is 238,900 miles away from the center of Earth. Write these miles in scientific notation.

Answer: __________

4. Put these numbers in numerical order starting with the smallest number.

$$3^2 \quad \pi \quad \sqrt{9} \quad 2^3 \quad \sqrt{3}$$

Answer: __________

5. What is 0.85 written as a fraction?

Answer: __________

6. On an official New Jersey map, the scale shows 0.5 inches for every 10 miles. How far apart are two towns if the map shows them as 2.5 inches apart?

Answer: __________

7. If your dad receives an average of 40 e-mails each day at work, how many e-mails can he expect to receive during a month? (For this example, we'll let 4 weeks = 1 month and each week = 5 workdays)

Answer: __________

Use the menu below to answer questions 8 and 9.

LUNCHEON MENU

(Special Today: NO SALES TAX)

Item	Price
Hamburger plain	\$2.50
Cheeseburger	\$3.00
Chicken nuggets	\$3.25
Hot dog plain	\$2.00
Hot dog with chili	\$2.50
French fried potatoes	\$1.50
Soda	\$1.00, \$1.50, \$2.00 (sm., med., large)
Juice	\$1.50, \$2.00 (sm., med.)

8. If Max had \$5.00, would he have enough money to buy a hamburger, fries, and a small soda?

   Answer: ___________

9. Jose has a \$10 bill. How much change would he receive if he bought chicken nuggets, fries, and a medium juice?

   Answer: ___________

10. Which is closet to 1?

    $\frac{1}{2}$ 40% $\frac{20}{40}$ $\frac{8}{9}$

    Answer: ___________

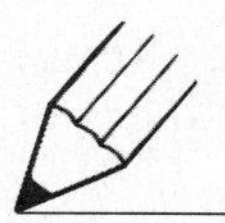

PRACTICE ECR QUESTIONS

(For answers, see page 63.)

1. A New York City parking garage charges $40.00 to park for the whole day, or $6.00 for the first hour and $4.00 for each additional half-hour.
   - If Bernice and her mom were taking a trip to the New York City Planetarium, they would need to park for 3 hours. Which plan should they use? Show your work and explain your answer.
   - Next month they plan to go to the theater and a dinner in New York City and expect to be there for 7 hours. Show how the $40.00 per-day rate would be the best choice.

2. Sam's brother, Daryl, works at an automobile factory and makes $12.00 per hour. Since the company was not selling as many cars lately, they decreased each person's wages by 8%.
   - What will Daryl's new hourly wage be?
   - If Daryl worked a typical 40-hour week, how much less will his weekly wages be with this 8% reduction?

3. Brett decided to have a big breakfast at the Ridge Diner. Below is a copy of the diner's breakfast menu.

THE RIDGE DINER MENU

Juice
(orange, grapefruit, tomato, or cranberry)
$1.20

Eggs any style
One $1.25 Two $2.25 Three $3.25

Toast or Plain Muffin
$1.20

Coffee or Tea with refill
$1.25

Milk
Small $.95 Large $1.25

Side Dishes
Sausage or Bacon $1.50
Homefries $1.00

Morning Special
(5:00 A.M.–11:00 A.M., MONDAY–FRIDAY)
Juice, toast or muffin, coffee or tea,
2 eggs any style, sausage or bacon,
and homefries
$6.95

   - How much money would Brett save if he ordered the Morning Special instead of buying each item separately?
   - How much would his total bill be with a 6% sales tax and a tip of $1.50? (Round your answer up to the nearest penny.)
   - How much change would he get from $10.00?

4. In a recent survey about how many people attended a local amusement park, it was determined that 3 out of 8 customers had also attended the park the week before.

   - What percentage of customers attended the park the week before?
   - If 4,000 people attended the park this week, how many of these people can we estimate also attended the week before?

5. Mr. B. takes excellent care of his car. He gets the oil changed every 5,000 miles, and has the tires rotated every 10,000 miles.

   - If he drove his car for 50,000 miles, how many trips would he make to the service station for the above two services?
   - If each of these trips to the service station is 8 miles, each round-trip would then be 16 miles. If Mr. B's car gets 20 miles per gallon and gasoline costs $2.45 per gallon, how much would he spend in gas for the above trips to the service station?

6. The Sweater Barn in an upstate New York mall and the Teen Outlet in New Jersey are both having big sweater sales.

Sweater Barn N.Y.	Sweater $24 less 15%	Matching cardigan is 50% off of the sweater sale price
Teen Outlet N.J.	Sweater $18 less 10%	Matching cardigan is the same price as the sweater

Mary wants to purchase three sets as birthday gifts for her granddaughters. She notices that the New York State store charges 8% sales tax on clothing and New Jersey has no sales tax on clothing.

   - What will the three sets cost at the Sweater Barn in New York?
   - What will the three sets cost at the Teen Outlet in New Jersey?
   - Which is the better buy? Explain.

SOLUTIONS TO PRACTICE QUESTIONS

COMPARE AND ORDER INTEGERS (page 14)

1. A $-2, 2, 6$ is in order from least to greatest
2. A $-4, -6, -8$ is in order from greatest to least
3. C $-3, 0, 2, 5$ is in order from least to greatest

POSITIVE AND NEGATIVE INTEGERS (page 15)

1. C -2
2. D $+10$
3. D $+1{,}200$
4. $+2$
5. $+45$
6. A Dan Dan had $100 - 24.50 = \$75.50$; Jay had $100 - 27 = \$73.00$.

INEQUALITIES AND EQUALS (page 16)

1. $-4 < 0$
2. $10 > 3$
3. $0 < 3$
4. $2 > -2$
5. $(5)(3) = (3)(5)$ or $15 = 15$
6. $16 - 3 < 13 + 6$ or $13 < 19$
7. $0 > -7$
8. $10 - 2 = 2^3$ or $10 - 2 = 8$ and $2^3 = 2 \times 2 \times 2 = 8$

ABSOLUTE VALUES (page 17)

1. A $|-16|$ $|-16| = 16$; $|5|=5$; $16 > 5$, 16 is the greater distance.
2 B $|-20|$ $|-20| = 20$; $|-6| = 6$; $20 > 6$; 20 is greater.
3. B $|-20|$ The -20 would be measuring beneath the water, and the absolute-value sign tells you it is distance with a positive value. This could represent 20 feet.

RATIONAL AND IRRATIONAL NUMBERS (page 18)

1. B $\sqrt{6}$ The calculator shows [2.44948743]. There seems to be no repeating pattern. This is an irrational number.

2. B $\sqrt{8}$ The calculator shows [2.828427125]. There seems to be no repeating pattern. This is an irrational number.

ADDING, SUBTRACTING, MULTIPLYING, AND DIVIDING INTEGERS (page 22)

1. B 72
2. B 20
3. C –15
4. D 10
5. B –10
6. C 5
7. A 9
8. B 1.35
9. B –15.6
10. D 10
11. –18
12. 8
13. 5
14. 5
15. –10

EXPONENTS (page 24)

1. B 6^2
2. C five cubed
3. D $3 \times 3 \times 3 \times 3$
4. D $8^2 - 4^2 = 8 \times 8 - 4 \times 4 = 64 - 16 = 48$ is the largest number.
   $4^2 = 16$; $(4)(2) = 8$; $2 \times 2 \times 2 \times 2 = 16$; $240 = 1$
5. C $(9)(3)(3) = (9)(9) = 81$ 9^2 also $= (9)(9) = 81$
6. 2^3
7. 8^2
8. 8^3

SQUARE ROOTS AND CUBE ROOTS (page 28)

1. 5 $5 \times 5 = 25$
2. 9 $9 \times 9 = 81$
3. 6
4. 12
5. 7
6. 10
7. B $5 + 10$
8. D $\sqrt{36} + \sqrt{25} = 6 + 5 = 11$
9. A $16+5$ $4^2 + \sqrt{25} = 16 + 5$
10. C $8 + 7$ $2^3 + \sqrt{49} = 2 \times 2 \times 2 + 7 = 4 \times 2 + 7 = 8 + 7$
11. D $\sqrt{100} - 1^2 = 10 - 1 = 9$

12. B A. $2^3 + (4)(2) = 2 \times 2 \times 2 + 8 = 8 + 8 = 16$
B. $\sqrt{4} = 2$
C. $5^2 - 3^2 = 25 - 9 = 16$
D. $6^2 - (10)(2) = 36 - 20 = 16$

ESTIMATING SQUARE ROOTS (page 29)

1. D $\sqrt{50}$ is between $\sqrt{49}$ and $\sqrt{64}$, $\sqrt{49} = 7$, and $\sqrt{64} = 8$.
2. B $\sqrt{15}$ is a little less than 4. $\sqrt{16}$ is 4 . $\sqrt{15}$ is about 3.87.
3. C $\sqrt{30}$ because $\sqrt{25} = 5$ and $\sqrt{36} = 6$; 30 is between 25 and 36.
4. C $\sqrt{115}$ is between $\sqrt{100}$ and $\sqrt{144}$. (It is between 10 and 12.)
5. B $\sqrt{100} = 10$ The only choice that does NOT equal 10 is B.
A. (2) (5) =10 B. 100/2 = 50
C. $\sqrt{4} + 2^3 = 2 + 8 = 10$ D. $4^2 - 6 = 16 - 6 = 10$

SCIENTIFIC NOTATION (page 31)

1. See the bold answers in the chart.

Standard Form	How Many Places Will You Move the Decimal Point?	Scientific Notation Form
650,000,000,000,000	14	**6.5×10^{14}**
2,000,000,000	**9**	2.0×10^9
45,000,000,000	**11**	4.5×10^{11}
3,250,000,000	9	3.25×10^9
12,000,000,000	**10**	**1.2×10^{10}**

2. A 9.45×10^{12} because you move the decimal point 12 places 9,450,000,000,000.
3. C 6.5×10^{11}
4. B 110,000,000
5. B 2.8×10^8 Barbara is correct. A number in scientific notation format has only one digit to the left of the decimal point.

ALGEBRAIC ORDER OF OPERATIONS (page 33)

1. B Divide (B) 24 because $27 - 3 = 24$
2. C Divide (D) 55 $36 + 9 \times 3 - 8 = 36 + 27 - 8 = 63 - 8 = 55$
3. A 12 $(72 \div 9 - 2) + 2 \times 3 = (8 - 2) + 6 = 6 + 6 = 12$
4. C 58 $(20 + 8 \div 4) + (3^2 \times 4) = (20 + 2) + (9 \times 4) = 22 + 36 = 58$
5. B 16 $[(2(4) + 3(4)] - 4 = 8 + 12 - 4 = 20 - 4 = 16$

6.	C	4	$(6 - 4)^3 - 4 = 2^3 - 4 = 8 - 4 = 4$
7.	D	2×4	Work inside Parenthesis first; Multiply before Subtracting.
8.	A		$(2 + 3) \times 2 \times 5 - 1 = 5 \times 2 \times 5 - 1 = 10 \times 5 - 1 = 50 - 1 = 49$
9.	D		$4 + 3 \times 2 \times (5 - 1) = 4 + 3 \times 2 \times (4) = 4 + 6 \times (4) = 4 + 24 = 28$
10.	Billy		In the expression $2(2 + 5 \times 2)$ you should work inside the Parentheses first, then Multiply before Adding: $2(2 + 10) = 2(12) = 24$
11.	A	−15	$(-3 - 4) + (-6 - 2) = -7 + -8 = -15$
12.	C	−32	$(-9 + 2) + (-30 + 5) = -7 + -25 = -32$
13.	B	−2	$(-8 + 2) + (-4 + 8) = -6 + 8 = -2$
14.	A	−28	$(-16 + 2)(2) = (-14)(2) = -28$
15.	B	−18	$(-2 - 4)(3) = (-6)(3) = -18$
16.	A	−3	$(-18) \div (10 - 4) = -18 \div 6 = -3$
17.	C	3	$(-16 + 4) \div (-2 - 2) = -12 \div -4 = 3$
18.	A		$(10 - 2) \div (2^2) = 8 \div 4 = 2$
19.	B		$2 + (8 - 6) = 2 + 2 = 4 = 2 \times 2 = 2^2$
20.	B		$-6 + 4 + (2)(5 - 3) = -6 + 4 + 2(2) = -6 + 4 + 4 = -6 + 8 = 2$

ADDING, SUBTRACTING, MULTIPLYING, AND DIVIDING FRACTIONS (page 40)

1.	C	$\frac{3}{4}$	
2.	A	$\frac{1}{12}$	
3.	C	11	
4.	B	$2\frac{3}{4}$	($2\frac{6}{8}$ reduces to $2\frac{3}{4}$)
5.	A	$\frac{2}{5}$	($\frac{4}{10}$ reduces to $\frac{2}{5}$)
6.	B	$1\frac{1}{4}$	($\frac{5}{4}$ simplifies to $1\frac{1}{4}$)
7.	A	$\frac{98}{9}$	
8.	A	$\frac{8}{30}$	($2 \times 1 \times 4 = 8$ for numerator and $3 \times 2 \times 5 = 30$ for denominator)

ORDERING DECIMALS BY SIZE (page 41)

1.	B	3.0, 0.3, 0.03, 0.00009 or 3, 3/10, 3/100, 9/100,000 rock concert, door slamming, normal conversation, falling leaves	
2.	D	air	0.025, which is 25/1000 (wool = 0.040, wood = 0.15, water = 0.6)

PERCENTS (page 42)

1. B Oxygen, Neon, Helium, Krypton (0.21, 0.0018, 0.0005, 0.00001)
2. B 0.83 65% + 18% = 83% Oxygen + Carbon
3. C 3.5% = 0.035 (as a decimal number) = 35/1,000 (as a fraction)

COMPARING AND ESTIMATING RATIONAL NUMBERS (page 46)

1. B $7.00 Estimate: 3 sodas ($3.00), 2 pretzels ($2.00), 1 bag chips ($2.00)
2. A Hot dog 1.50 Choice B = $3.90; Choice C = $4.90; Choice D = $3.70
   Burger 2.50
   Soda 1.10
   5.10
3. B $18.00 To estimate the tax you can use 10% (or $1.60); $16.40 + $1.60 = $18.00
4. C (5 + 5) 3 + (7 + 3) First round off to nearest whole numbers, then group the numbers so you can add 10s.
5. Change all to decimals (use your calculator), or make common denominators and compare the numerators.

Tires	$\frac{3}{4}$	0.75	=	0.750	or	$\frac{30}{40}$
Glass	$\frac{2}{5}$	0.40	=	0.400	or	$\frac{16}{40}$
Cans	$\frac{5}{8}$	0.625	=	0.625	or	$\frac{25}{40}$
Paper	$\frac{7}{10}$	0.70	=	0.700	or	$\frac{28}{40}$

a. A tires

b. B glass $\frac{2}{5}$ is less than $\frac{1}{2}$ $(\frac{2}{5} = \frac{4}{10}, \frac{1}{2} = \frac{5}{10})$ or $(\frac{2}{5} = 0.40, \frac{1}{2} = 0.50)$

c. D glass/tires: glass, cans, paper, tires; 0.40, 0.625, 0.70, 0.75; or $\frac{16}{40}, \frac{25}{40}, \frac{28}{40}, \frac{30}{40}$

6. B $\frac{5}{32}$ $\frac{1}{4}$ $\frac{3}{8}$ $\frac{7}{16}$ $\frac{9}{16}$ or $\frac{5}{32}$ $\frac{1}{4}$ $\frac{3}{8}$ $\frac{7}{16}$ $\frac{9}{16}$

   0.15625 0.25 0.375 0.4375 0.5625 $\frac{5}{32}$ $\frac{8}{32}$ $\frac{12}{32}$ $\frac{14}{32}$ $\frac{18}{32}$

7. D $\frac{9}{16}$ $\frac{9}{12} > \frac{1}{2}$ because $\frac{1}{2} = \frac{8}{16}$ or $\frac{9}{16} = 0.5625 >$ because $\frac{1}{2} = 0.5000$

8. A $\frac{1}{4}$ $\frac{1}{3}$ $\frac{2}{5}$ or $\frac{1}{4}$ $\frac{1}{3}$ $\frac{2}{5}$

$\frac{20}{60}$ $\frac{15}{60}$ $\frac{24}{60}$ 0.250 0.333 0.400

9. B Just over 5 miles. First add the miles run by Carlos, Jim, and Terrell

$$4\frac{1}{4} + 5\frac{1}{3} + 3\frac{1}{8}$$

You can use 24 as a common denominator

$$4\frac{6}{24} + 5\frac{8}{24} + 3\frac{3}{24} = 12\frac{17}{24} \text{ miles}$$

Next, subtract from the total to get what is left over for Zak.

$$18 \text{ miles less } 12\frac{17}{24} = 5\frac{7}{24} \text{ miles or just over 5 miles}$$

or use decimals $4.25 + 5.33 + 3.125 = 12.705$ miles

$$18.00 - 12.71 = 5.29 \text{ miles left over for Zak}$$

10. B 100 Novels borrowed = 40% of 450 = $0.40 \times 450 = 180$
Biographies borrowed = 18% of 450 = $0.18 \times 450 = 81$
$180 - 81 = 99$ or about 100

RATIO AND PROPORTION (page 51)

1. B girls/total = $\frac{14}{25}$
2. B 3,012
3. B $\frac{60}{100}$
4. C 40 cm
5. D 9 feet $\frac{3}{4} = \frac{x}{12}$; therefore, $4x = (3)(12)$, $4x = 36$, $x = 9$
6. B
7. A $m = 10$ $2m = (2.5)(8)$, $2m = 20$, $m = 10$
8. C 180 $0.3c = (16)(9)$, $0.3c = 54$, $c = \frac{54}{0.3}$, $c = 180$
9. B 1.35″ $\frac{24}{5.4} = \frac{6}{x}$, $(6)(5.4) = 24x$, $32.4 = 24x$, $1.35 = x$
10. B 1.25 or $1\frac{1}{4}$ $\frac{5}{j} = \frac{12}{3}$, $12j = 15$ $j = \frac{15}{12}$ or $\frac{5}{4}$ or $1\frac{1}{4}$ or 1.25

SCR NONCALCULATOR QUESTIONS (page 53)

1. (5 – 2) — Remember **Pemdas** (*Please Excuse My Dear Aunt Sally*); work inside Parentheses first
2. Sarah
3. 2.389×10^5 miles away
4. $\sqrt{3}$ $\sqrt{9}$ π 2^3 3^2 — $\sqrt{3} \sim 1.7$ $\sqrt{9} = 3$ $\pi \sim 3.14$ $2^3 = 8$ $3^2 = 9$
5. $\frac{85}{100}$
6. 50 miles — $\frac{0.5}{10} = \frac{2.5}{50}$ or think that $0.5 \times 5 = 2.5$ miles then $10 \times 5 = 50$ miles
7. 800 e-mails — (5 days)(4 weeks) = 20 days; (40 e-mails)(20 days) = (40)(20) = 800
8. yes — burger = \$2.50, fries =\$1.50, small soda = \$1.00, for a total of \$5.00
9. \$3.25 — chicken nuggets = \$3.25, fries = \$1.50, medium juice = \$2.00, for a total of \$6.75; \$10.00 – \$6.75 = \$3.25
10. 8/9 — 1/2 = 0.5 or 50%, which is halfway to number 1
    40% = is less than 50%, it is less than halfway to number 1
    20/40 = 2/4 = 1/2 = 50%, it also is halfway to number 1
    8/9 is only 1/9 away from the number 1, so it is the closest

ECR QUESTIONS (page 55)

1. ▪ They should *not* use the whole-day rate. The whole-day rate would cost \$40.00; the other option would cost only \$30.00 [6 + 3(8)].

   ▪ The total with the second option would be \$6.00 + \$48.00 = \$54.00.

   The first hour would cost \$6.00.
   The next 6 hours would cost \$8.00 each, so (8)(6) = \$48.00.

2. ▪ His new hourly rate would be $11.04 instead of $12.

   (12)(0.08) = $.96, then $12.00 – $.96 = $11.04, or (12)(.92) = $11.04.

   ▪ His weekly salary would be $38.40 less with the 8% reduction.

   Original weekly salary: (12)(40 hrs.) = $ 480.00
   Reduced weekly salary: (11.04)(40 hrs.) = $441.60
   The difference is $480 – 441.60 = $38.40

3. ▪ He would save $2.45 buying the *Morning Special* instead of buying each item separately.

   1.20 + 1.20 + 1.25 + 3.25 + 1.50 + 1.00 = $9.40 if items are purchased separately.
   $9.40 – $6.95 (*Morning Special*) = $2.45 difference.

   ▪ His total bill, including 6% sales tax and a $1.50 tip, would be $8.87.

   ($6.95)(1.06) = 7.367 (round up to nearest penny); $7.37 + $1.50 = $8.87

   ▪ His change from $10.00 would be $1.13. 10.00 – 8.87 = 1.14

4. ▪ 37.5% (or $35\frac{1}{2}\%$) also attended the park the week before. $\frac{3}{8}$ = 0.375 or 37.5%

   ▪ Approximately 1,500 people who attended the park this week also attended the park the previous week. (0.375)(4,000 people) = 1,500 people

5. ▪ He would make 10 trips in all.

Oil changes at miles	500,	**1,000,**	1500,	**2,000,**	2,500,	**3,000,**	3,500,	**4,000,**	4,500,	**5,000**
Tires rotated at miles		**1,000,**		**2,000,**		**3,000**		**4,000**		**5,000**

   Notice that there are 5 times when he gets only his oil changed and there are 5 more times when he gets both an oil change and the tires rotated at the same time.

   ▪ He would spend $19.60 on gasoline for the 10 round trips to the service station.

   (10 trips)(16 miles each round trip) = 160 miles total
   (160 miles)(20 miles per gallon) = 8 gallons of gasoline used
   (8 gallons)($2.45 per gallon) = $19.60 on gasoline

6. ■ Sweater Barn at the upstate New York mall where there is an 8% sales tax on clothing:

   Cost for sweater: \$24 less 15% is same as \$24 × 0.85 = \$20.40 per sweater
   Cost for cardigan: \$20.40 × 50% is (1/2 of \$20.40) = \$10.20 per cardigan
   Total cost for one sweater and one cardigan before taxes = \$30.60 per set
   Three sets = (3)(\$30.60) = \$91.80
   Total for 3 sets with sales tax = (1.08)(\$91.80) = **\$99.15** (\$99.14 also accepted as correct) total with sales tax

   ■ Teen Outlet in New Jersey (where there is no sales tax on clothing):
   Cost for sweater: \$18 less 10% = \$16.20 for each sweater
   Cost for cardigan: Each cardigan is the same price = \$16.20 per cardigan
   Total cost for one sweater and one cardigan = (2)(\$16.20) = \$32.40 per set
   Total for 3 sets = (3)(\$32.40) = **\$97.20** total This would be the final price because there is no sales tax on clothing in New Jersey.

   ■ The Teen Outlet store in New Jersey would be a slightly better buy. The cost at the New Jersey store would be **\$1.95** (or \$1.94) less than the New York store.

Name: ______________________________ Date: ________________

Cluster I Test

35 minutes
(Use the *NJ ASK 8 Mathematics Reference Sheet* on page 265.)

SHORT CONSTRUCTED RESPONSE AND MULTIPLE-CHOICE QUESTIONS

DIRECTIONS FOR QUESTIONS 1 THROUGH 12: Each of the questions or incomplete statements below is followed by four suggested answers. Select the one that is the best in each case, and fill in the corresponding lettered circle. Be sure the circle is filled in completely so you cannot see the letter. Unless you are told to do so in the question, do NOT include sales tax in your answer to questions involving purchases.

1. Evaluate the following expression:

$$(-6)(-2) + 4^2$$

Answer: __________

2. Evaluate the following expression:

$$\sqrt{36} + 4^2 + |-20|$$

Answer: __________

3. Which rational number could be the value of x on this number line?

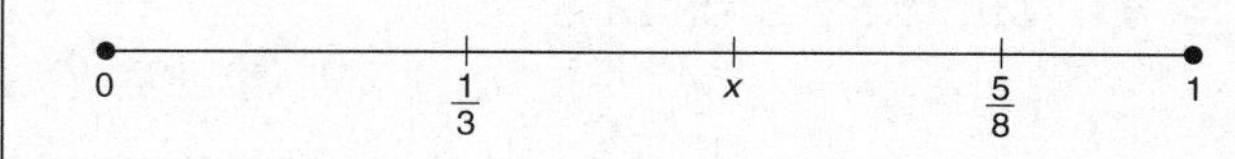

A. $\frac{1}{4}$

B. $\frac{2}{3}$

C. 0.7

D. 0.45

Ⓐ Ⓑ Ⓒ Ⓓ

4. The $\sqrt{115}$ is between

A. 8 and 9
B. 9 and 10
C. 10 and 11
D. 11 and 12

Ⓐ Ⓑ Ⓒ Ⓓ

5. The large number 68,500,000,000 has the same value as the following number written in scientific notation form.

A. 0.685×10^{11}
B. 6.85×10^{10}
C. 68.5×10^{9}
D. 685×10^{8}

Ⓐ Ⓑ Ⓒ Ⓓ

GO ON TO THE NEXT PAGE ➡

6. Which operation should you do first to evaluate the following expression?

$$16 \times 4 \div 2 + (8 - 2)^2 + 10^2$$

A. square the 10
B. divide 4 by 2
C. multiply 16 by 4
D. subtract 2 from 8

Ⓐ Ⓑ Ⓒ Ⓓ

7. Arrange the following integers in order from least to greatest:

$$\frac{1}{3} \quad 0.60 \quad \frac{2}{5} \quad 0.50$$

A. $\frac{1}{3}, \frac{2}{5}, 0.50, 0.60$

B. $0.60, \frac{2}{5}, \frac{1}{3}, 0.50$

C. $\frac{2}{5}, 0.60, \frac{1}{3}, 0.50$

D. $0.50, \frac{1}{3}, 0.60, \frac{2}{5}$

Ⓐ Ⓑ Ⓒ Ⓓ

8. The circle graph below shows the composition of Earth's crust. What fraction is equivalent to the amount of *other elements* in Earth's crust?

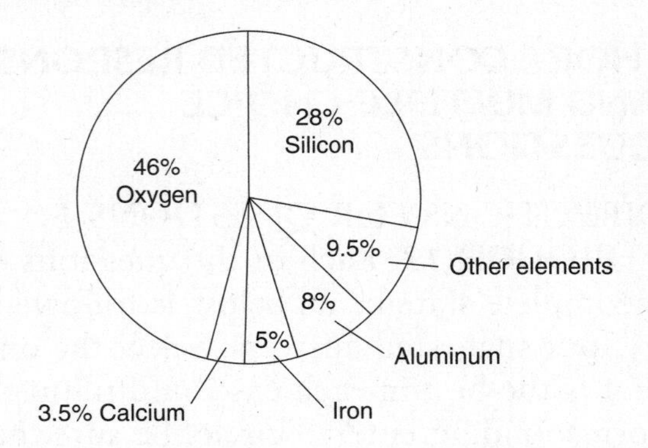

A. $\frac{95}{10}$

B. $\frac{95}{100}$

C. $\frac{95}{1,000}$

D. $\frac{95}{10,000}$

Ⓐ Ⓑ Ⓒ Ⓓ

GO ON TO THE NEXT PAGE

9. The smaller the U-value, the better the material is for insulation. Use the chart below and select the material that is the best insulator.

Material	U-value
Roof with no insulation	2.2
Insulated roof	0.3
Single brick wall	3.6
Double brick wall, air cavity between	1.7
Double brick wall filled with foam	0.5
Single-glazed window	5.6
Double-glazed window with air gap	2.7
Floor without carpets	1.0
Floor with carpets	0.3

A. double brick wall filled with foam
B. roof with no insulation
C. double-glazed window with air gap
D. floor with carpets

Ⓐ Ⓑ Ⓒ Ⓓ

10. A used car was priced at $7,000. The salesperson then offered a discount of $350. This means the salesperson offered a discount of ____ %.

A. 5
B. 20
C. 80
D. 95

Ⓐ Ⓑ Ⓒ Ⓓ

11. Roni is making a large poster for the hall bulletin board. She is enlarging the figure drawn below so the side corresponding to the right side will be 9 inches long instead of 3 inches. What will be the distance around this enlarged shape (the perimeter)?

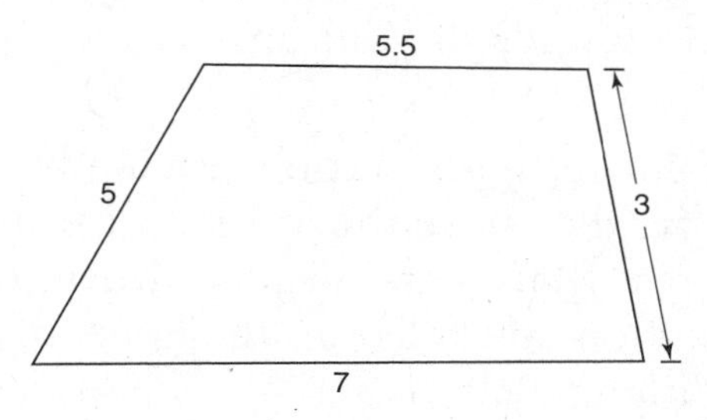

A. 61.5 inches
B. 41 inches
C. 20.5 inches
D. 189 inches

12. There are 650 students who were surveyed. Use the chart below to determine how many students selected pizza or chili as their favorite.

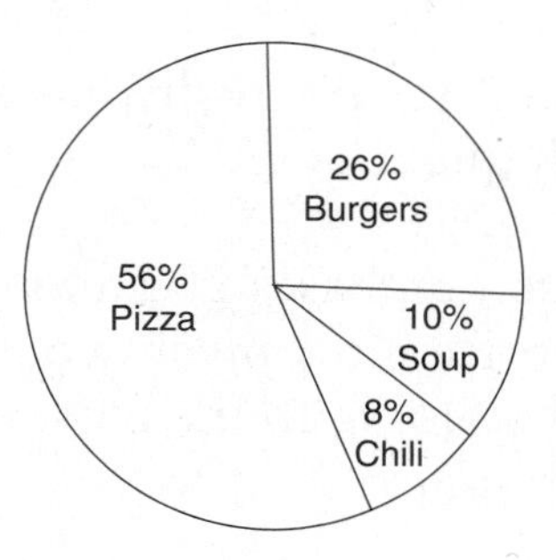

A. 429 students
B. 416 students
C. 364 students
D. 64 students

Ⓐ Ⓑ Ⓒ Ⓓ

GO ON TO THE NEXT PAGE

EXTENDED CONSTRUCTED RESPONSE QUESTIONS

DIRECTIONS FOR QUESTIONS 13 AND 14: Respond fully to the ECR questions that follow. Show your work and clearly explain your answer. You will be graded on the correctness of your method as well as the accuracy of your answer.

13. Four local gas stations each advertise that they have the lowest prices in the area. Marco and his dad drove to each station and filled up the tanks of five different cars.

Use the chart below as a guide:

Gas Station	Total Cost	Number of Gallons
Alco	$18.24	9.5
Bright	$18.31	9.2
Custom	$17.02	8.3
Dixon	$16.80	8
Extra	$18.45	9

- Which gas station really has the lowest price? Explain.
- If Marco purchases 20 gallons at the lowest price, how much would he save instead of going to the most expensive gas station?

14. There is a big sale at the local department store. Every hour their winter jackets get reduced by 5%. Susan wants to buy the jacket that is originally priced at $89.50. The sale begins at 1:00 P.M.

- If Susan has only $70.00, will she be able to buy the jacket today?
- If so, what will it cost?
- What time will she be able to buy it?

Explain or show your work to describe how you arrived at your answer.

SOLUTIONS TO CLUSTER I TEST

SHORT CONSTRUCTED RESPONSE AND MULTIPLE-CHOICE QUESTIONS

1. 28 $(-6)(-2) + 4^2 = 12 + 16 = 28$
2. 42 $\sqrt{36} + 4^2 + |-20| = 6 + 16 + 20 = 42$
3. D
4. C $\sqrt{115}$ is between numbers that I know are perfect square numbers.
   $\sqrt{115}$ is between $\sqrt{100}$ and $\sqrt{121}$.
   Since $\sqrt{100}$ is 10, and $\sqrt{121}$ is 11, then $\sqrt{115}$ is between 10 and 11.
5. B 6.85×10^{10} To change a number in *standard form* to a number in *scientific notation form*, you move the decimal point from the right until you have only one digit in the ones place. To change the number 68,500,000,000 to 6.85 we moved the decimal point 10 places. Therefore, $6.85 \times 10^{10} = 68{,}500{,}000{,}000$.
6. D Subtract 2 from 8 (8 – 2)
   Use the correct *Algebraic Order of Operations*. Work inside parentheses first. Remember, *PEMDAS* or *Please Excuse My Dear Aunt Sally.*
7. A $\frac{1}{3}$ $\frac{2}{5}$ 0.50 0.60
   First change each fraction to a decimal. Use your calculator to divide easily.
   $\frac{1}{3} = 0.33$ $\frac{2}{5} = 0.40$
   Now you can see that the order from lowest to highest is 0.33 0.40 0.50 0.60
   Rewrite with the original equivalent numbers: $\frac{1}{3}$ $\frac{2}{5}$ 0.50 0.60
8. C $\frac{95}{1{,}000}$ 9.5% as a decimal is written as 0.095, which reads as 95 thousandths, or $\frac{95}{1{,}000}$.
9. D Floor with carpets. Look at the given choices in A, B, C, and D and write them down or circle their values.
   A. double brick wall filled with foam 0.5
   B. roof with no insulation 2.2
   C. double-glazed window with air gap 2.7
   D. floor with carpets 0.3
   Then, reread the question to see that you need the *smallest U-value* (the smallest number), which is 0.3.
10. A 5 In this case it is easiest to use the choices given and plug in the numbers. To find 5% of \$7,000, multiply (\$7,000)(.05) = \$350
11. A 61.5 First, notice that each side of the large shape is 3 times larger than the smaller shape. You can find the perimeter of the large shape two ways.

1. Find the perimeter of the smaller shape and multiply by 3:

   $3 + 5 + 5.5 + 7 = 20.5; (20.5)(3) = 61.5$

2. Multiply each side of the small shape by 3 to get the dimensions of the large shape, and then add them together:

   $9 + 15 + 16.5 + 21 = 61.5$

12. 416 students

Add the percentage of students who select pizza or chili: 56% + 8% = 64%.
Next, find 64% of the total number of students: (.64) (650) = 416 students.

EXTENDED CONSTRUCTED RESPONSE QUESTIONS

QUESTION 13 (3 points if all parts are answered correctly)

- Alco is the least expensive. Gas at Alco costs $1.92 per gallon.

Total price ÷ Number of gallons = Price per gallon

18.24 ÷ 9.5 =	$1.92 per gallon	Alco
18.31 ÷ 9.2 =	$1.99 per gallon	Bright
17.02 ÷ 8.3 =	$2.05 per gallon	Custom
16.80 ÷ 8 =	$2.10 per gallon	Dixon
18.45 ÷ 9 =	$2.05 per gallon	Extra

- Marco would save $3.60.

20 gal × 2.10 =	$42.00 most expensive
20 gal × 1.92 =	– 38.40 least expensive
savings:	$3.60

QUESTION 14 (3 points if all parts are answered correctly)

Again, it is easy to solve this problem if you organize your information in a table or chart. Show your work.

If on sale for 5% off, then she would pay 95% or 0.95 times previous price		Price Each Hour
Before 1:00 P.M. Original price		$89.50
Price at 1:00 P.M. less 5% of 89.50	89.50 × 0.95	$85.03
Price at 2:00 P.M. less 5% of 85.03	85.03 × 0.95	$80.78
Price at 3:00 P.M. less 5% of 80.78	80.78 × 0.95	$76.74
Price at 4:00 P.M. less 5% of 76.74	76.74 × 0.95	$72.90
Price at 5:00 P.M. less 5% of 72.90	72.90 × 0.95	$69.26

Answer

- Yes, Susan will be able to buy the jacket today.
- It will cost her $69.26.
- She will be able to buy the jacket at 5:00 P.M.

Chapter 2

Cluster II: Spatial Sense and Geometry

WHAT DO ASK 8 SPATIAL SENSE AND GEOMETRY QUESTIONS LOOK LIKE?

MULTIPLE-CHOICE QUESTION (MC)

Example 1: You are given a square with a side measuring 8 inches and a triangle with a base of 8 inches and a height of 16 inches. Which of the following is true?

A. Area of the square is > area of the triangle.
B. Area of the square is < area of the triangle.
C. Area of the square = area of the triangle.
D. There is not enough information to answer the question.

Example 1: Strategies and Solutions
Draw and label diagrams.

Show all work, even work done with a calculator.

Remember symbols (< means less than; > means greater than).

Area square = $s^2 = 8^2 = (8)(8) = 64$ sq. in.

Area triangle = $\frac{bh}{2} = \frac{(8)(16)}{2} = 64$ sq. in.

The correct answer is C.

SHORT CONSTRUCTED RESPONSE QUESTION (SCR)

Example 2: (No calculator permitted.)

What is the area of the *smallest* polygon described below?

- A rectangle 18 ft long by 2 ft wide
- A square that is 6 ft long on each side
- A triangle that has a base of 9 ft and is 4 ft tall

Answer: ______

Example 2: Solution

- Area rectangle = (length)(width) = (18)(2) = 36
- Area square = (length)(width) = (6)(6) = 36
- Area triangle = $\frac{(\text{base})(\text{height})}{2} = \frac{(9)(4)}{2} = \frac{36}{2} = 18$

The triangle has the smallest area, 18 sq. ft.

Answer: 18 sq. ft

Example 3: (No calculator permitted.)

What is the area of an *isosceles trapezoid* with one base 6 cm long, the other base 10 cm long, and a height of 4 cm?

Answer: ______

Example 3: Solution

- Area any trapezoid = $\frac{(\text{base}_1 + \text{base}_2)(\text{height})}{2}$
- Area this trapezoid = $\frac{(6+10)(4)}{2} = \frac{(16)(4)}{2} = (8)(4) = 32$

Answer: 32 cm

You'll notice that in the SCR (short constructed response) questions you are NOT permitted to use a calculator, so the numbers used are not complicated.

EXTENDED CONSTRUCTED RESPONSE QUESTION (ECR)

Example 4: John has an odd-shaped backyard. He wants to plant grass in the entire area, and he also wants to put fencing around all sides.

1. What is the area he will cover with grass seed?

2. If grass seed costs $0.05 per square foot, what would it cost him to buy enough seed for the backyard?

3. How much fencing will he need to purchase?

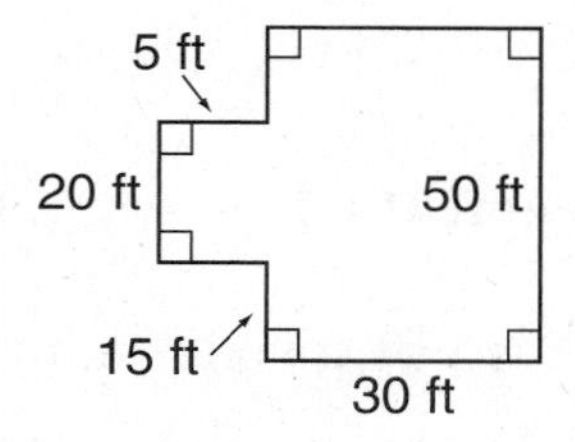

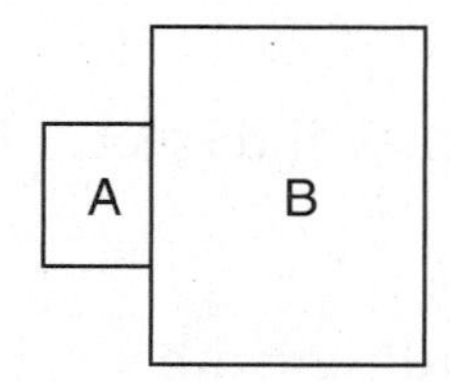

Example 4: Strategies and Solutions
Underline or circle important information in the question.

Write formulas.
Use *NJ ASK 8 Mathematics Reference Sheet on page 265.*

Make a chart or table to help you organize data. Remember, almost every open-ended NJ ASK 8 question can be organized with a chart or table.

Check computation. Does it make sense?

Think	Plan	Do
Draw and divide shape	Rectangles labeled	See diagram
Area of A and B rectangles	Area = $l \times w$ $A = 20 \times 5 = 100$ $B = 30 \times 50 = 1{,}500$	Total area $A + B =$ 1,600 sq. ft
Cost of grass seed	Total Area × Cost per square foot $1{,}600 \times .05 =$	$80.00
Perimeter for fence	Add all sides $50 + 50 + 35 + 35$	170 feet

Final Answers:
1. He will need to cover an area of 1,600 square feet with grass seed.
2. It will cost $80 to by the grass seed.
3. He will need 170 feet of fencing.

Important Reminder: The people who score these extended constructed response questions must work very fast. They spend only a short time reading each one. Therefore, it is very important that your work be neat, clear, and easy to read. On the math portion of the test, it is better to use a chart or a table to show your work than to write a complicated paragraph.

Geometry: Geometry really means measuring the earth. First, we'll review some basic terms that will be used throughout this section.

POINTS, LINES, ANGLES, AND PLANES

POINTS AND LINES

Definitions you should know:

- A *point* is a location in space. It doesn't have any height or width.
- If there are at least two points on a plane, there is a *line*. A line contains an indefinite number of points and continues indefinitely in both directions.
- A portion of a line that has two endpoints is a *line segment*.
- A portion of a line that has only one endpoint is a *ray*.

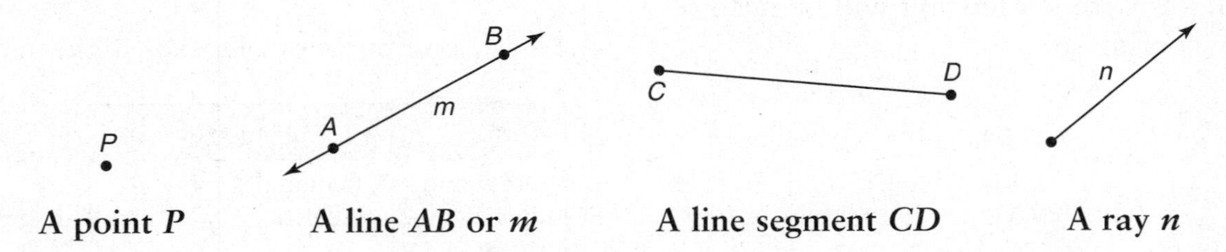

A point *P* **A line *AB* or *m*** **A line segment *CD*** **A ray *n***

The points are written as uppercase letters, but a line can also be given a name using a single lowercase letter (line *AB* can also be called line *m*).

ANGLES

When lines, segments, or rays meet or when they intersect they form angles.

Some different types of angles you should recognize and remember:

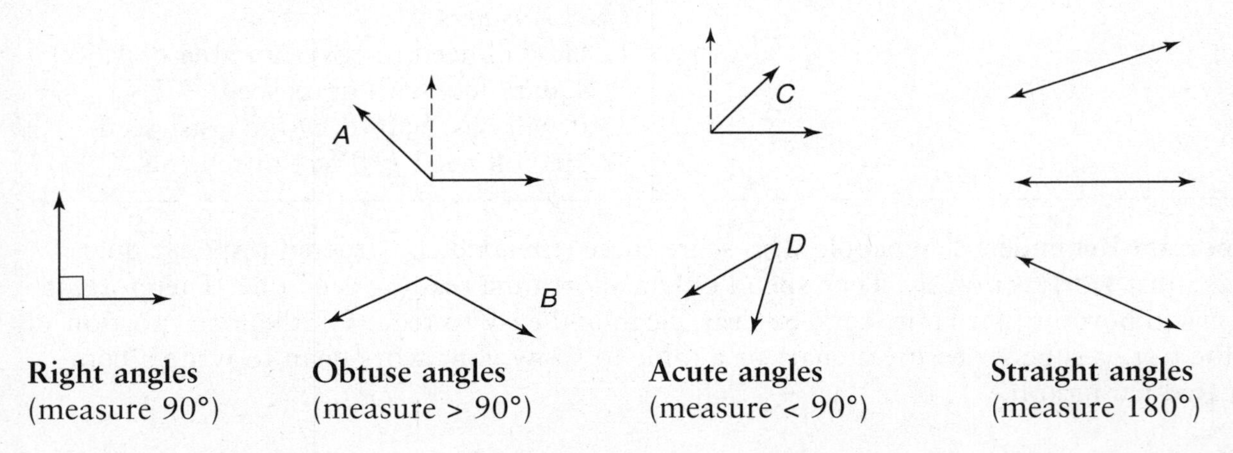

Right angles (measure 90°) **Obtuse angles** (measure > 90°) **Acute angles** (measure < 90°) **Straight angles** (measure 180°)

The term "vertex" is often used in geometry. A *vertex* is the endpoint that is shared by two segments that form an angle. The little symbol for a right angle is the little square near the vertex of the angle.

Angles and Pairs of Angles with Special Names

Complementary Angles When the sum of two angles is 90°, the angles are called *complementary angles.*

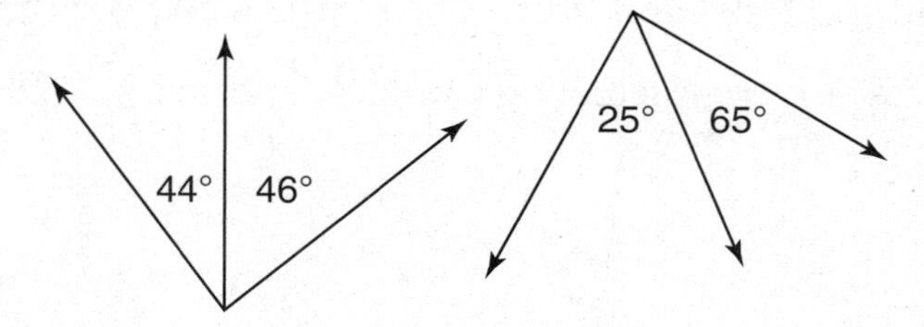

Hint: You can remember the name "complementary" by thinking of "corner," (complementary angles look like the corner of a room). You can also write the word "complementary" and draw a line next to the C. You will see the number 90: Complementary.

Vertical Angles When lines, rays, or segments intersect they form four angles; their *vertical angles* are equal, ($m\angle c = m\angle e$) and ($m\angle d = m\angle f$).

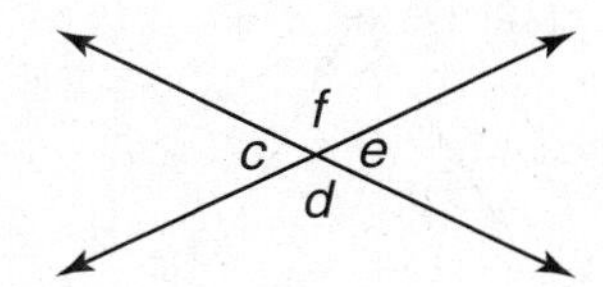

Hint: Vertical angles look like two letter *X*s.

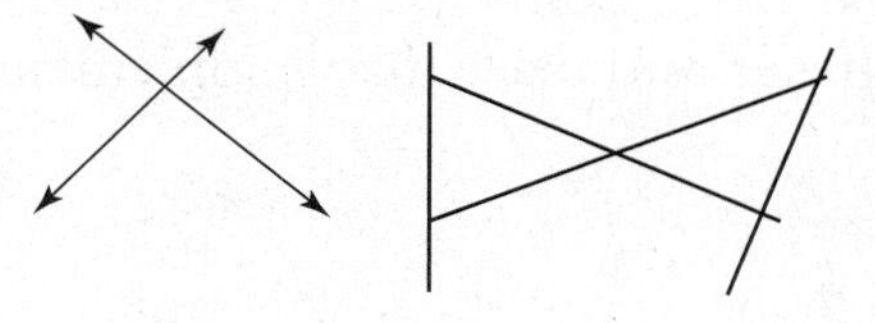

When we write "the measure of an angle is," we use symbols instead of words. The example $m\angle 5 = 210°$ is read as "the measure of angle five is two hundred ten degrees." A 50° angle and a 40° angle are complementary angles (50° + 40° = 90°). A 33° and a 57° angle are also complementary angles (33° + 57° = 90°). If you are told that two angles are complementary and you know the measure of one angle, you can find the measure of the other. Here are two examples.

Examples

A. Given: $\angle 5$ *and* $\angle 6$ are complementary angles, and $m\angle 5 = 10°$. What is $m\angle 6$?

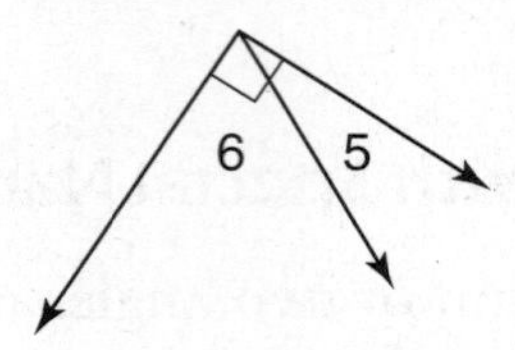

Since they should total 90° and $90° - 10° = 80°$, we know that $m\angle 6 = 80°$.

B. Given: $\angle 3$ *and* $\angle 4$ are complementary angles, and $m\angle 4 = 78°$. What is $m\angle 3$?

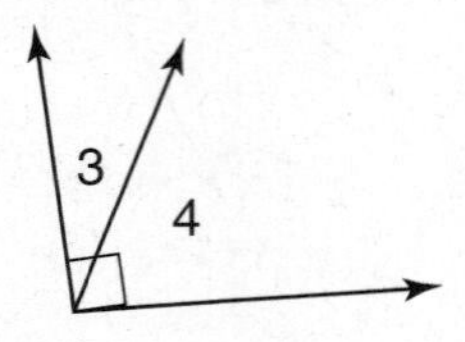

Since they should total 90° and $90° - 78° = 12°$, we know that $m\angle 3 = 12°$.

C. If $\angle a$ and $\angle b$ are complementary, and $m\angle a = 35°$, then $m\angle b$ must = _____° because their sum is 90°. ($m\angle b = 55°$)

D. If $m\angle d = 55°$ and $m\angle e = 35°$, they are complementary angles because their sum is ____. ($55° + 35° = 90°$)

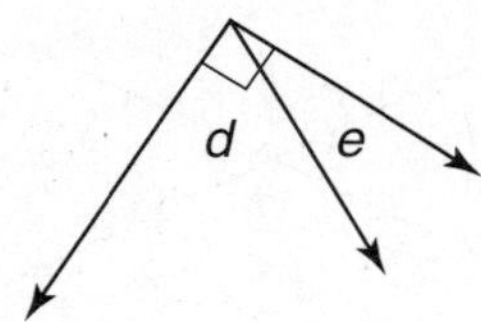

Supplementary Angles When the sum of two angles combine to make a straight line, they are called *supplementary angles* (their sum = 180°). In the following diagrams, angles *a* and *b* are supplementary. Angles *d* and *e* are also supplementary.

Hint: You can remember the word "supplementary" by remembering s̲upplementary angles make a s̲traight line.

Examples (Refer to the diagrams at the bottom of page 78.)

A. In the diagram above, if m$\angle b$ = 40°, then m$\angle a$ would = 180° – 40° or 140°.

B. In the diagram above, if m$\angle d$ = 52°, then m$\angle e$ would = 180° – 52° or 128°.

C. If m$\angle a$ = 35°, then m$\angle b$ must = _____ because their sum is 180°. (m$\angle b$ = 145°)

D. If m$\angle d$ + m$\angle e$ = 180°, then they are called ___________ angles. (supplementary)

RELATIONSHIPS OF LINES

- Lines are *parallel* if they are on the same plane and are the same distance apart (equidistant) from each other.
- When two lines cross over each other they are *intersecting lines.*
- When two lines cross over each other at right angles they are *perpendicular* to each other.

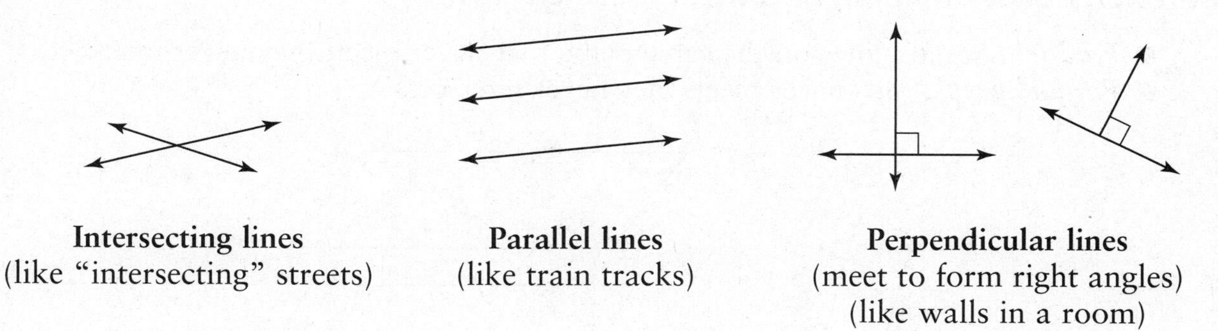

Intersecting lines (like "intersecting" streets)

Parallel lines (like train tracks)

Perpendicular lines (meet to form right angles) (like walls in a room)

Rays with Special Names and Characteristics

An *angle bisector* divides an angle into two equal angles. Ray *AB* is an angle bisector. Ray *EF* is also an angle bisector.

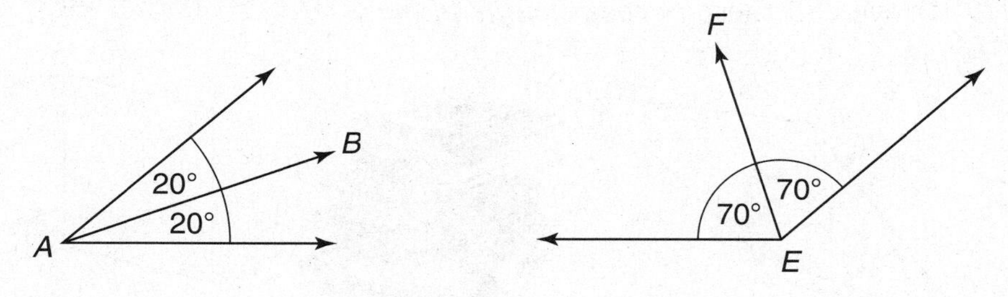

PLANES

Besides points, lines, and angles, you should be familiar with *planes*. When we write on a sheet of paper we are writing on one plane, but when we fold that paper, or make a three-dimensional form, then we have more planes. Above you see two different *planes*. Notice that when two planes intersect they intersect *at a line*. (See dotted line.)

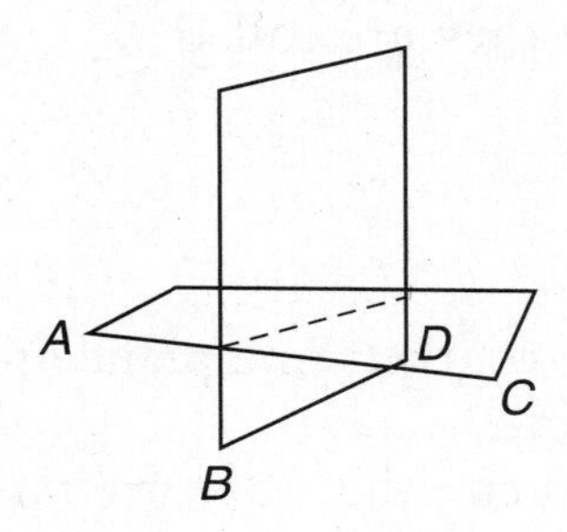

Parallel, Perpendicular, and Intersecting Planes

- *Parallel lines* are lines on the same plane that never meet; they never intersect.
- *Parallel planes* also never meet; they never intersect.

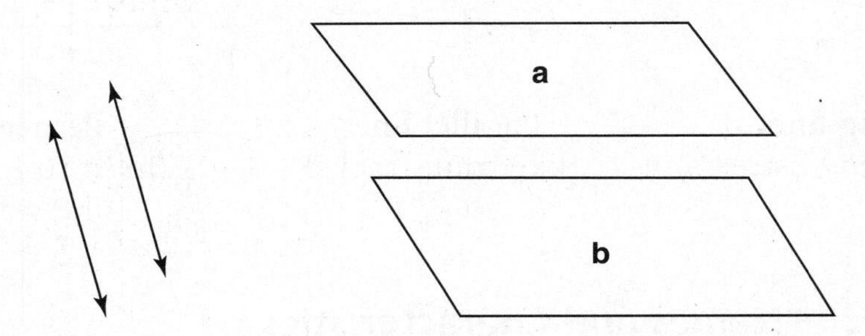

- Pages in a book look like *intersecting planes*.
- Walls in a room look like intersecting planes; since most walls meet at right angles (90°), we call these *perpendicular planes*.

Three-Dimensional Forms

Planes can combine to make three-dimensional forms. Three-dimensional forms have length, width, and depth.

- A tissue box is a three-dimensional form.

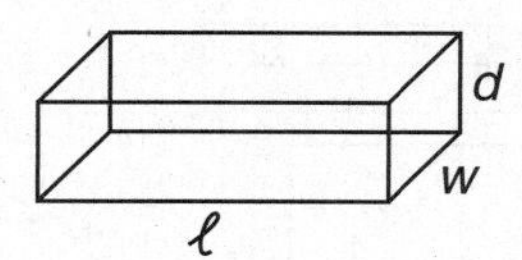

- A special rectangular form (or solid) that has six identical surfaces is called a *cube*. It is also called a *regular solid* because all of its faces have the same size and shape. (In a cube: Area of top = Area of one side = Area of front = Area of back = Area of bottom)

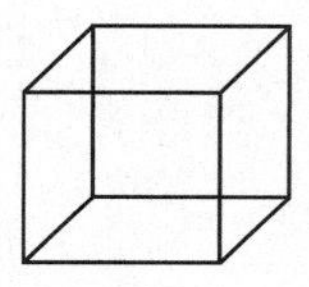

There are other three-dimensional solids. Some are drawn below. Look at these figures and answer the questions below.

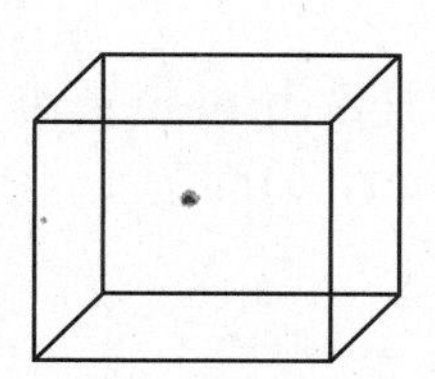
Rectangular prism

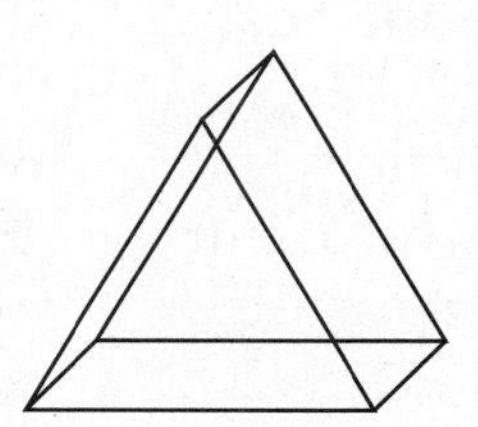
Triangular prism with a rectangular base

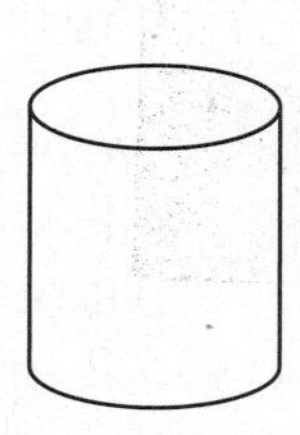
Cylinder

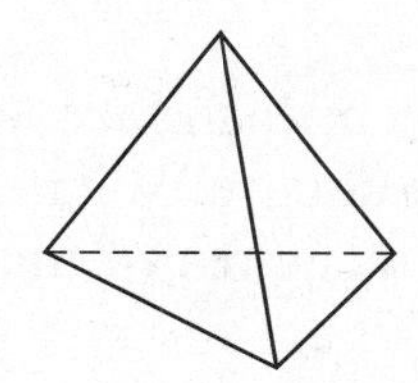
Triangular prism with a triangular base

A. How many faces does a rectangular prism have? It has a top, bottom, front, back, and two sides. Therefore it has __________ faces. (six)

B. How many faces does a triangular prism with a rectangular base have? (five)

C. How many faces does a triangular prism with a triangular base have? (four)

D. If you unrolled the cylinder you would have the following flat shapes: one __________ and two _________. (rectangle; circles)

E. How many faces does a cube have? (six)

Now let's work in reverse. Look at the flat shapes below and see if you can tell what three-dimensional forms they would make when folded. These flat shapes are called *nets*.

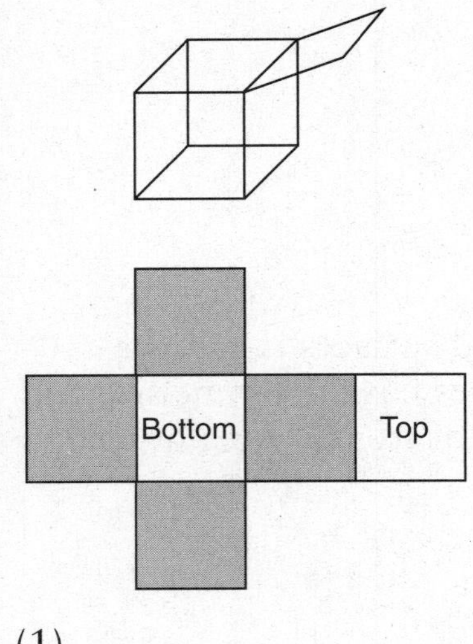

(1)

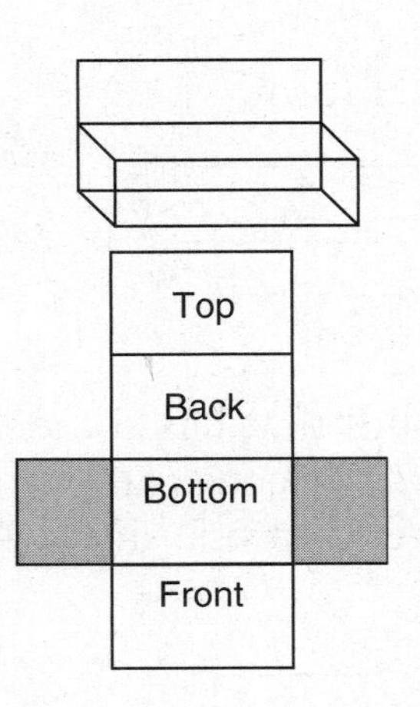

(2)

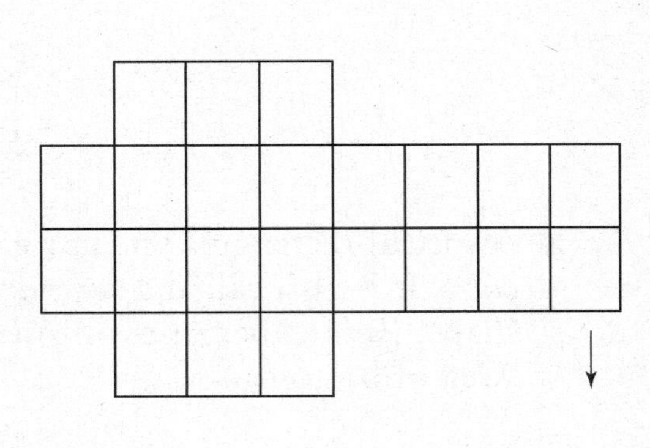

(3) Draw the three-dimensional box you could make from this net.

Number (1) is a cube with a top cover that flaps over from the right side.

Given: each face is an 8 ft by 8 ft square. What is the total surface area of this cube?

Think:

a. How many faces? ____
b. Are all faces the same size? ___
c. What is the area of one face? ____
d. What is the sum of the area of all faces? (Total surface area) = _______sq. ft

Number (2) is a rectangular box with a top cover that flaps over from the back.

Given: The two sides measure 5.2 in. by 5.2 in. All other faces measure 5.2 in. by 10.4 in.

What is the total surface area of this box?
____ sq. in.

Describe box number (3):

What are the dimensions of box number (3)?

What is the total surface area of box number (3)?

____

Answers (1)

a. 6
b. yes
c. $8 \times 8 = 64$ sq. ft
d. $6 \times 64 = 384$ sq. ft

Answers (2)

$A_{sides} = 2(5.2)(5.2) = 54.08$

$A_{all\ other\ faces} = 4(5.2)(10.4) = 108.16$

Total surface area = 162.24 sq. in.

Answers (3)

Answers will vary.

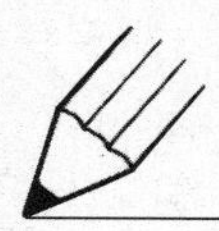

PRACTICE: Points, Lines, and Planes

(For answers, see page 145.)

1. If $\angle A$ is a right angle, and $m\angle 1 = 43°$, then $m\angle 2 =$

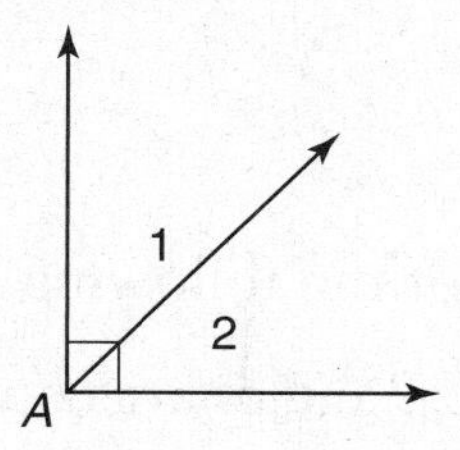

A. 43°
B. 47°
C. 57°
D. 86°

2. If $m\angle 1 = 35°$ and $m\angle 2 = 145°$, we say these angles are

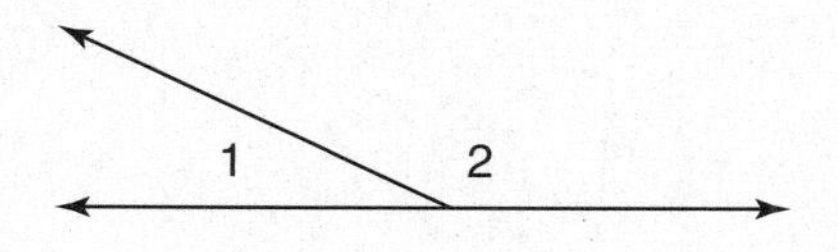

A. complementary
B. supplementary
C. perpendicular
D. equivalent

3. If the bolder-colored ray AB is an angle bisector, and $m\angle 2 = 48°$, then we know that $m\angle 1 =$ ________°.

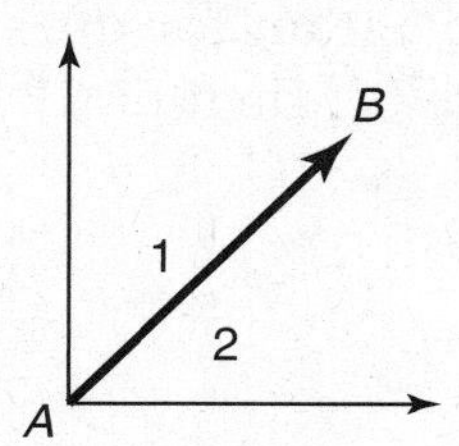

4. If an angle measures 23°, then we know its supplement will measure

Answer: ________°.

5. If an angle measures **60°**, then we know its complement will measure

Answer: ________°.

6. If one angle measures 10° and another measures 80°, we call these angles

A. complementary
B. congruent
C. vertical
D. supplementary

7. In triangle ABC, there is a perpendicular bisector (**bold** line segment BD) from vertex B to line segment AC; this tells us that

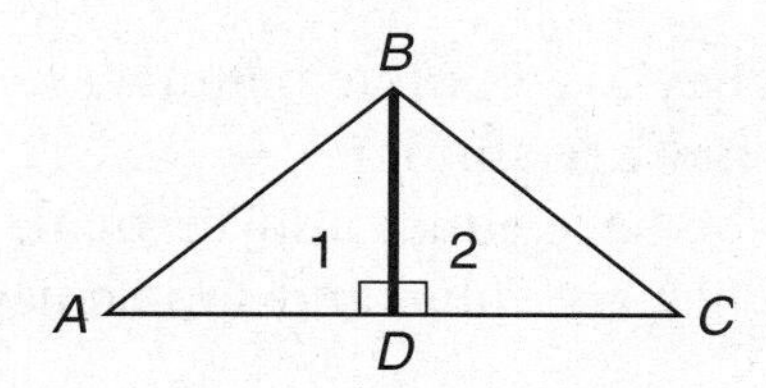

A. $m\angle A = 60°$
B. $m\angle A = 90°$
C. $m\angle 1 = 90°$
D. $m\angle 1 + m\angle 2 = 90°$

8. In $\triangle ABC$, ray BD is an angle bisector of $\angle B$; therefore, we know that

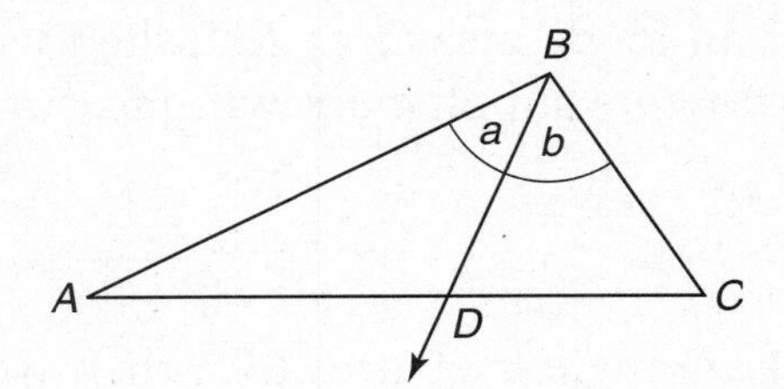

A. $m\angle A = m\angle C$
B. line segment $AD \cong$ line segment DC
C. $m\angle A = 90°$
D. $m\angle a = m\angle b$

9. Given a four-sided figure with diagonal EG intersecting diagonal FH, what can you say about angles a and b?

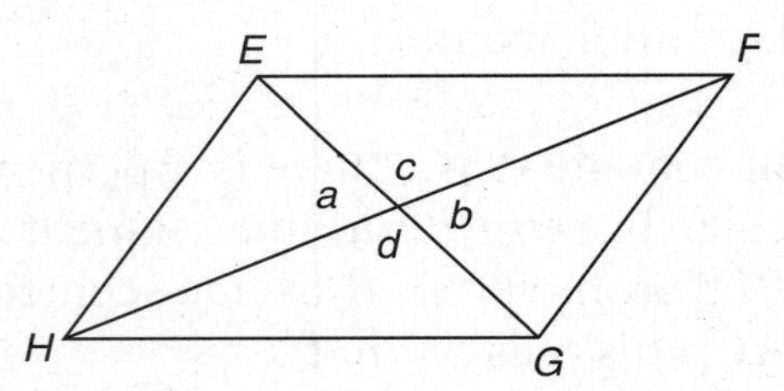

A. They are complementary.
B. They are similar.
C. They are equal and vertical.
D. They are equal and supplementary.

10. Use the same diagram as in question 9. If $m\angle a = 78°$, then $m\angle b =$

A. 78°
B. 12°
C. 102°
D. 90°

11. Use the same diagram as in question 9. If $m\angle a = 78°$, then $m\angle c =$

A. 78°
B. 12°
C. 102°
D. 112°

12. Which of the following is true?

A. Vertical angles are complementary.
B. Supplementary angles total 90°.
C. Complementary angles are equal.
D. Vertical angles are equal.

13. In the diagram shown, if ray NM is an angle bisector, and $m\angle 2 = 46°$, then

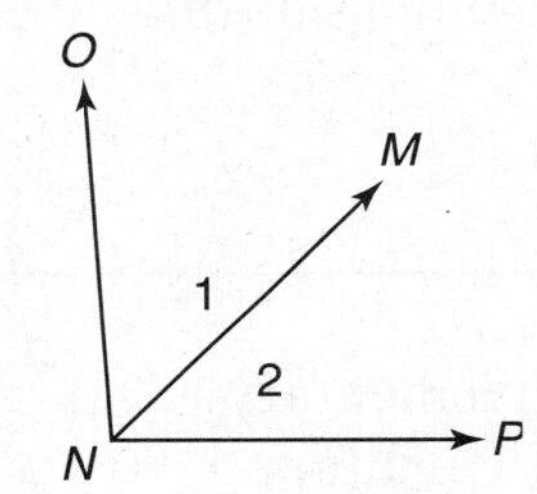

A. $m\angle 1 = 34°$
B. $m\angle 1 = 46°$
C. $m\angle 1 + m\angle 2 = 90°$
D. $m\angle 1 + m\angle 2 = 180°$

14. In the diagram for question 13, if $m\angle 1 = 70°$, then $m\angle 1 + m\angle 2 =$

A. 180°
B. 140°
C. 170°
D. 90°

15. In the diagram below, if $m\angle 1 + m\angle 2 = 90°$, then we could say that

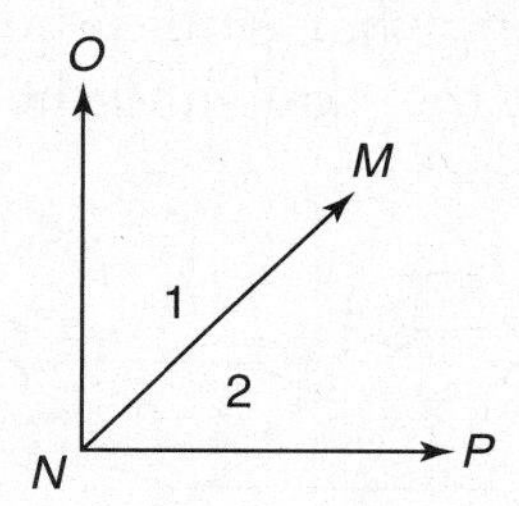

A. line segment $ON \cong$ line segment NP
B. line segment ON is perpendicular to line segment NP
C. ray NM is perpendicular to line segment NP
D. $m\angle 1 = m\angle 2$

16. In the diagrams below, you see two rectangular prisms. Each three-dimensional form has six planes. Each plane intersects another plane at a

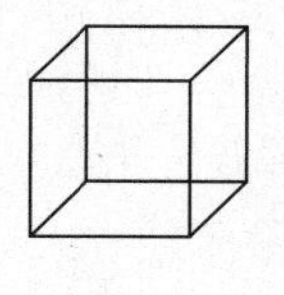

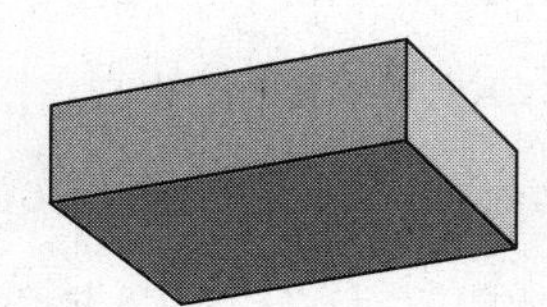

A. line
B. point
C. ray
D. bisector

17. If this is a rectangular prism, you know that

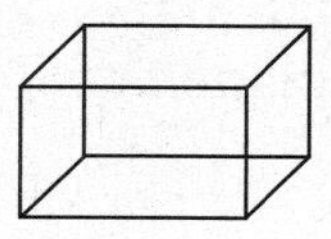

A. it is a cube
B. it has six faces
C. it is larger than a square prism
D. all sides are equal

18. Looking at this figure you can tell that it

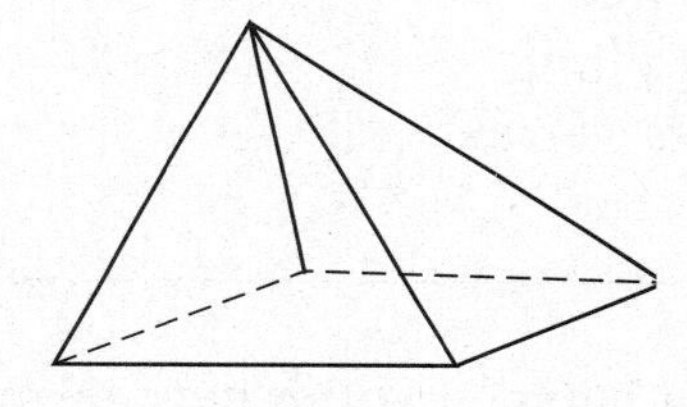

A. is a rectangular prism
B. is a triangular prism with a rectangle base
C. is a six-sided figure
D. has only one face that is triangular

19. Use the diagram below. Look at the line going through this two-dimensional shape. This line intersects this plane at

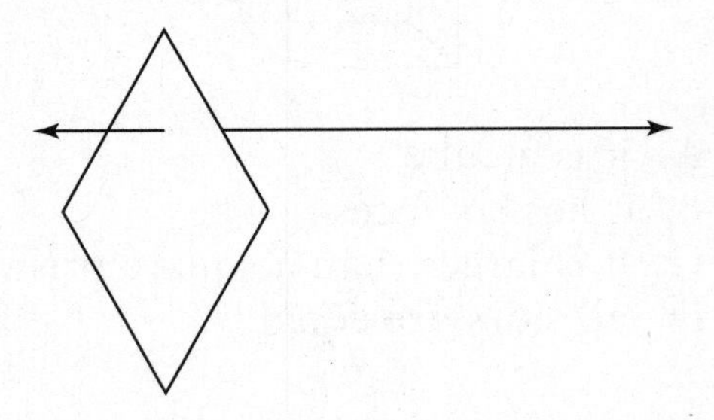

A. a point
B. a line
C. a plane
D. its vertex

20. Here is a cylinder on a table. If a horizontal plane cut through this cylinder, what would the intersection of the plane and the cylinder look like?

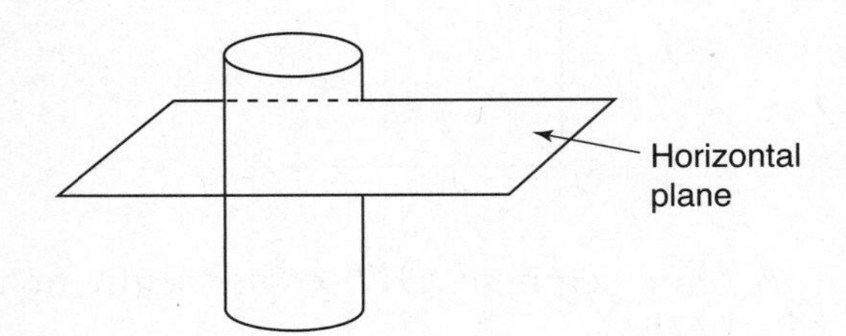

A. a square
B. an oval
C. a rectangle
D. a circle

21. You learned some new terms in this section. Match each term on the left with its description on the right to be sure you understand what each one means.

1. A triangle
2. A cube
3. A vertex
4. A regular shape
5. A three-dimensional form
6. A plane
7. Parallel planes
8. Complementary angles

A. All sides are the same size and all angles measure the same, like a square
B. A two-dimensional flat shape
C. It has length, width, and depth
D. A point where two lines meet and form an angle
E. They never meet; like a ceiling and a floor in a room
F. Two angles that add up to 90°
G. Two angles that add up to 180°
H. A flat surface like a table top
I. A regular three-dimensional solid

POLYGONS: AREA AND VOLUME

Polygons *Polygon* really means many angles (*poly-* is Greek for many and *-gon* is Greek for knee or bend or angle). *Polygons* are figures with many angles and many sides. We usually identify them by the number of their sides. Some polygons are illustrated on the following pages.

Regular Polygons If the lengths of the sides of a particular polygon are the same and the angles measure the same, that figure is called a *regular polygon.* Some regular polygons are shown below with their special names.

A *regular triangle* is called an *equilateral triangle.* (All sides are the same length and all angles have the same measure.)

A *regular rectangle* is called a *square.* (All sides are equal and all angles are equal.)

Parallelogram *A parallelogram* is a four-sided polygon with *opposite* sides equal and parallel.

Trapezoid Another polygon you should know about is a trapezoid. A *trapezoid* has four sides, but it is different from a rectangle, square, or parallelogram. A *trapezoid* has only *one pair of parallel sides.* A trapezoid can be a *right trapezoid* (with one or two right angles), or it can be an *isosceles trapezoid* (with two equal nonparallel sides and equal base angles).

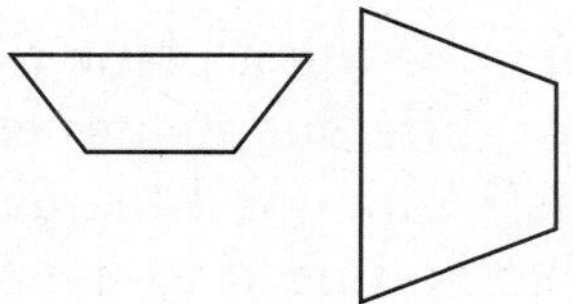

Drawings of polygons	Name of polygon	Angles	Sides
	Triangle	3	3
	Quadrilateral (When sides are even, it could look like a square or like a diamond.)	4	4
	Pentagon (When sides are the same length, it sometimes looks like a house.)	5	5
	Hexagon	6	6
	Octagon (When sides are the same length, it looks like a stop sign.)	8	8
	Decagon (This polygon is difficult to draw; all of its angles are obtuse.)	10	10

AREA OF FLAT SHAPES (Two-Dimensional Shapes)

When you are finding area, think of *covering* a plane figure in small squares. There are formulas to help you find the area of different shapes easily. Remember to always label your answers in *square units* such as 14 in.^{2} or 14 square inches. You are working with two dimensions (length and width) so you need to use square units.

Examples

A. You can count the boxes in the rectangle on the next page and see that the area of this rectangle is 35 square units. A faster way would be to multiply the base (7 units) by the height (5 units).

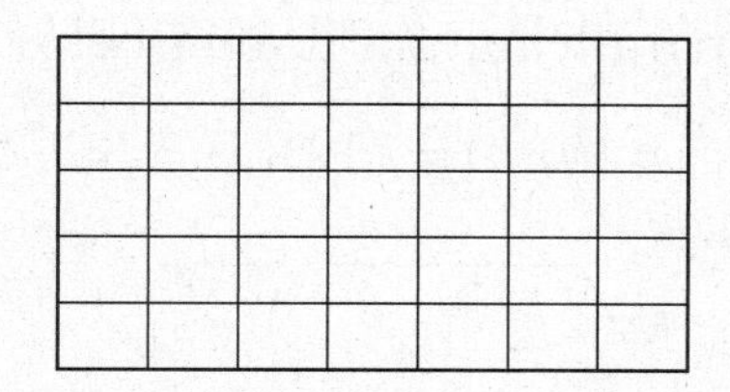

Base × Height = Area of rectangle
$7 \times 5 = 35$ square units

B1. The square to the left is 6 units long and 6 units high. The area of this square is (Base)(Height) or $(b)(h) = (6)(6)$ or 6^2, which is 36 square units.

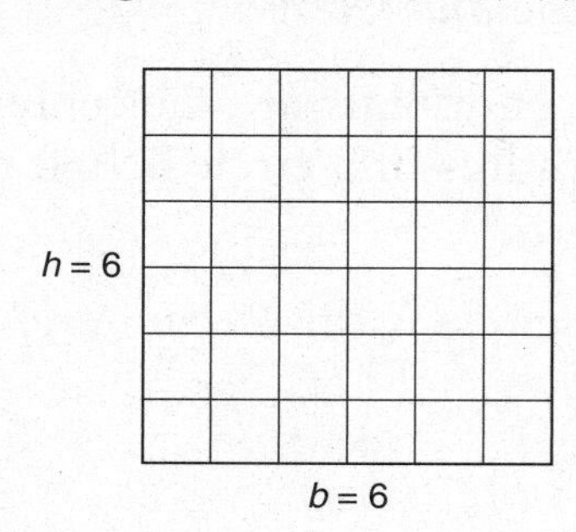

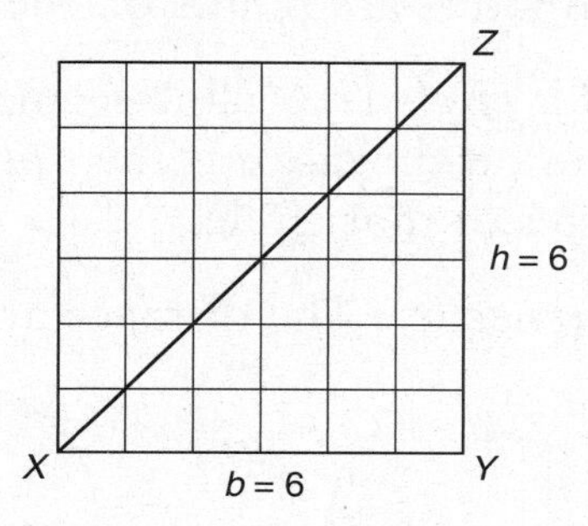

B2. $\triangle XYZ$, to the right, is half the size of the square.

$$A_{\text{triangle}} = \frac{bh}{2} = \frac{(6)(6)}{2} = \frac{36}{2} = 18 \text{ square units}$$

C. What is the area of a rectangle with one side 12 inches and one side 3 inches?

$$A_{\text{rectangle}} = \text{Base} \times \text{Height} = (b)(h) = 12 \times 3 = 36 \text{ square inches}$$

D. What is the area of a triangle with a base of 5 feet and a height (altitude) of 12 feet?

$$A_{\text{triangle}} = \frac{\text{Base} \times \text{Height}}{2} = \frac{(b)(h)}{2} = \frac{(5)(12)}{2} = \frac{(60)}{2} = 30 \text{ square feet}$$

E. What is the area of the obtuse triangle drawn below? (*Hint:* 4 is the height.)

$$A_{\text{triangle}} = \frac{(b)(h)}{2} = \frac{(5)(4)}{2} = \frac{(20)}{2} = 10 \text{ square units}$$

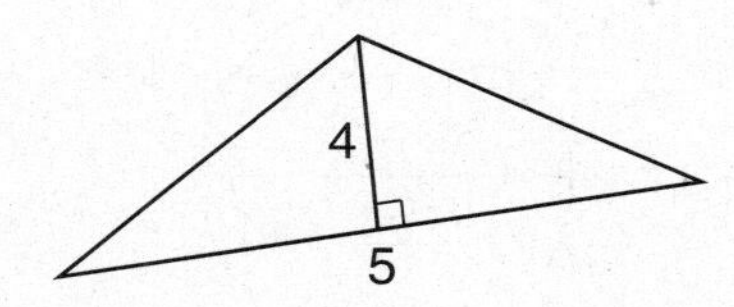

F. What is the area of the parallelogram drawn below? (*Notice:* 5 is the height.)

$$A_{\text{parallelogram}} = (b)(h) = (12)(5) = 60 \text{ square units}$$

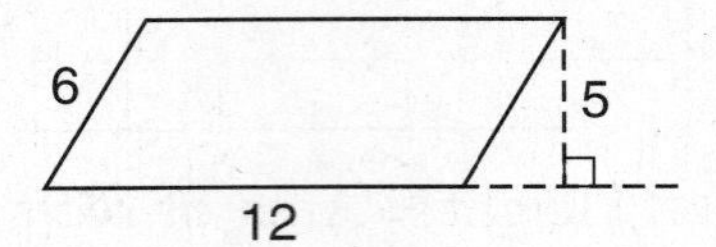

G. What is the approximate area of a circle with radius (r) = 5 cm?

Diameter of a circle **(d):** A line segment that contains the center of the circle and whose endpoints are points on the circle is the *diameter*.

Radius of a circle **(r):** A line segment whose endpoints are the center of the circle and a point on the circle is the *radius*. The radius of a circle is half the length of the diameter of that circle.

Circumference **(c):** The distance around the circle is the *circumference*.

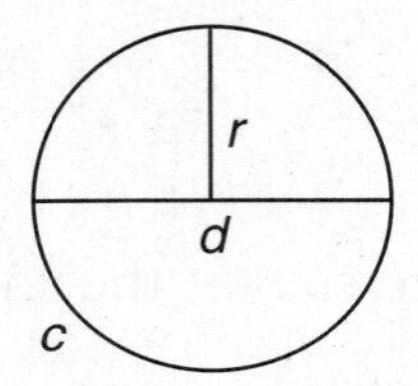

$$A_{\text{circle}} = \pi r^2 = (3.14)\ (5^2) = (3.14)(25) = 78.5 \text{ or } 79 \text{ sq. cm}$$

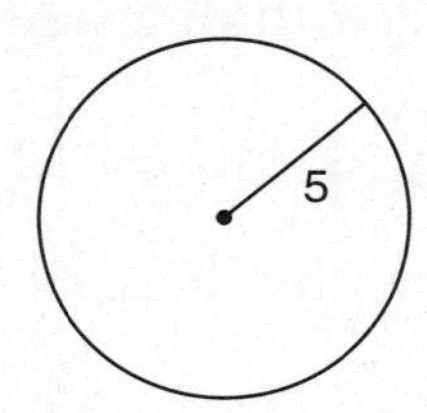

If you use a calculator, press [3.14] [×] [5^2], and you will see [78.5]. Or press [second] [π] [×] [5] [x^2]. You will see 3.141592654 × 25 = [78.53981634].

H. What is the approximate area of a circle with diameter (d) = 20 feet?

$$A_{\text{circle}} = \pi r^2 = (3.14)(10^2) = (3.14)(100) = 314 \text{ square feet}$$

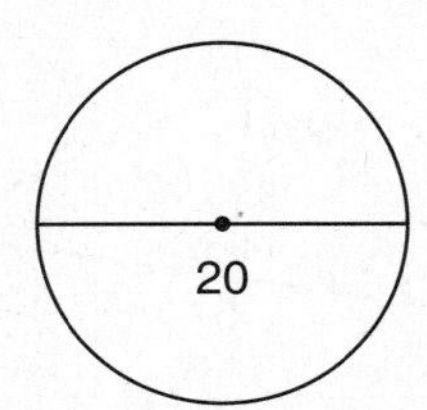

If you use a calculator, press the following keys: [3] [.] [1] [4] [×] [1] [0] [x^2].

You'll see 314. However, if you press π instead of 3.14 you will get a more accurate answer: [second] [π] [×] [1] [0] [x^2] [=]. You will see 3.141592654 × 100 = [314.1592654].

I. What is the area of this trapezoid?

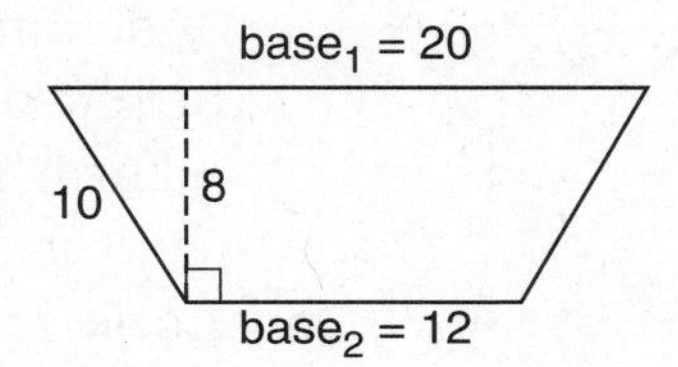

The area of a trapezoid is the average of the two bases times the height

$$\text{Area of a trapezoid} = \frac{\text{Base}_1 + \text{Base}_2}{2}(\text{Height})$$

Remember to use the altitude for the height. In this figure the height is 8.

$$\text{Area of this trapezoid} = \left(\frac{12+20}{2}\right)(8) = \left(\frac{32}{2}\right)(8) = (16)(8) = 128 \text{ square units}$$

If you can't remember the formula, (and don't have a reference sheet), use your problem-solving techniques and divide a big problem into smaller ones. Trapezoids can be divided into rectangles and triangles.

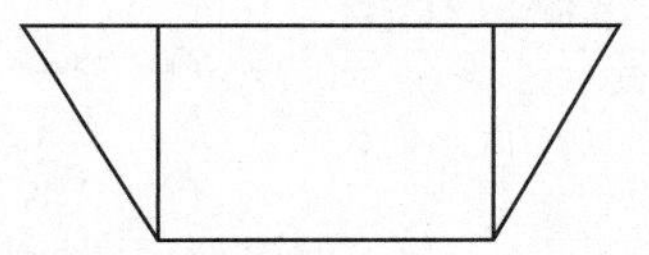

Find the area of each smaller figure and then add up the numbers to get the area of the entire trapezoid.

Formulas to Use

Area of a rectangle = Base × Height

Area of a triangle = $\frac{1}{2}$ Base × Height

Area of a circle = $\pi \times r^2$ (r = radius)

Use 3.14 for π

(Remember to use *radius*, not diameter, when finding the area of a circle.)

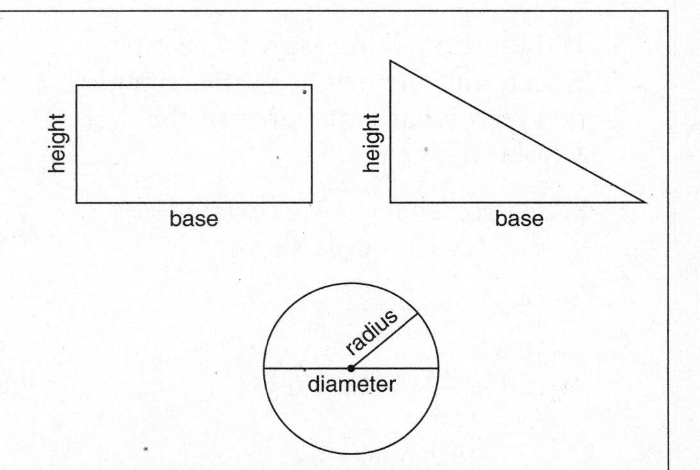

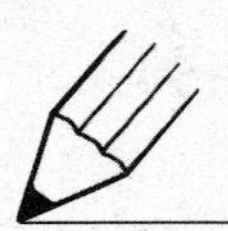

PRACTICE: Area of Flat Shapes

(For answers, see page 145.)

Refer to the triangle below to answer questions 1 and 2.

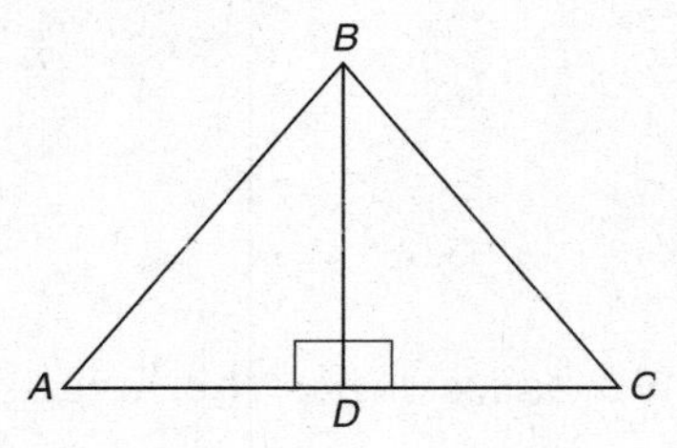

1. If AC = 12 inches and BD = 14 inches, what is the area of triangle ABC?
_____ square inches

2. If AC = 14.5 feet and BD = 8.2 feet, what is the area of triangle ABC?
_____ square feet

The figure below shows a rectangle connected to a triangle (shape *ABCDE*). Use this shape to answer questions 3 and 4.

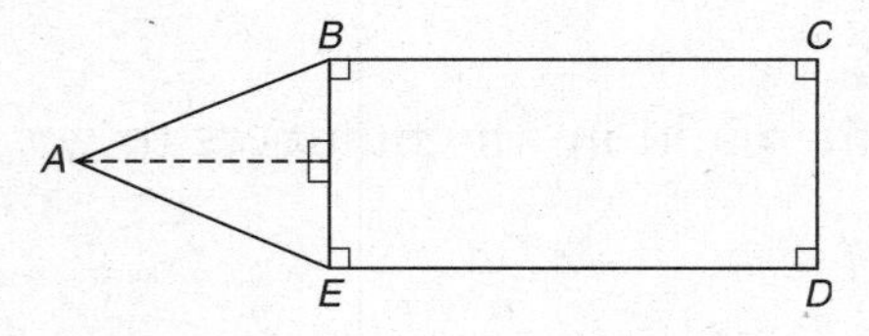

3. If the rectangle measures 4 feet by 6 feet, and the height of the triangle is 3 feet, what is the area of the whole shape?

Area rectangle + Area triangle = Area whole shape

$$4 \times 6 + \frac{(3)(4)}{2} =$$

$$24 + \frac{12}{2} = 24 + 6 =$$

_____ square feet

4. If the rectangle measures 8 inches wide and 6 inches high, and the height of the triangle is 4 inches, what is the area of that whole shape?

5. Look at the circle next to "Formulas to Use" on page 81 to help you answer the following questions:

a. If the diameter of a circle is 15 inches, what is the *radius* of that circle?

A. 15
B. 7.5
C. 8
D. 30

b. If the radius of a circle is 4.3 yards, what is the *diameter* of that circle?

A. 4.3
B. 2.15
C. 2.6
D. 8.6

c. If the radius of a round garden is 6 feet, what is the *approximate area* of that garden?

A. 36 sq. ft
B. 19 sq. ft
C. 38 sq. ft
D. 113 sq. ft

d. If the diameter of a circular cover for a pool is 18 feet, what is the *approximate area* of this cover?

A. 19 sq. ft
B. 9 sq. ft
C. 113 sq. ft
D. 255 sq. ft

e. If I had a rectangular room that measured 12 feet wide and 18 feet long, what is the *approximate area* of the largest round rug I could fit in this room? (*Hint:* Draw and label a diagram first.)

A. 38 sq. ft
B. 113 sq. ft
C. 254 sq. ft
D. 452 sq. ft

6. Find the area of a circle whose radius is 2.5 cm. Use 3.14 as an approximation for π. Round your answer to the nearest tenth. Show your work.

7. Find the length of the radius of a circle whose *circumference* and *area* measure the same. (*Remember*: Circumference is the distance around a circle.) Explain your answer.

8. A square has an area of 256 square meters. To the nearest whole number, what would the radius of a circle be so that the circle would have approximately the same area?

A. ~ 6 m
B. ~ 7 m
C. ~ 8 m
D. ~ 9 m

9. A triangle has an area of 72 square inches. What could the height and length be in this triangle? List as many integer values as you can.

10. A rectangle has an area of 36 square yards. What could the height and length be in this rectangle? List at least three different possibilities.

AREA OF IRREGULAR SHAPES

Sometimes you need to find the area of shapes that are really combinations of other geometric shapes. In Figure A, you see a shape that seems to be made up of different rectangles. Since you know how to find the area of one rectangle, you can easily find the area of the entire shape.

Examples

A. Figure A is the figure given. Figures B and C show different ways you could divide this irregular shape into rectangles. Now, just find the area of each rectangle and add them to find the total area of the irregular shape.

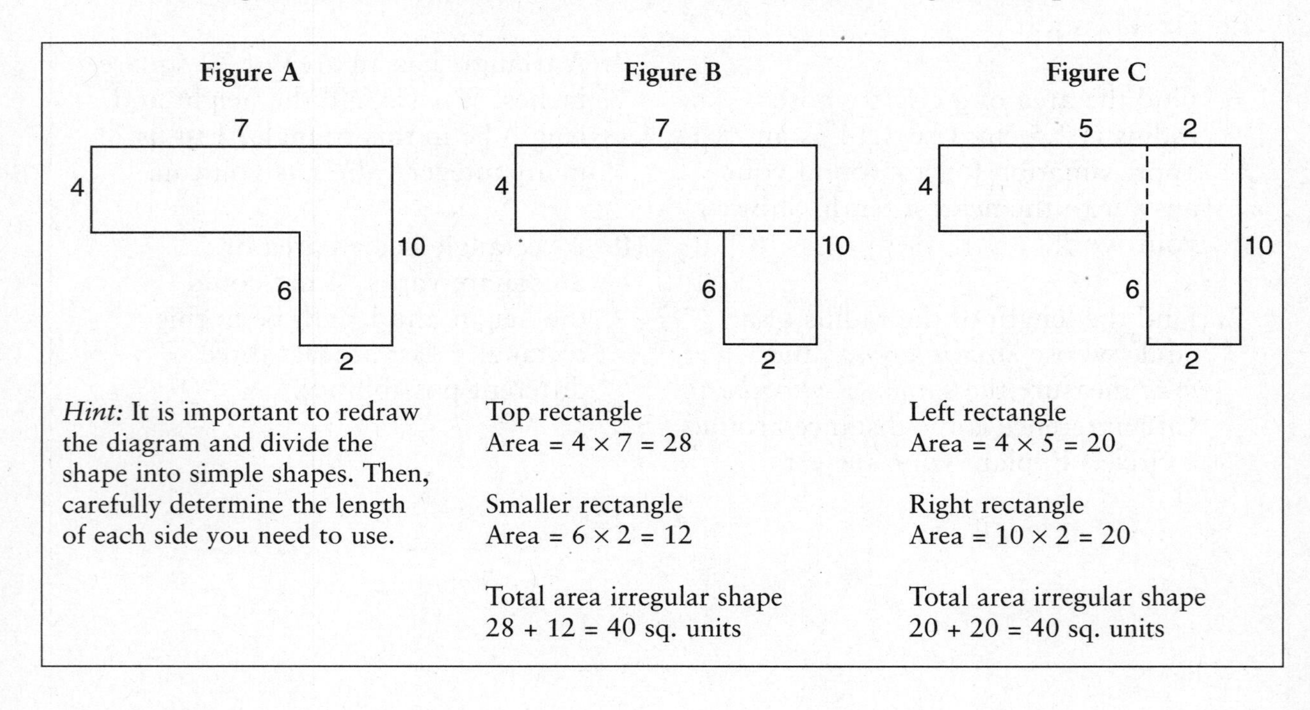

Hint: It is important to redraw the diagram and divide the shape into simple shapes. Then, carefully determine the length of each side you need to use.

Top rectangle
Area = 4 × 7 = 28

Smaller rectangle
Area = 6 × 2 = 12

Total area irregular shape
28 + 12 = 40 sq. units

Left rectangle
Area = 4 × 5 = 20

Right rectangle
Area = 10 × 2 = 20

Total area irregular shape
20 + 20 = 40 sq. units

B. Find the area of this irregular shape (composed of a rectangle and a triangle). Given: The height of the triangle is 4 feet, the base of the rectangle is 6 feet, and the height of the rectangle is 5 feet.

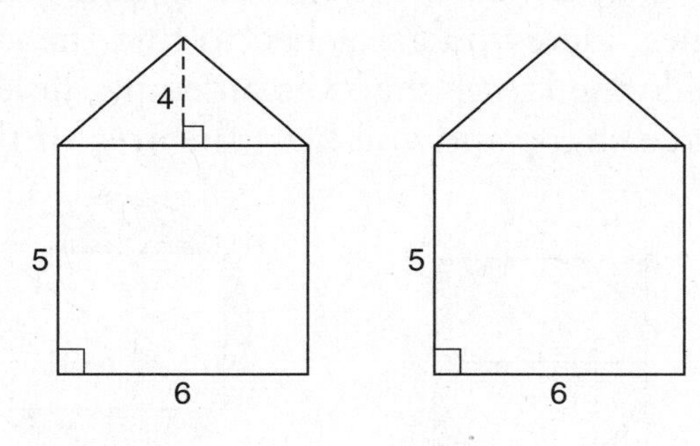

$$A_{\text{triangle}} = \frac{(b)(h)}{2} = \frac{(6)(4)}{2} = \frac{(24)}{2} = 12 \text{ square units}$$

$$A_{\text{rectangle}} = (b)(h) = (6)(5) = 30 \text{ square units}$$

$$A_{\text{total figure}} = A_{\text{triangle}} + A_{\text{rectangle}} = 42 \text{ square units}$$

C. Find the approximate area of this irregular shape (composed of a rectangle and half a circle). Given: The width of the rectangle is 6 inches, and the height of the rectangle is 4 inches. From this information you can see that the *diameter* of the circle is also 4 inches (so the radius is 2 inches). Now, we'll find the area of the entire shape.

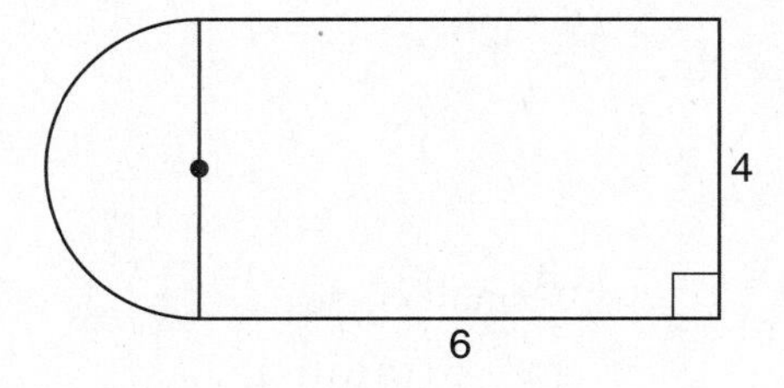

$$A_{\text{rectangle}} = (b)(h) = 6 \times 4 = 24.00 \text{ square inches}$$

$$A_{\text{half circle}} = (\pi r^2) \div (2) = \frac{(3.14)(4)}{2} = \underline{6.28 \text{ square inches}}$$

$$A_{\text{entire shape}} = A_{\text{rectangle}} + A_{\text{half circle}} = 30.28 \text{ square inches}$$

AREA OF THE SHADED REGION

Other times you may need to find the area of a shape within a shape. You often are asked to find the area of part of the shape. Below are four diagrams with overlapping shapes of circles, rectangles, and triangles. Here you are asked to find the area of the shaded region. Think of the white shape inside the larger shape as an empty hole (like a donut hole). Just subtract the hole from the large shape and you have the area of the shaded region.

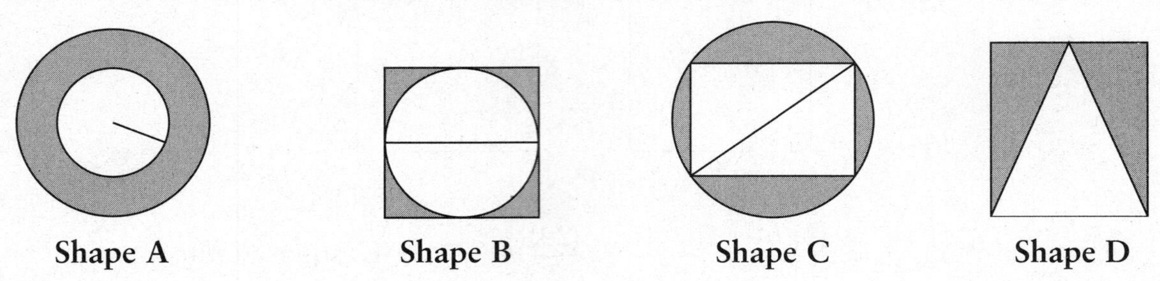

For Shape A, use the area of the larger circle minus the area of the smaller circle. For Shape B, use the area of the square minus the area of the circle. For Shape C, use the area of the circle minus the area of the rectangle. (The rectangle might also be a square.) For Shape D, use the area of the rectangle minus the area of the triangle.

Example

Use Shape A. Diameter of large circle = 14, diameter of small circle = 10. What is the *approximate area* of the shaded region? (Use $\pi = 3.14$.)

A. 232
B. 154
C. 75
D. 6

$A_{\text{large circle}}$	−	$A_{\text{small circle}}$	$= A_{\text{shaded region}}$
$(\pi) \times (\text{radius})^2$		$(\pi) \times (\text{radius})^2$	=
3.14×7^2		3.14×5^2	=
3.14×49		3.14×25	=
153.86	−	78.50	= 75.36 square units

If you use a calculator, press [π] [×] [7] [x^2] [–] [π] [×] [5] [x^2] [=]. You will see [75.9822369].

Notice that when we used 3.14 as π, our answer was only 75.36. But the calculator used a more accurate estimate for π; it used 3.141592654; therefore, the final answer with a calculator is a little higher and more accurate. However, we were able to select the closest multiple-choice answer, C.

Note: Formulas for areas of two-dimensional shapes are on the official *NJ ASK 8 Mathematics Reference Sheet* in this book (page 265).

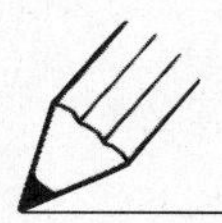

PRACTICE: Area of Shaded Region

(For answers, see page 146.)

1. Use Shape B from page 96. One side of the square = 12 centimeters. What is the area of the shaded region?

_____________ – _____________ = _________________

(Area of square) (Area of circle) (Area of shaded region)

2. Use Shape C from page 96. The square measures 6 feet on each side, and its diameter measures approximately 8.5 feet. What is the approximate area of the shaded region?

_____________ – _____________ = _________________

(Area of circle) (Area of square) (Area of shaded region)

A. 4–5 square feet
B. 6–8 square feet
C. 15–16 square feet
D. 20–21 square feet

3. Use Shape D from page 96. The rectangle has a base of 5 inches and is 8 inches tall. What is the area of the shaded region?

_____________ – _____________ = _________________

(Area of rectangle) (Area of triangle) (Area of shaded region)

4. Use Shape A from page 96. If the area of the large circle = 36 square units, and the area of the shaded region is 14 square units, what is the area of the small circle?

_____________ – _____________ = _________________

(Area of large circle) (Area of small circle) (Area of shaded region)

5. Which figure below does NOT have an area of 100?

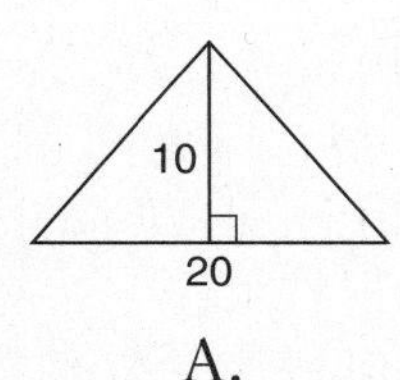

A.

4
25

B.

25
25

C.

40
5

D.

6. Find the area of the irregular shape below. (These are sometimes called *composite* shapes because they are *composed* of two or more geometric shapes.)

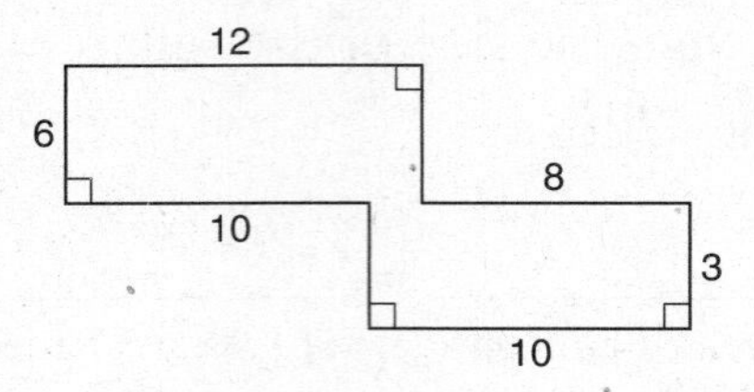

A. 96 square units
B. 90 square units
C. 102 square units
D. 55 square units

7. You are given a rectangle connected to a semicircle. If the height of the rectangle is 6 cm and the width is 4 cm, what is the approximate area of the whole shape?

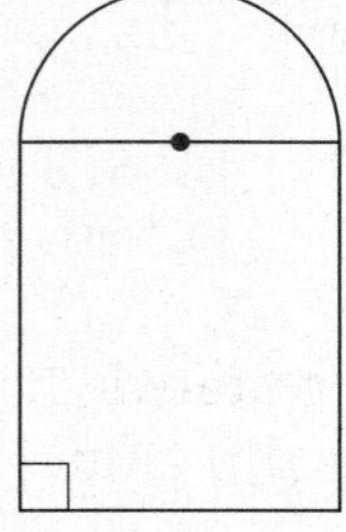

A. 24 square cm
B. 30 square cm
C. 18 square cm
D. 50 square cm

8. There is a rectangular yard that is 30 feet long and 20 feet wide. In the yard there is a circular fountain with a radius of 2 feet, and a path (as shown) that is 3 feet wide. The area left over will be covered with grass seed.

What is the area of the grass seed surface? Round your answer to the nearest whole number. (*Hint:* Label all dimensions on the diagram first.)

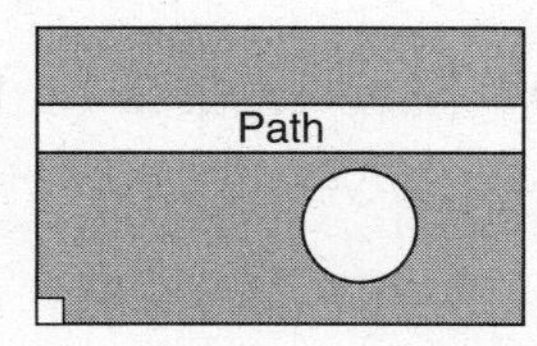

A. 497 square feet
B. 510 square feet
C. 534 square feet
D. 527 square feet

9. There is a 3-foot-wide walkway around a small park. What is the area of the park? See diagram.

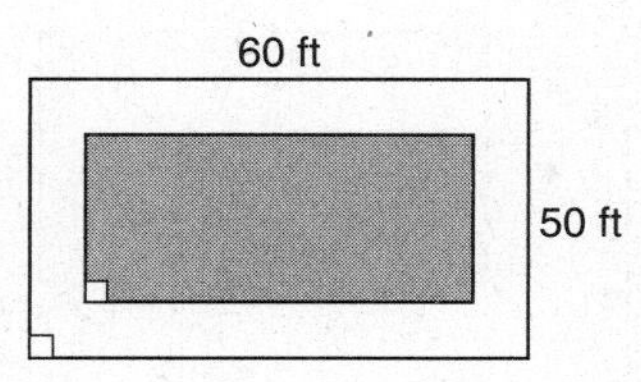

A. 220 square feet B. 2,376 square feet
C. 2,679 square feet D. 3,000 square feet

10. Which one does not have an area of 36 sq. units?

A.

8

9

A right triangle

B.

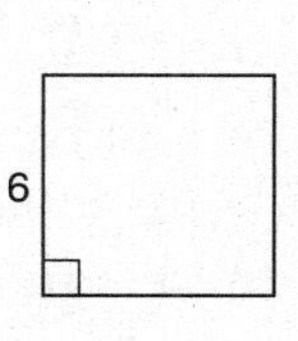

A square

C.

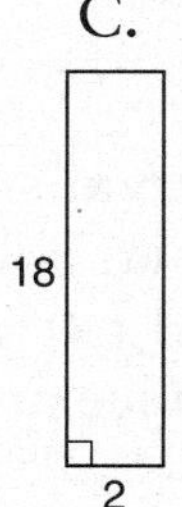

A rectangle

D.

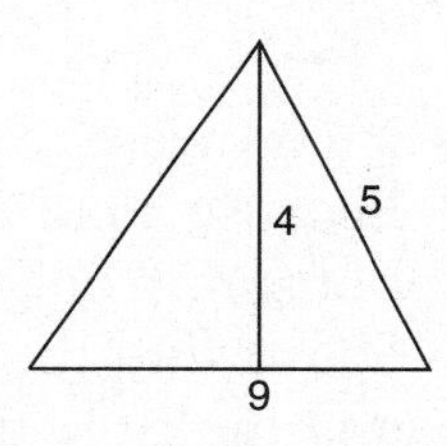

An acute triangle

11. What is the area of the shaded portion inside the rectangle?

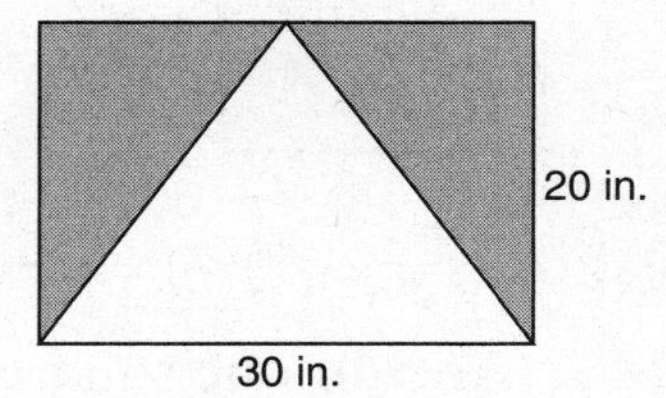

SURFACE AREA OF THREE-DIMENSIONAL FORMS

Sometimes you are asked to think about *flat surfaces* even when you have a three-dimensional form. Don't get this confused with volume. You are still looking for the amount of *surface area*. When you find the surface area of a three-dimensional form, you just find the area of each face (top, bottom, and sides) and then add them together.

Examples

A. If you were to paint the outside of this rectangular box, what is the *total surface area* you would be painting?

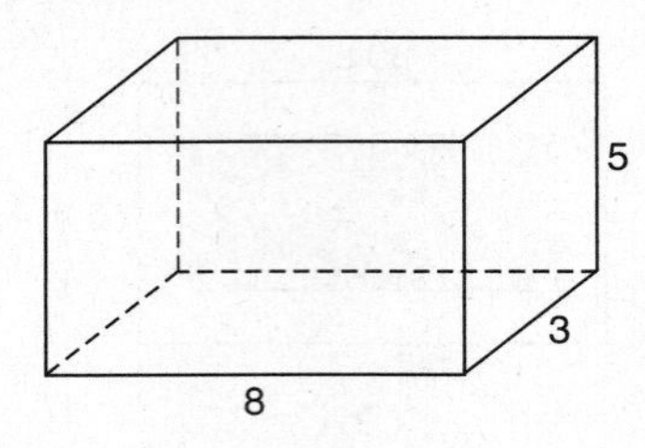

Area of top and bottom	$(b)(h) = (3 \times 8)(2)$	=	48
Area of two sides	$(b)(h) = (3 \times 5)(2)$	=	30
Area of front and back	$(b)(h) = (8 \times 5)(2)$	=	80
Total surface area			158 sq. units

B. If you were to find the total surface area of a cube, it could be much less work. Because a cube is a regular solid, it is made up of six identical squares (top, bottom, front, back, and two sides), and these squares all have the same area. So, to find the total surface area, you just need to find the area of one square and multiply that by 6.

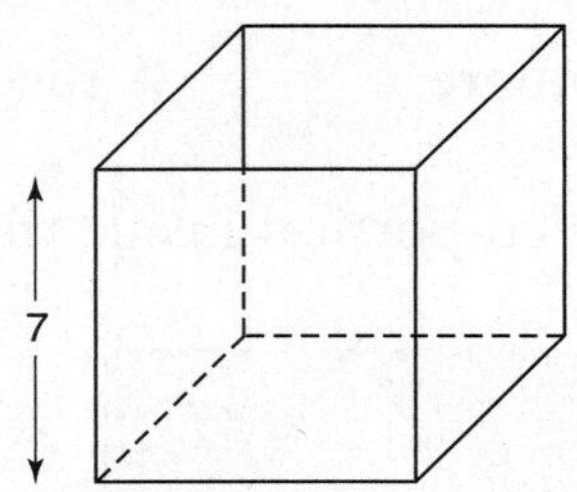

Area of one square = $(s)(s)$ or $(s^2) = (7)(7) = 49$

Surface area of entire cube = (Area of one square)(6) = (49)(6) = 294 square units

C. If you were to paint the outside of this triangular box, what is the total surface area you would be painting?

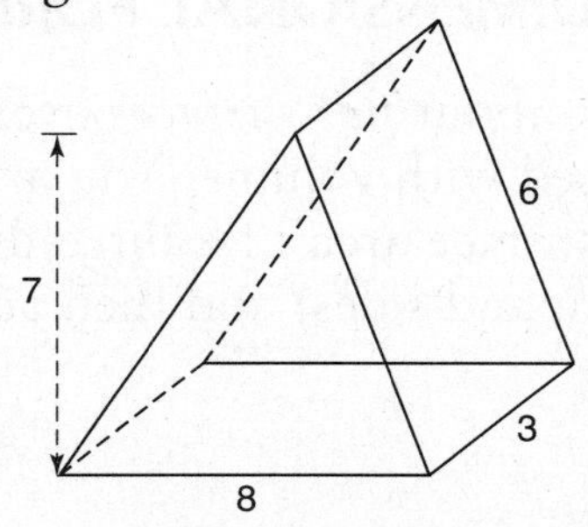

Area of bottom (1 rectangle) $(b)(h) = (3 \times 8)$ $= 24$

Area of sides (2 rectangles) $(2)(b)(h) = (2)(6)(3)$ $= 36$

Area of front and back (2 triangles) $(2)\,\dfrac{(b)(h)}{2} = (8)(7)$ $= \underline{56}$

Total surface area $= 24 + 36 + 56$ $= 116$ sq. units

D. Extended-Response Question: A nursery school teacher asks you to paint a set of wooden blocks. The set contains three different sizes of blocks. There are ten blocks of each size. The blocks are all cubes. She asks you to paint the small and the medium blocks yellow, and the large blocks red. The length of the edge of each small block is 1 inch, the edge of each medium block is 3 inches, and the edge of each large block is 4 inches.

The teacher gives you one can of yellow paint and one can of red paint. Each can will cover 1,000 square inches. Will you have enough paint for all the blocks as she asked? Explain!

This sounds complicated. But if you organize your information carefully, and label all the information, it can be easy. It is best to use a table, label each section, and show all your work.

Blocks (Cubes) Used	Area of 1 Face	Area of 1 Cube (add top, bottom, front, back, 2 sides = 6 faces)	Area of 10 Cubes (with work shown)	Area of 10 Cubes
Small blocks (1 in. each edge)	(1)(1) = 1 sq. in.	(6)(1) = 6 sq. in.	(10)(6) = 60 sq. in.	60 sq. in.
Medium blocks (3 in. each edge)	(3)(3) = 9 sq. in.	(6)(9) = 54 sq. in.	(10)(54) = 540 sq. in.	540 sq. in.
Total surface area of all yellow cubes				**700 sq. in.**
Large blocks (4 in. each edge)	(4)(4) = 16 sq. in.	(6)(16) = 96 sq. in.	(10)(96) = 960 sq. in.	960 sq. in.
Total surface area of all red cubes				**960 sq. in.**

Since each can of paint can cover 1,000 square inches, you would have enough paint. I need 700 square inches of yellow paint and 960 square inches of red paint.

E. Find the approximate surface area of the can below to the nearest square inch. The circumference is about 6.25 inches, and the height is 5.5 inches.

First you should sketch the parts of a can the way they would look if you could lay them flat. A can (really a cylinder) is made of two circles and a rectangle.

$$A_{\text{rectangle}} = l \times w = 6.25 \times 5.5 = 34.375 \text{ square inches}$$

(**Note:** The circumference of the circle is really the same as the length of the rectangle.)

$$A_{\text{circle}} = \pi r^2 = (3.14)\ (?)$$

Even though you don't know the radius, you do know the circumference and the value of π. That can help you find the radius.

C	$= \pi d$	Formula for circumference of a circle.
6.25	$= (3.14)(d)$	Use the values that you know.
$\frac{6.25}{3.14}$	$= \frac{(3.14)(d)}{3.14}$	Divide both sides by 3.14 to solve for d.
1.99	$= d$ (diameter)	Since the radius is half the diameter, the radius = about 1.

$$A_{\text{circle}} = \pi r^2 = (3.14)(1)(1) = 3.14 \text{ square inches}$$

The total area of the can is the area of the rectangle + the area of two circles.

$$34.375 + 2(3.14) = 40.65 \text{ or approximately } 41 \text{ in.}^2$$

VOLUME OF 3-DIMENSIONAL FORMS

Other times you need to think about the total space *inside* a three-dimensional form. You then will be thinking about its *volume*. Since you are working with three dimensions (width, length, and depth), you need to use *cubic units*.

Examples

A. If you were to fill this rectangular solid with water, how many cubic units of water would it hold?

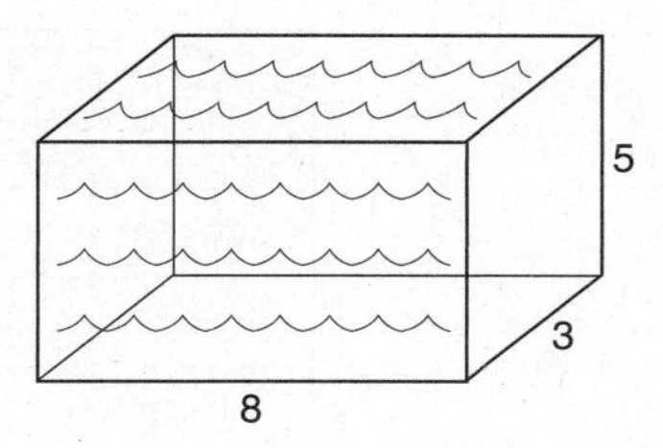

$$\text{Base} \times \text{Height} \times \text{Depth} = \text{Volume}$$
$$b \times h \times d = V$$
$$V = 8 \times 5 \times 3 = 120 \text{ cubic units}$$

B. If you were to fill this rectangular cube with water, how many cubic feet of water would it hold? (*Remember*: A cube is a regular solid; all of its faces are identical.)

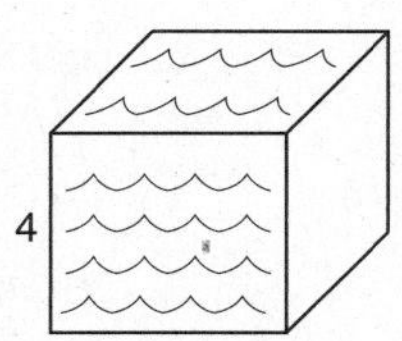

$$\text{Base} \times \text{Height} \times \text{Depth} = \text{Volume}$$
$$b \times h \times d = V$$
$$V = (b)(h)(d) = (4)(4)(4) = 64 \text{ cubic feet}$$

IMPORTANT REMINDERS

There are many other solid forms (for example, cylinders, triangular prisms, and spheres); however, for the ASK 8 test, we will concentrate only on the ones we have shown you here.

Remember the difference between *surface area* and *volume*!

- To find *surface area*, you find the area of each face and add them together.
- To find *volume*, you multiply Length × Width × Depth to see how much will fit inside the solid.
- In a *regular solid*, all faces are identical (Length = Width = Depth).

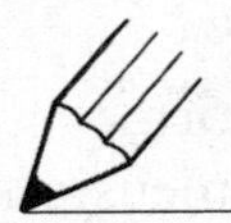

PRACTICE: Surface Area and Volume

(For answers, see page 146.)

1. Given: A cube with each length equal to 3 inches. Find the *surface area* of this form (the area I would paint if I painted the exterior of the cube).

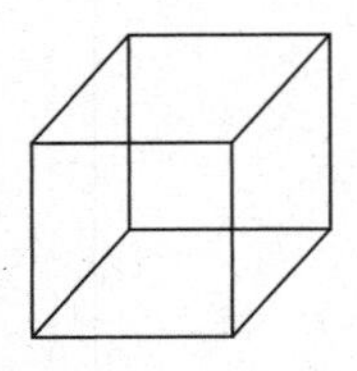

A. 27 sq. in.
B. 36 sq. in.
C. 54 sq. in.
D. 63 sq. in.

2. Given: A box with the following dimensions: Length = 12, Height = 4, Depth = 2

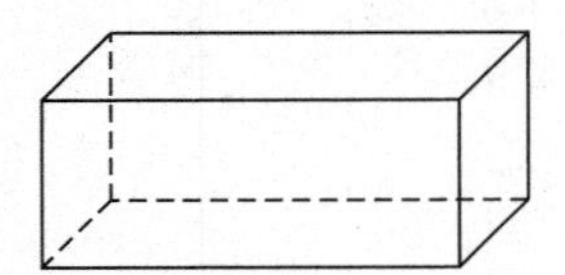

a. What is the surface area (the area I would cover with wrapping paper without overlapping)? (*Hint:* First draw and label the box with the dimensions you are given.)

b. What is the volume of this box? For example, how many 1-inch cube-shaped ice cubes would fit inside this box?

c. Draw another box with the *same volume* but a *different surface area.* What are the dimensions of the new box? Insert your information in the table below.

	Show work	Original box	New box
Length		12 units	___ units
Width		4 units	___ units
Depth		2 units	___ units
Volume		___ cubic units	___ cubic units
Surface area		___ square units	___ square units

3. Two boxes have the same volume.

Box A: Height = 3 in.
Width = 2 in.
Length = 6 in.

Box B: Height = 4 in.
Width = 3 in.
Length = __ in.

What is the length of Box B?

A. 2 in.
B. 3 in.
C. 4 in.
D. 9 in.

4. If a cube has a volume of 64 cubic centimeters, the length of one edge would =

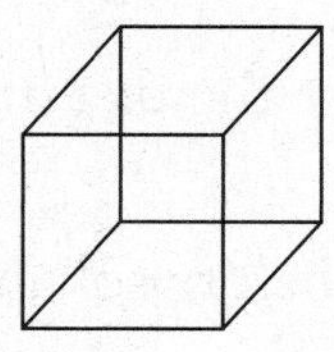

A. 6 cm
B. 4 cm
C. 8 cm
D. 16 cm

5. If a cube has a volume of 27 cubic feet, then the area of one face would =

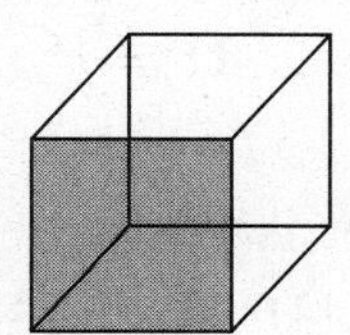

A. 3 sq. ft
B. 6 sq. ft
C. 9 sq. ft
D. 12 sq. ft

6. If a cube has a volume of 8 cubic feet, then what is the perimeter of one of its faces?

A. 4 feet
B. 8 feet
C. 12 feet
D. 16 feet

7. If the perimeter of one face of a cube is 20 inches, what is the volume of that cube?

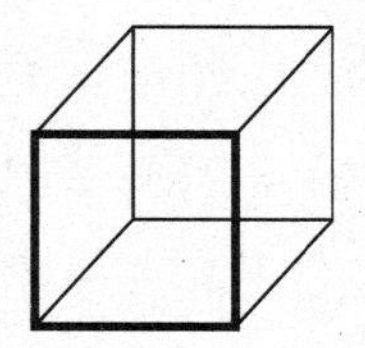

A. 40 cu. in.
B. 60 cu. in.
C. 26 cu. in.
D. 125 cu. in.

PERIMETER

Think *distance around* when you see the word "perimeter." When you are looking for the perimeter of a two-dimensional or plane figure, you are adding the lengths of the sides of the figure. Formulas help us do the arithmetic. Here is the formula for the perimeter of a rectangle.

$$P_{rectangle} = 2(l + w) = 2l + 2w \ (l = \text{length and } w = \text{width})$$

Examples

A. Judy and Dee take a walk around their neighborhood a few times a week. They walk east 900 feet, south 1,200 feet, and then continue in a rectangular shape until they return home again. How far do they walk?

Add the four sides.

$900 + 1{,}200 + 900 + 1{,}200 = 4{,}200$ feet　　　　$2(900 + 1{,}200) = 4{,}200$ feet

Length + Width + Length + Width = Perimeter　**or**　2(Length + Width) = Perimeter

$2l + 2w = P$　　　　$2(l + w) = P$

This is a typical perimeter problem where we measure the distance *around* a shape.

Hint: It is always helpful to draw and label a diagram before solving perimeter problems.

B. Find the perimeter of a rectangle whose width is 3 cm and whose length is twice the width.

Draw a rectangle. Label the lengths: Width (w) = 3 and Length (l) = 6

$$2(l) + 2(w) = P_{rectangle}$$
$$2(6) + 2(3) = P_{rectangle}$$
$$12 + 6 = P_{rectangle}$$
$$18 = P_{rectangle}$$

or

$$2\,(l + w) = P_{rectangle}$$
$$2(6 + 3) = P_{rectangle}$$
$$2(9) = P_{rectangle}$$
$$18 = P_{rectangle}$$

C. Find the length of a side of a square whose perimeter is 64 inches. Remember, in a square, all sides are the same length. Instead of saying Perimeter = Side + Side + Side + Side, we can just say 4 times the length of one side.

Using the formula, substitute 64 for *P*.

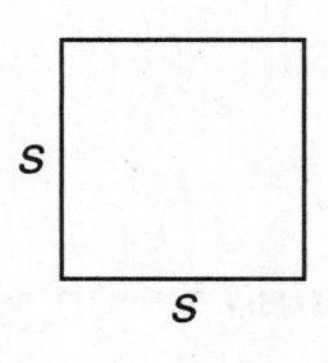

Perimeter of a square

$$P_{square} = 4s$$
$$64 = 4s$$
$$16 = s$$

D. To find the perimeter of a triangle, add the lengths of the three sides together.

$$P_{triangle} = s_1 + s_2 + s_3$$

If $s_1 = 4$ and $s_2 = 5$ and $s_3 = 8$, then　$P_{triangle} = 4 + 5 + 8 = 17$

E. Sometimes you are asked to find the perimeter of *special* triangles. The following is information you should remember about two special triangles. We use tick marks to show which sides are the same length (which sides are congruent).

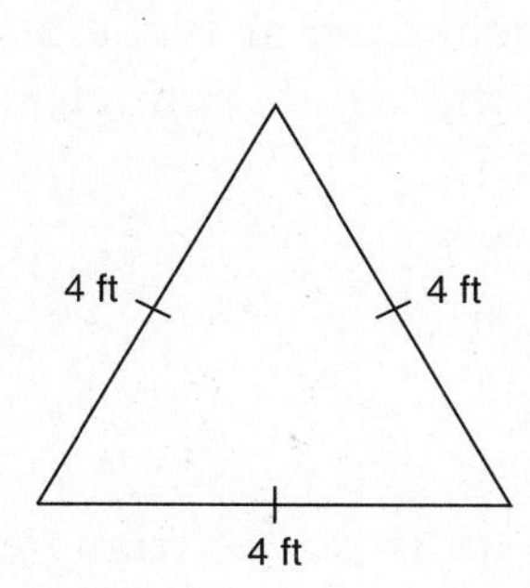

Equilateral triangle

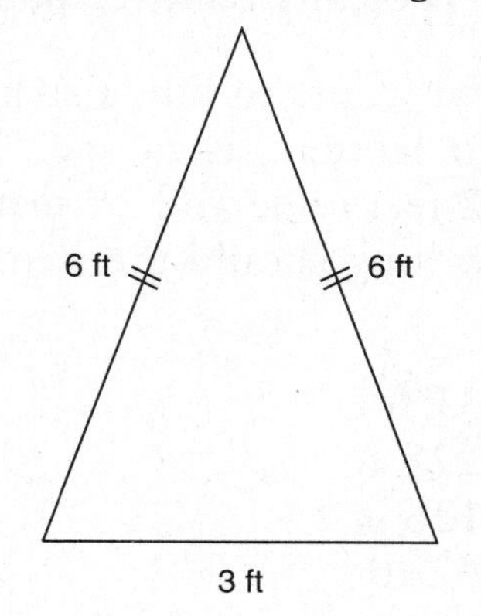

Isosceles triangle

This first triangle is an ***equilateral*** triangle.
All sides are the same length.
If one side = 4 ft, Perimeter = 4 + 4 + 4 = 12 ft
or Perimeter = 3(4) = 12 ft

This is an ***isosceles*** triangle.
Two sides are the same length.
If base = 3 ft and side a = 6 ft,
$P = 3 + 6 + 6 = 15$ or $P = 3 + 2(6) = 15$

F. Circles use different vocabulary. The perimeter of a circle is called its *circumference.* (See the *NJ ASK 8 Mathematics Reference Sheet* on page 265.) The formula used to calculate the circumference is

$$\textit{Circumference} = (\pi)(\textit{diameter})$$

$$\mathbf{C = \pi d}$$

The approximation for π (pi) is 3.14 or 22/7. Find the approximate circumference of a circle whose radius is 4.5 feet. Round to nearest whole number. (*Remember:* We know the diameter is twice the radius, so the diameter is 9 feet.)

$C = \pi d$ or $(3.14)(d)$

$C = (3.14)(9)$

$C = 28.26$ feet

$C = 28$ feet (rounded to nearest whole number)

If you use a calculator, press [π] [×] [9] [=] and see [28.27433388]. Remember that the calculator uses a more accurate estimate for π; it uses 3,141592654 not just 3.14, so the calculator answer will be a little higher.

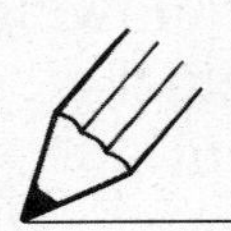

PRACTICE: Perimeter

(For answers, see page 146.)

1. David wants to buy a string of lights to go around his garage. His garage is 12 feet wide and 19 feet deep. How long should the string of lights be?

   A. 31 ft
   B. 228 ft
   C. 126 ft
   D. 62 ft

2. If one side of a regular pentagon measures 8 inches, what is the perimeter of this shape? (*Note*: Remember that, in regular geometric shapes, all sides are the same length.)

   A. 16 in.
   B. 30 in.
   C. 40 in.
   D. 64 in.

3. **a.** If we were to fence off either garden shown below, how much fencing should we buy?

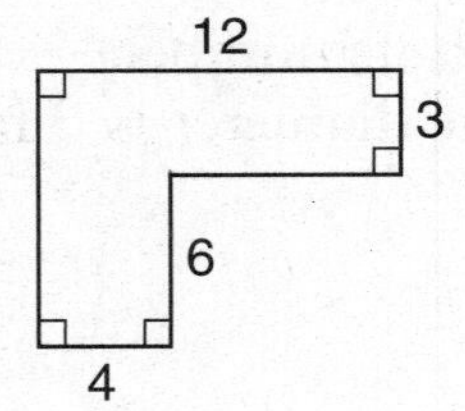

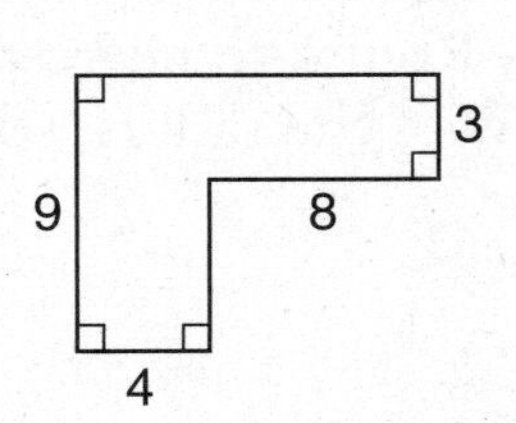

   Answer: ______ feet

   b. What is the area of this garden? Divide each diagram differently and explain why the total areas are still the same. Show all work!

4. If one side of an equilateral triangle measures 9 inches, what is the distance around the whole triangle?

   A. 24 inches
   B. 27 inches
   C. 32 inches
   D. 36 inches

5. If the base of an isosceles triangle measures 10 cm and one side measures 8 cm, what is the distance around the whole triangle?

   A. 18 cm
   B. 26 cm
   C. 40 cm
   D. 80 cm

6. The perimeter of the figure shown is 70 feet. What is the measure of line segment *AB*? (*Hint*: Be sure ALL sides show measurements.)

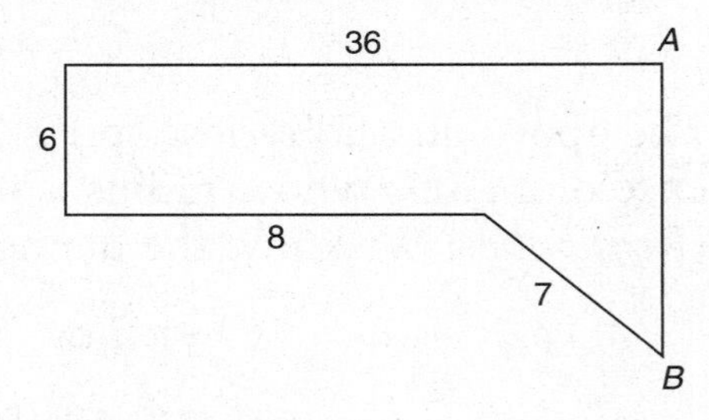

7. The diameter of a circle is 12.6 feet. What is its circumference?

8. The radius of a circle is 8 mm. What is the circumference of this circle?

9. Tom has 100 feet of fencing. He wants to use it to make a rectangular play area for his dog, but he wants to be sure his dog will have the largest area possible. (*Hint*: Be sure to answer all parts of the question completely.)

a. Should it be a long and narrow area or more like a square?

____

b. What should the dimensions of his rectangle be?

____

c. Give examples to support your answer. Explain your answer.

10. A regular pentagon has the same perimeter as this rectangle. How long is each side of the pentagon?

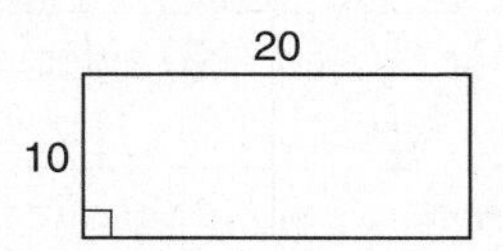

A. 6 ft
B. 7.5 ft
C. 12 ft
D. 15 ft

11. The triangle shown is isosceles. What is the perimeter of this isosceles triangle?

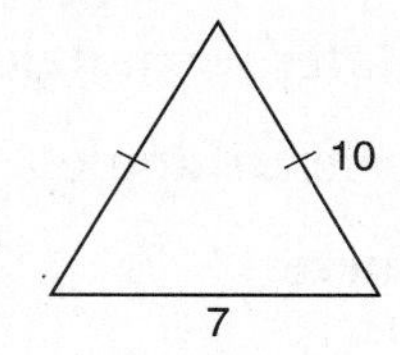

A. 23 inches
B. 27 inches
C. 20.5 inches
D. 25.5 inches

12. The triangle shown is equilateral. What is the perimeter of this equilateral triangle?

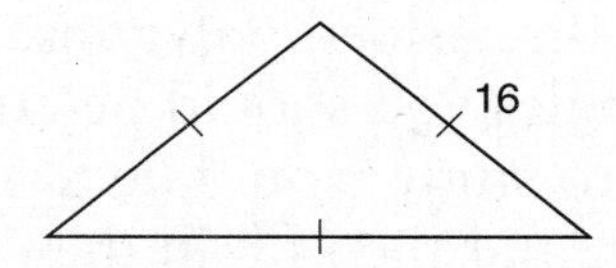

A. 144 cm
B. 24 cm
C. 48 cm
D. 36 cm

13. You reviewed some terms in this section that you should remember when finding perimeter. Match each term on the left with its description on the right.

1. Equilateral triangle
2. Isosceles triangle
3. Perimeter
4. Circumference
5. A regular shape

A. Two sides are the same length
B. The distance around a circle
C. All sides are the same length
D. Any shape where each side measures the same length
E. The distance around a shape

TRIANGLES AND OTHER POLYGONS

TRIANGLES

Triangles are three-sided polygons. Review the following chart to remind yourself of the different triangles you should be familiar with.

One way to study would be to make your own table, draw each triangle (or name each triangle), and then fill in the other columns without looking at the chart.

Drawing of Triangle	Name of triangle	Angles	Sides*
	Scalene triangle	No angles are =	No sides are ≅
	Isosceles triangle	Base angles are =	Opposite sides are ≅
	Equilateral triangle	All angles are =	All sides are ≅
	Obtuse triangle	One angle is greater than 90° (> 90°)	No sides are ≅
	Acute triangle	All angles are less than 90° (< 90°)	
	Right triangle	One angle is 90° (= 90°)	The *hypotenuse* (the side opposite the right angle) is always the longest side.

*The symbol ≅ means "are congruent," same size and shape.

Remember:

- The *sum* of the three angles in any triangle is always 180°.
- The *shortest* side of a triangle is always opposite the *smallest* angle.
- The *longest* side of a triangle is always opposite the *largest* angle.

Examples

A. If one angle of a triangle is 20°, and another is 50°, then the third angle = ?

$$20 + 50 = 70, \qquad 180 - 70 = 110°$$

B. If one angle of a triangle is 120°, and another is 35°, then the third angle =?

$$120 + 35 = 155,\ 180 - 155 = 25°$$

C. If you have a right triangle, and one angle = 40° then the third angle =?

$$90\text{ (right angle)} + 40 = 130, \qquad 180 - 130 = 50°$$

D. In the triangle below, which angle is the smallest? (Do not just guess because it looks the smallest. Use what you know about the relationship of angles to sides in a triangle.)

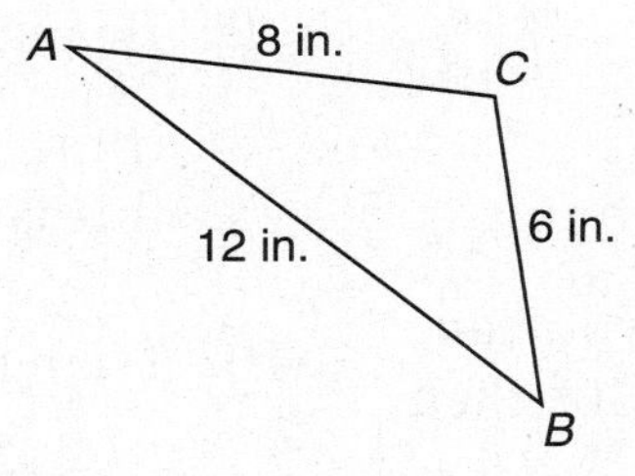

A. $\angle A$
B. $\angle B$
C. $\angle C$
D. not enough information

The answer is $\angle A$ because $\angle A$ is opposite the shortest side, which measures 6 inches.

E. Mark says that a triangle can have two obtuse angles; Aisha says that it cannot. Who is correct? Explain your answer.

If a triangle had two obtuse angles, then it would have two angles greater than 90°. They would add up to more than 180°; however, we know that the three angles in a triangle must add up to 180°, not just two. Aisha is correct.

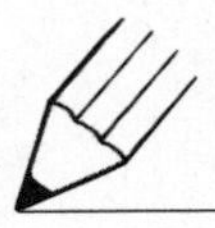

PRACTICE: Triangles

(For answers, see page 147.)

1. Given: A triangle ABC, $m\angle A = 40°$, $m\angle B = 50°$, then $m\angle C =$ _____ .

2. Given: An isosceles triangle DEF; each base angle = 30°, then the vertex angle = _____ .

3. Given: A right triangle, one angle = 20°, then the third angle must = _____ .

4. See the diagram below. Remember, vertical angles are = ; therefore, the arrow is pointing to an angle that measures _____ .

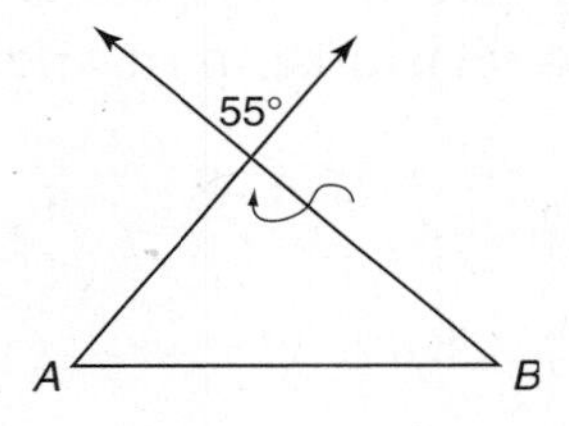

5. Using the same diagram, you can determine that if $m\angle A = 30°$ then $m\angle B =$ _____ .

For questions 6–8 you are given information about the sides of a triangle; classify each of the following triangles as equilateral, isosceles, or scalene.

6. Triangle XYX

   $XY = 12$
   $YZ = 12$
   $XZ = 10$

   ____

7. Triangle UVW

   $UV = 15$
   $VW = 15$
   $UW = 15$

   ____

8. Triangle DEF

   $DE = 9$
   $EF = 7$
   $FD = 12$

   ____

For questions 9–11 you are given information about the angles of each triangle; classify each of the following triangles as acute, right, or obtuse.

9. Triangle ABC

   If $m\angle A = 82°$, $m\angle B = 56°$, then this is ______________ triangle.

10. Triangle MNP

    If $m\angle M = 102°$, then this is ________________ triangle.

11. Triangle QRS

    If $m\angle Q = 50°$ and $m\angle R = 40°$, then this is ______________ triangle.

12. In the triangle below, which angle is the largest angle?

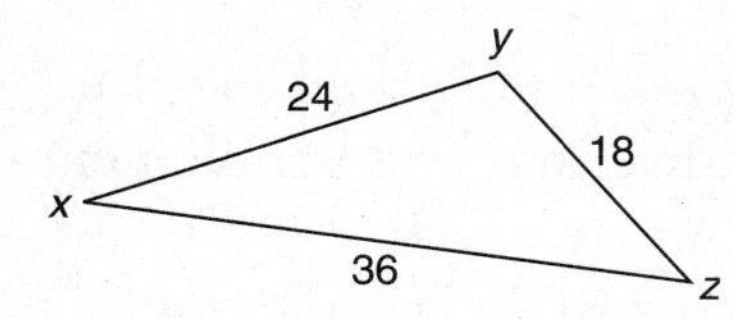

13. Area of triangle *A* is approximately ___ sq. units; area of triangle *B* is about ___ sq. units.

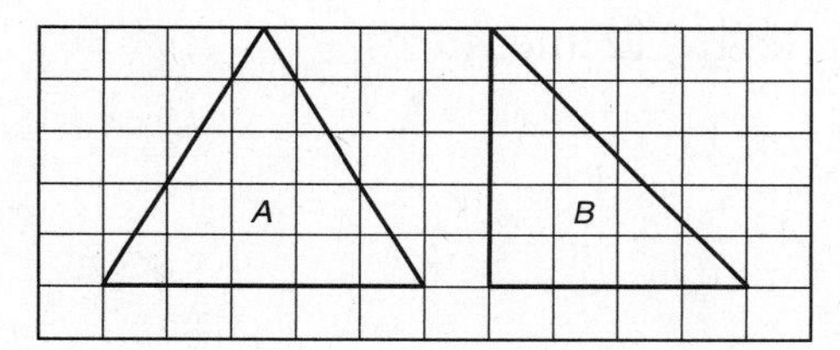

Remember:

- A straight line = 180°
- Vertical angles are equal
- The sum of all angles in a triangle = 180°
- Supplementary angles = 180°
- Complementary angles = 90°

Also remember to look for *overlapping* triangles. Do you see 8 triangles in this rectangle?

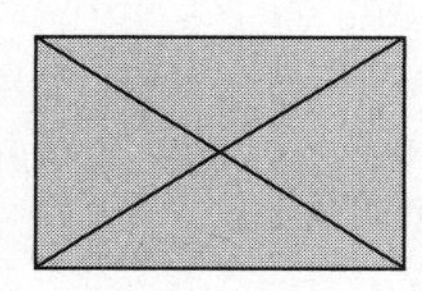

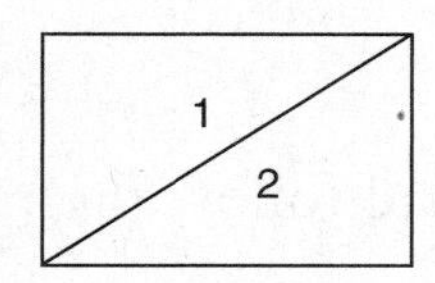

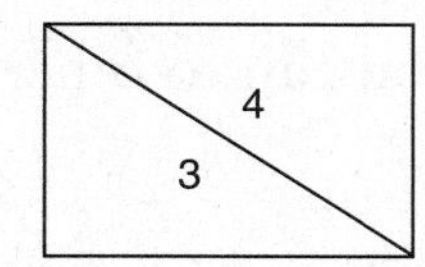

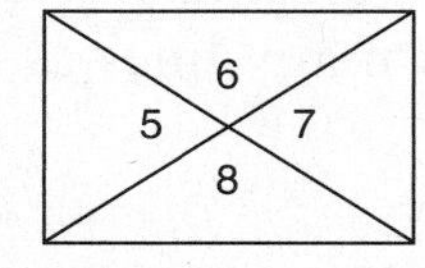

With this knowledge you should be able to answer the following questions.

Use the diagram below. Copy the diagram below onto your own sheet of paper. Label the diagram. As you find each measurement, write it on the diagram.

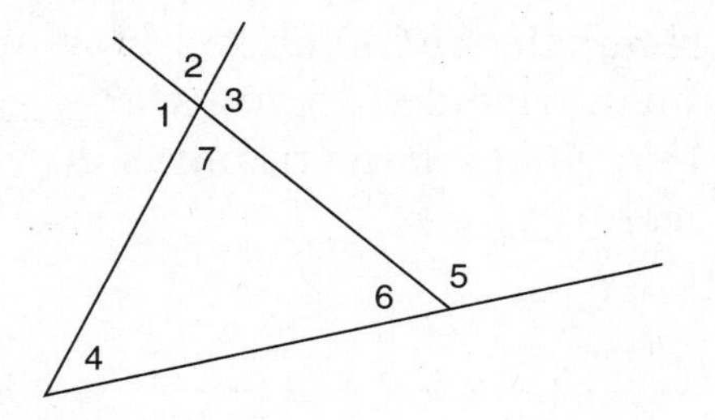

14. If m∠1 = 80°, then m∠3 = _____ because they are _____________ angles.

15. If m∠1 = 80°, then m∠2 = _____ because together they add up to _____ .

16. If m∠4 = 30°, then m∠6 = _____ because the three angles in a triangle add up to _____ . (*Hint*: Find m∠7 first.)

17. Now you can determine that m∠5 = _____ .

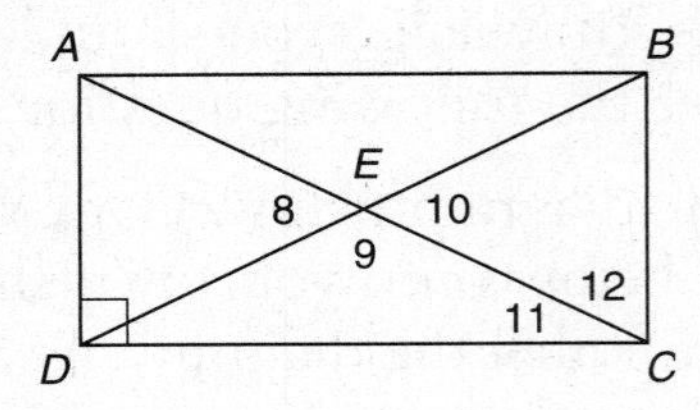

18. Look at the overlapping triangles in rectangle $ABCD$ above. How many small triangles do you see? _____ How many large triangles do you see? _____

19. In the rectangle shown, $m\angle 8 = 60°$; therefore, $m\angle 10 =$ _____ .

20. If $m\angle 8 = 60°$, then $m\angle 9 =$ _____ , because they are supplementary.

21. Since $ABCD$ is a rectangle, $m\angle C = 90°$. Name three other 90° angles: ____________ , ____________ , and ____________ .

22. If angle 11 measures 54°, then angle 12 measures _____ .

OTHER POLYGONS

Find the Sum of the Interior Angles of a Polygon

Because we know that the sum of the interior angles of a triangle = 180°, we can now find the sum of the interior angles of other polygons.

Example

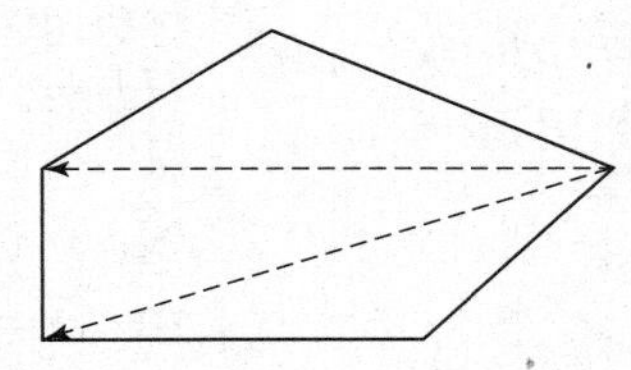

Use one of the basic polygons shown on page 88 and follow these steps.

1. Start at one vertex and draw a straight line to another vertex.
2. Continue from that same vertex and draw as many lines as you can to other vertices.
3. Now, count the number of triangles you have.
4. Notice that if you had chosen a five-sided polygon you would have been able to draw three triangles. Since one triangle has 180°, then three triangles would have $180 \times 3 = 540°$. This is the *sum of the interior angles* of that five-sided polygon.

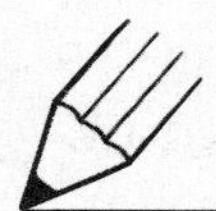

PRACTICE: Other Polygons

(For answers, see page 148.)

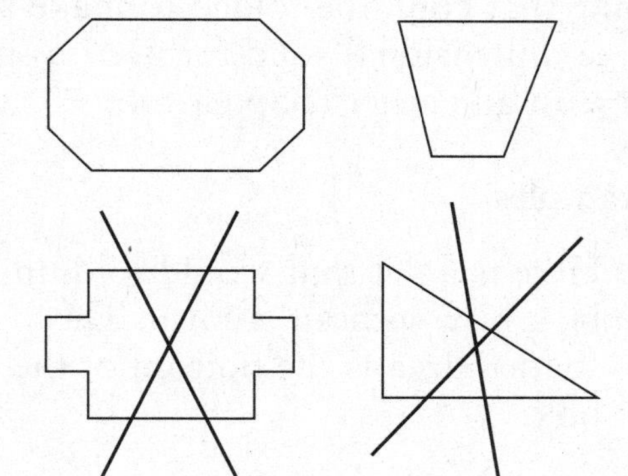

1. Draw a polygon with more than three sides. Use a ruler.

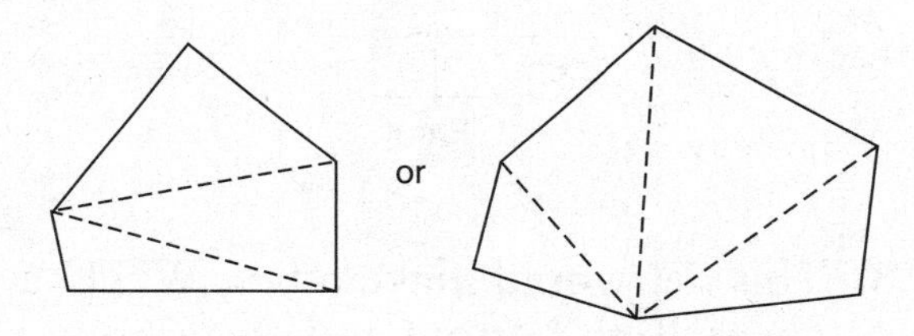

2. Use your drawing and start at one *vertex* (one point) and draw a line to every other vertex. Notice you are dividing the polygon into triangles. Use a ruler please.

3. How many sides does your polygon have?

4. How many triangles were you able to draw?

5. How many degrees in the sum of the interior angles of one triangle?

    _____

6. How many degrees in the sum of the interior angles of your polygon?

    _____

7. Some shapes are really different shapes put together. We call the new shape

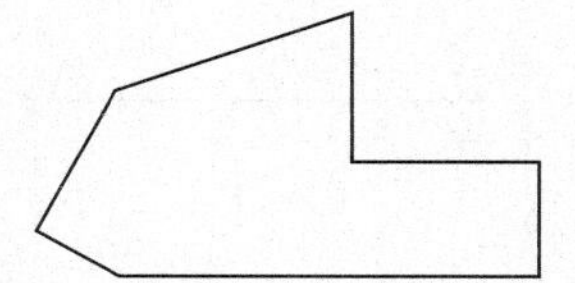

A. a regular shape
B. a new shape
C. a crooked shape
D. a composite shape

8. How many sides does a *pentagon* have? It has _____ sides.

A. 4
B. 5
C. 6
D. 8

9. If I had a hexagon and all sides were congruent and all angles were equal then it would be

A. an isosceles hexagon
B. a regular hexagon
C. a right-angled hexagon
D. an irregular hexagon

10. What is the area of a square that measures 6 feet on each side?

A. 36 sq. ft
B. 24 sq. ft
C. 30 sq. ft
D. 12 sq. ft

11. What is the perimeter of a rectangle that measures 3 cm long and 6.2 cm wide?

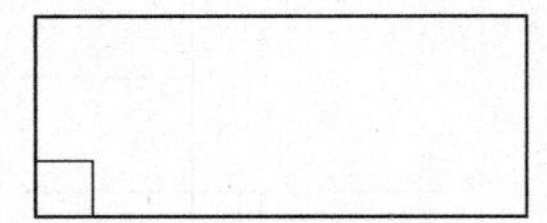

A. 13 cm
B. 18.4 cm
C. 12 cm
D. 9.2 cm

12. What is the area of the following shape that is made up of different rectangles? (*Hint:* Divide the shape into rectangles first.)

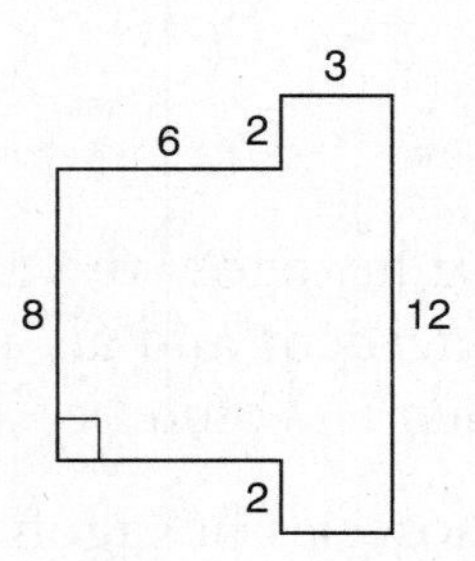

A. 42 sq. units
B. 78 sq. units
C. 84 sq. units
D. 92 sq. units

A *net* is another term you should know. A net is just a plan (a *pattern*) on a flat plane that could be folded to make a three-dimensional solid form. You saw these in an earlier chapter, too.

Examples

A. Here is a net that would fold up to be a box without a cover. The shaded area is the bottom of the box.

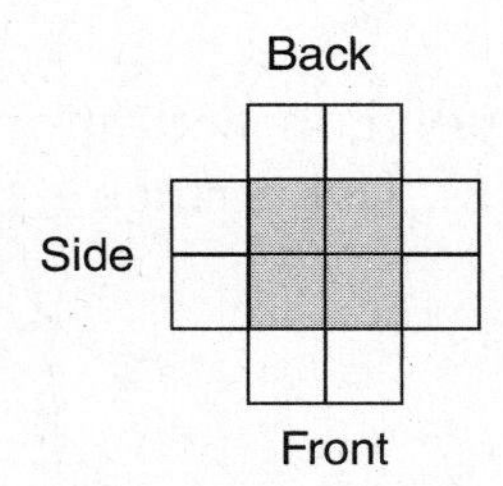

B. This net would make a box WITH a cover. Both bottom and cover are shaded.

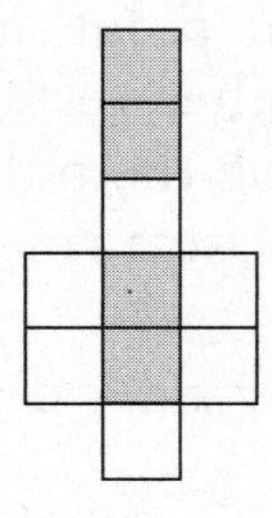

13. Which net will make a rectangular covered box (without overlapping)? The bottoms of the boxes have been shaded in to help you see the form.

A.

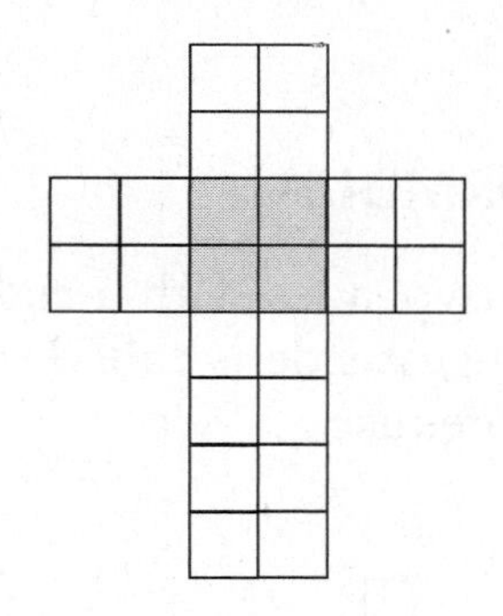

B.

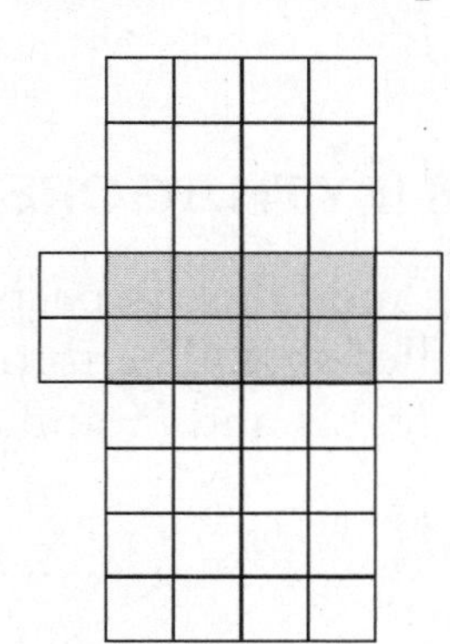

C.

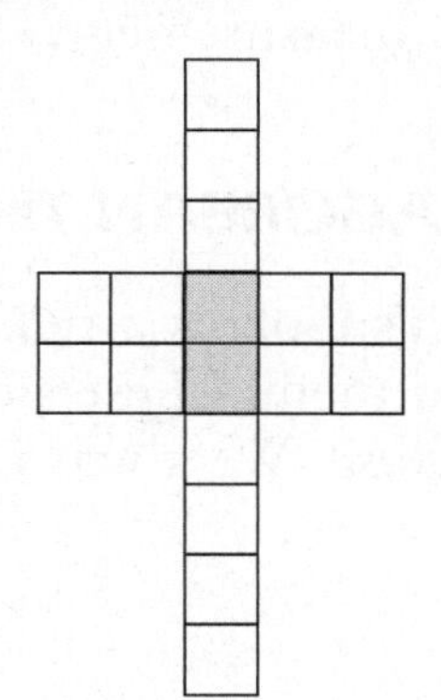

14. All of the nets drawn below are incorrect. Select one and correct it so it can be cut and folded into a box with a cover. Shade in the bottom and top of your corrected net. Please use a ruler and pencil.

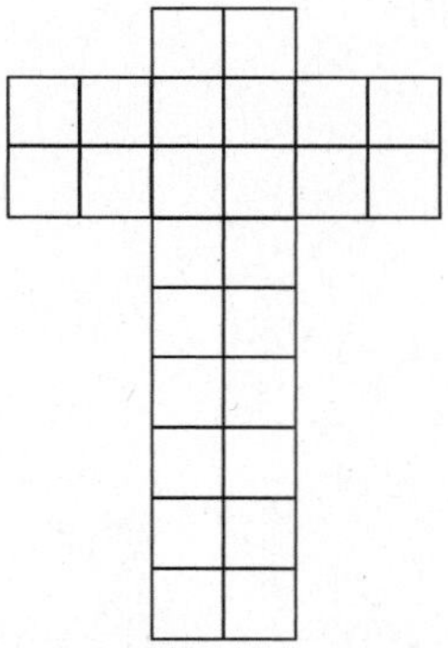

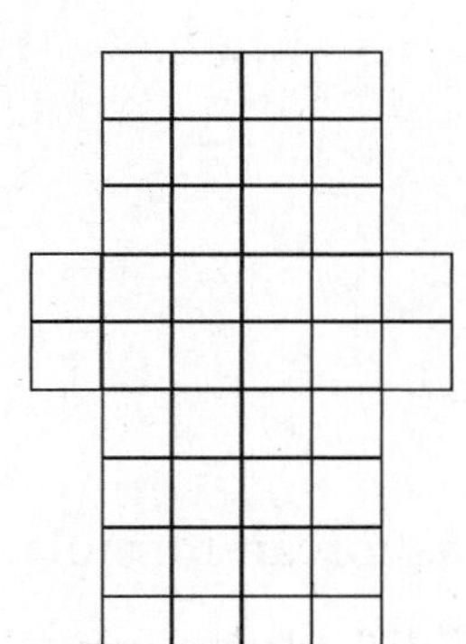

RIGHT TRIANGLES

When we study right triangles, we study the relationships between their angles and their sides. This study is called *trigonometry.* The word comes from a Greek word that means triangle measurement.

PYTHAGOREAN THEOREM (PYTHAGOREAN FORMULA)

First, let's look at a right triangle and label its parts. It has two sides with a right angle between them. These sides are called *legs*. The third and longest side is called the *hypotenuse*. We always label the legs a and b, and the hypotenuse c.

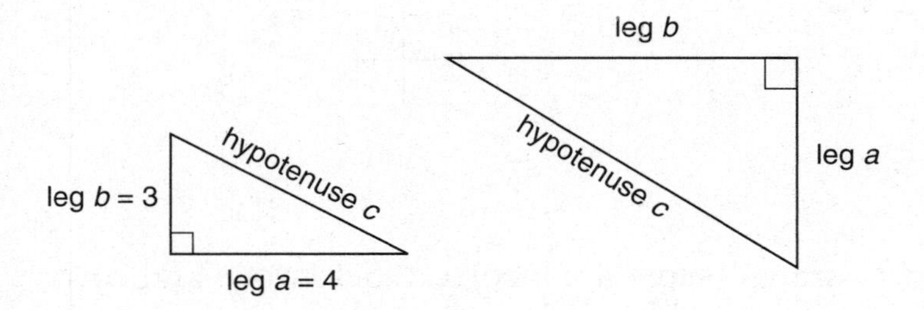

The theorem says that

$$(\text{Leg } a)^2 + (\text{Leg } b)^2 = (\text{Hypotenuse})^2$$

$$a^2 \quad + \quad b^2 \quad = \quad c^2$$

Examples

A.

- Write the general Pythagorean formula. $a^2 + b^2 = c^2$
- Replace the variables with the numbers you are given. $(3)^2 + (4)^2 = (c)^2$
- Do the computation (square the given numbers). $9 + 16 = (c)^2$
- Combine like terms (add the numbers). $25 = (c)^2$
- Simplify (take the square root of both sides). $\sqrt{25} = \sqrt{c}^{\,2}$
- Remember, c is the hypotenuse of the right triangle. $5 = c$

B. If one leg of a right triangle is 6, and the hypotenuse is 10, how long is the other leg? Notice that the hypotenuse is always the longest side and is labeled c in the formula.

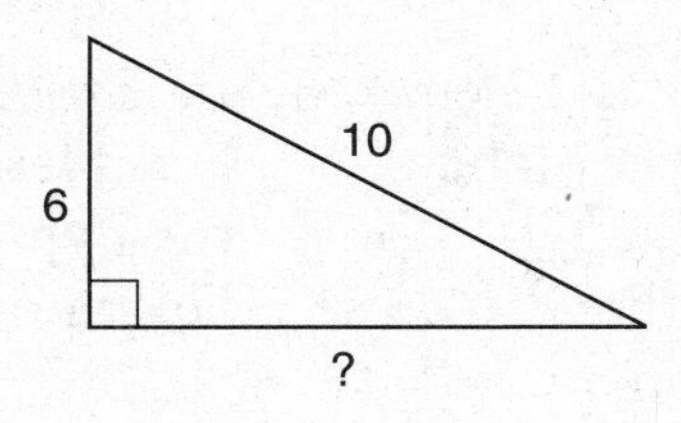

$$a^2 + b^2 = c^2$$
$$6^2 + b^2 = 10^2$$
$$36 + b^2 = 100$$
$$-36 \qquad\qquad -36$$
$$b^2 = 64$$
$$\sqrt{b^2} = \sqrt{64}$$
$$b = 8 \text{ (length of other leg)}$$

C. If one leg of a right triangle is 5, and the other leg is 4, how long is the hypotenuse? Here the answer is not a whole number. It may be easiest to use your calculator.

$$c^2 = a^2 + b^2$$
$$c^2 = 5^2 + 4^2$$
$$c^2 = 25 + 16$$
$$c^2 = 41$$

You should first *estimate* the answer.

Since $6^2 = 36$, and $7^2 = 49$, we know that the square root of 41 is between 6 and 7. To get an exact answer we can use a calculator.

Press [41] [second] [$\sqrt{x}$] [=]. You will see [6.4], the length of c (the hypotenuse).

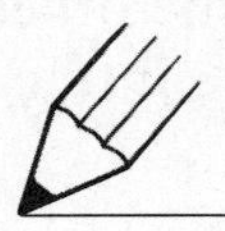

PRACTICE: Right Triangles

(For answers, see page 148.)

1. If one leg of a right triangle is 12 cm long, and the other leg is 9 cm long, how long is its third side (the hypotenuse)? (*Hint*: Label the diagram, write the Pythagorean formula, show all steps, and use your calculator when appropriate.)

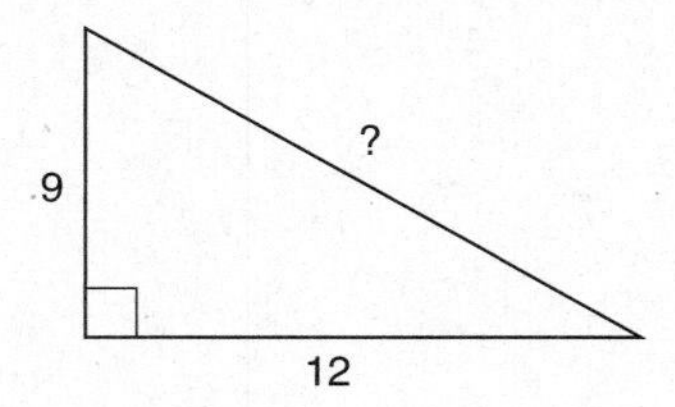

A. 10 cm long
B. 54 cm long
C. 15 cm long
D. 21 cm long

2. If leg *a* in a right triangle measures 6 inches, and leg *b* measures 5 inches, how long is the third side (the hypotenuse)? (*Hint*: Label the diagram, write the Pythagorean formula, show all steps, and use your calculator when appropriate.)

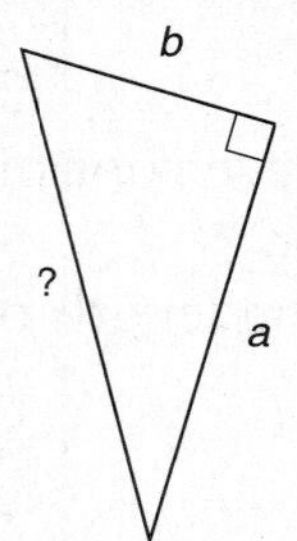

A. 8 inches long
B. 5 inches long
C. 11 inches long
D. 7.8 inches long

3. *Be careful, this one is a little different.* Here you are given the length of the hypotenuse and have to find the length of one of the legs.

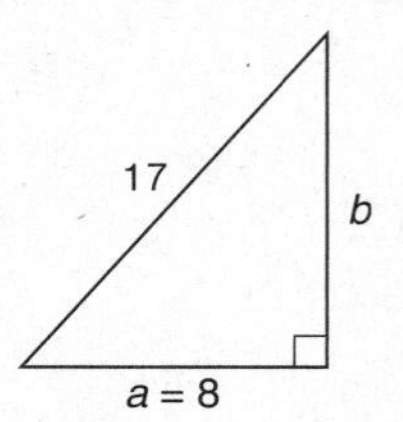

Given: a right triangle, leg $a = 8$ cm and the hypotenuse = 17 cm, how long is leg *b*? (*Hint*: Label the diagram, write the Pythagorean formula, show all steps, and use your calculator when appropriate.)

A. 15 cm
B. 12 cm
C. 9 cm
D. 6 cm

4. The window of a burning building is 24 feet above the ground. The base of the ladder is leaning against the wall and is 10 feet from the bottom of the wall. How tall must the ladder be to reach the window?

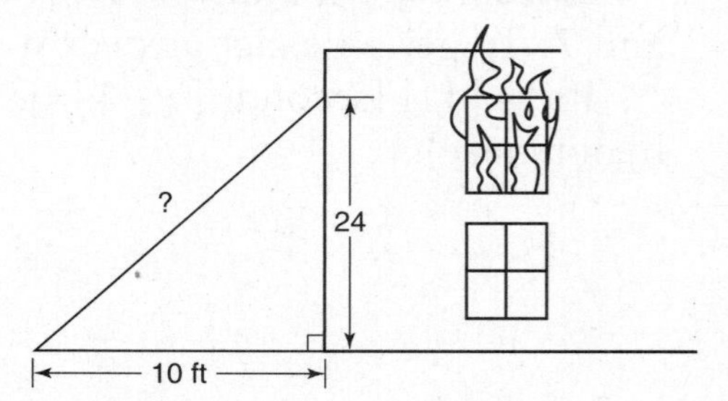

A. 30 feet tall
B. 26 feet tall
C. 24 feet tall
D. 20 feet tall

5. Jessica is in her backyard at *A* and her house is at *B*. How much shorter is her path along the diagonal compared to the walk along the sides of the rectangle?

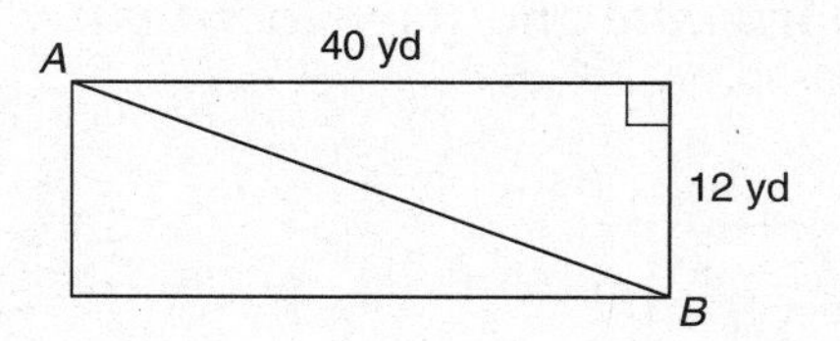

6. An isosceles right triangle has sides of length 5 meters each. Find the length of the hypotenuse to the nearest whole number. Show your work. (*Hint*: Use your calculator to find the square root.)

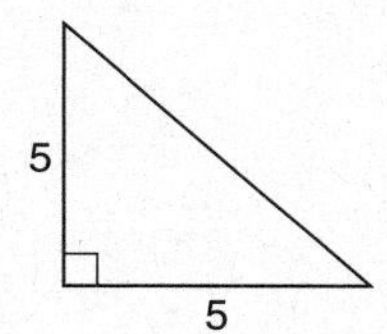

7. A 15-foot ladder is 6 feet away from the wall. How far up the wall does the ladder go? Show your work. (*Hint:* Label the diagram first.)

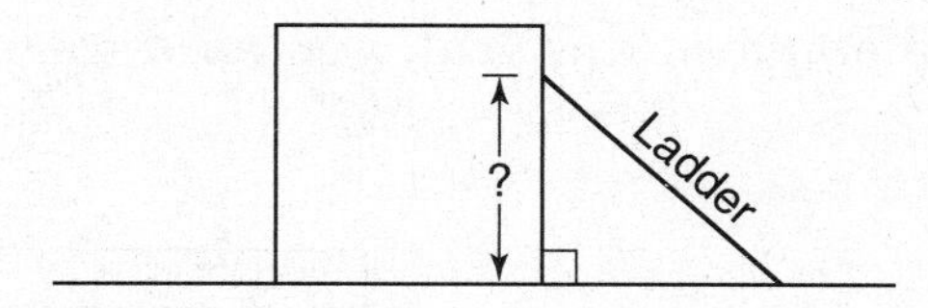

8. A square's sides are 8 cm long. Approximately, how long is its diagonal (the dotted line)? (*Hint*: Use your calculator to find the square root, or estimate.)

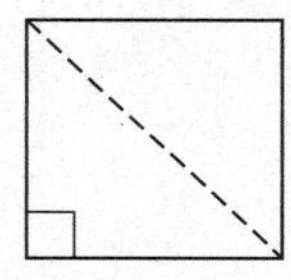

A. 9
B. 10
C. 11
D. 15

COORDINATE GEOMETRY

PLOTTING POINTS

The grid below is called a *coordinate plane.* A coordinate plane is formed by two number lines called axes. The horizontal number line is called the *x-axis* and the vertical number line is called the *y-axis.* The point where the number lines meet is called the *origin.* Each section of the coordinate plane is called a *quadrant.* To plot a point or to locate a point on this grid, you need to give its *x-coordinate* and its *y-coordinate.*

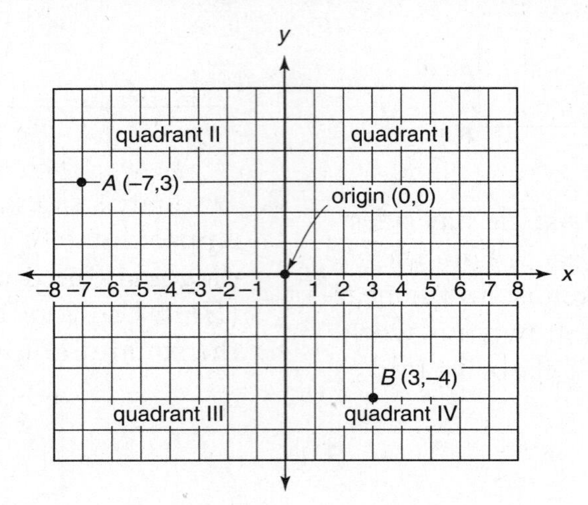

Example

To help us locate a point we give it an *ordered pair* of numbers. The first number gives you the x-coordinate and the second number gives you the y-coordinate. Notice that the x-coordinate tells you how to move left or right on the axis; the y-coordinate tells you how to move up or down on the axis. Always begin at the origin (0, 0) when you are counting left or right, up or down.

$B(3, -4)$ $A(-7, 3)$

(x, y) (x, y)

Steps to Graphing Any Point (x, y)

1. Start at the origin (0, 0).
2. Move x units right or left along the horizontal x-axis.
3. Then move y units up or down along the vertical y-axis.
4. Draw the point and label it with a capital letter.

AREA AND PERIMETER ON THE COORDINATE PLANE

Now that we have reviewed how to plot points on the coordinate plane, it will be very easy to connect the points to form different geometric shapes. It also will be easy to find the area or perimeter of these geometric shapes.

Examples

A. Find the area of triangle ABC. See the triangle on the coordinate plane below.

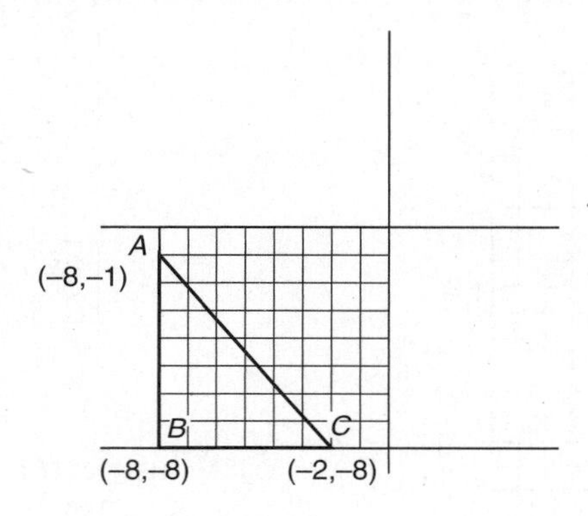

$$A\ (-8, -1) \quad B\ (-8, -8) \quad C\ (-2, -8)$$

BC (base) = 6 units long and AB (height) = 6 units long,

$$\text{Area triangle} = \frac{bh}{2} = \frac{(6)(6)}{2} = \frac{36}{2} = 18 \text{ square units}$$

B. Find the area and perimeter of the rectangle. Draw rectangle at $EFGH$ on the grid below. Start with E at (0,0), F at (0, 4), G at (7, 4), and H at (7, 0). Label these points. Now you can see that EF = 4 units and EH = 7 units.

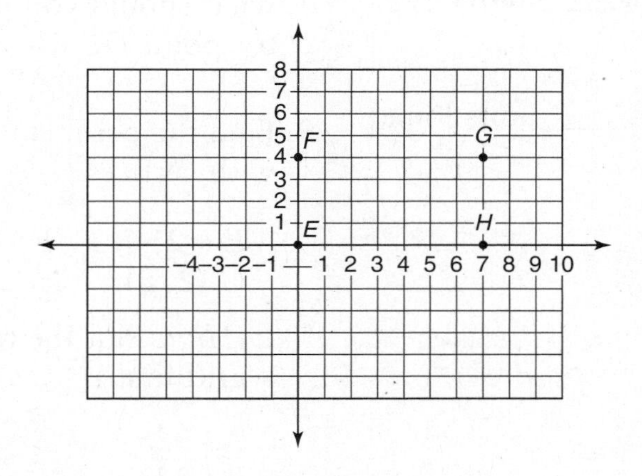

$$\text{Area} = bh = (7)(4) = 28 \text{ sq. units}$$

$$\text{Perimeter} = 2(w) + 2(l) = 2(4) + 2(7) = 8 + 14 = 22 \text{ units}$$

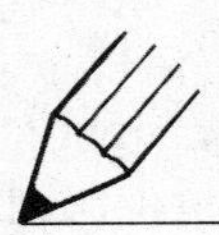

PRACTICE: Coordinate Geometry

(For answers, see page 149.)

Find the length of the line segments in the following four examples. Plot the points and connect the coordinates to help you see the length of each line.

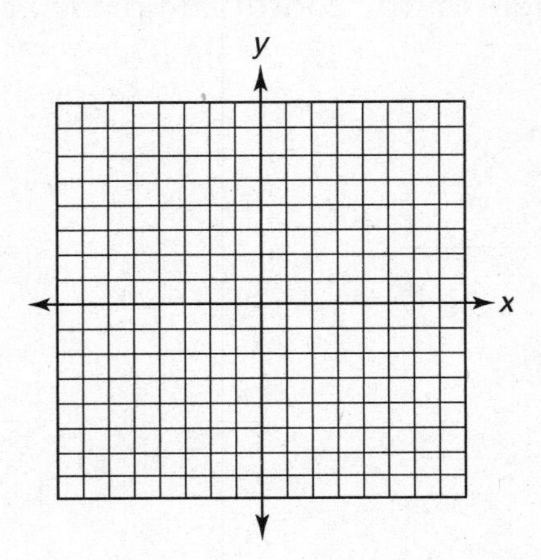

1. *A* (0, 0), *B* (6, 0)
   Line segment *AB* is ____ units long.

2. *F* (5, 5), *G* (5, 0)
   Line segment *FG* is ____ units long.

3. *C* (–4, 2), *D* (6, 2)
   Line segment *CD* is ____ units long.

4. *R* (4, –3), *S* (4, 6)
   Line segment *RS* is ____ units long.

Refer to the graph below to answer questions 5–8.

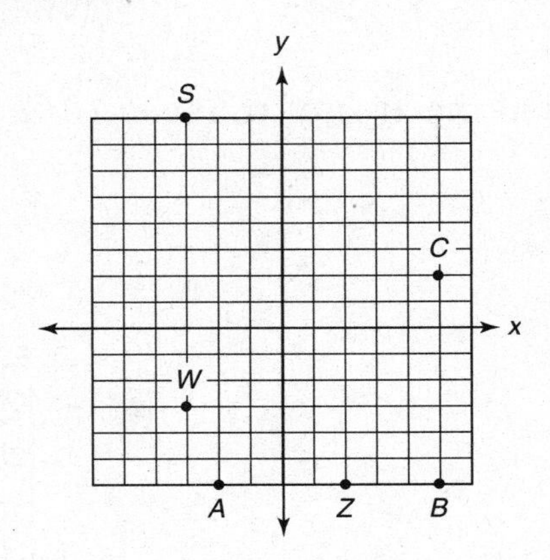

5. What letter is located at coordinates (–3, –3)?

6. What letter is located at (5, 2)?

7. What are the coordinates of the letter *B*?

8. If we wanted to connect *A*, *B*, *C*, with a point *D* to make a rectangle, what should you use as coordinates for point *D*?

9. Find the perimeter of a rectangle with vertices

   E (–2, 0) *F* (–2, 8)
   G (2, 8) *H* (2, 0)

   (*Hint*: Plot the rectangle on the grid first.)

   A. 32
   B. 22
   C. 12
   D. 24

10. Find the area of a rectangle with vertices P (–5, –2), Q (–5, 8), R(4, 8), and S (4, –2).

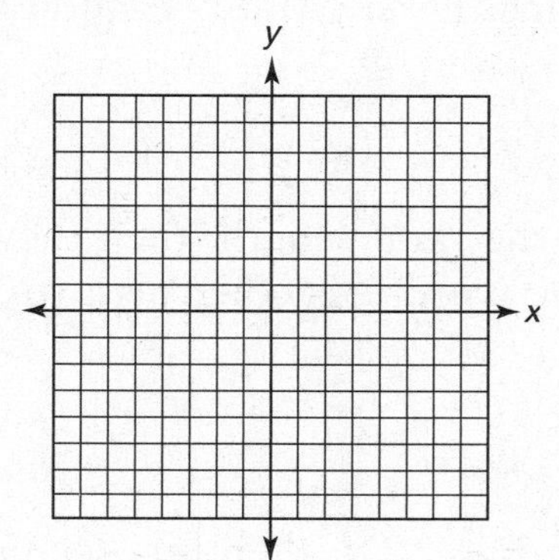

A. 90
B. 36
C. 38
D. 19

11. Remember to use a ruler to draw lines, and to show all work.

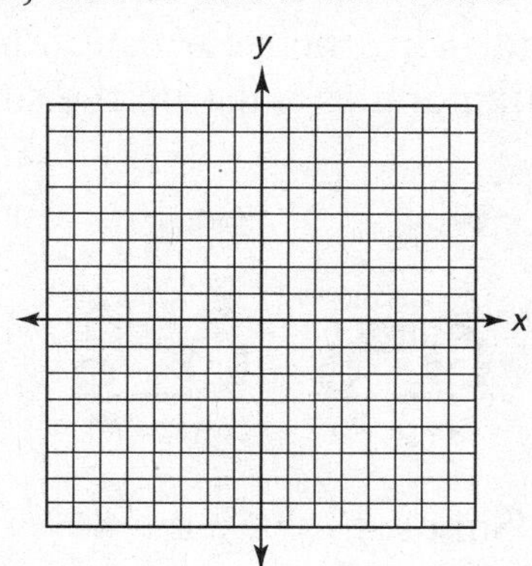

- Plot the following points on the grid: A (2, –2), B (2, 4), C (6, 4), D (6, 1), E (4, 1), F (4, –2)
- Connect the points from A to B, from B to C, from C to D, from D to E, from E to F.
- What is the perimeter of this shape?
- What is the area of this shape?

12. Find the area of a rectangle with vertices P (–5, –2), Q (–5, 6), R (4, 6), and S (4, –2). (*Hint*: Plot the rectangle on the grid first.)

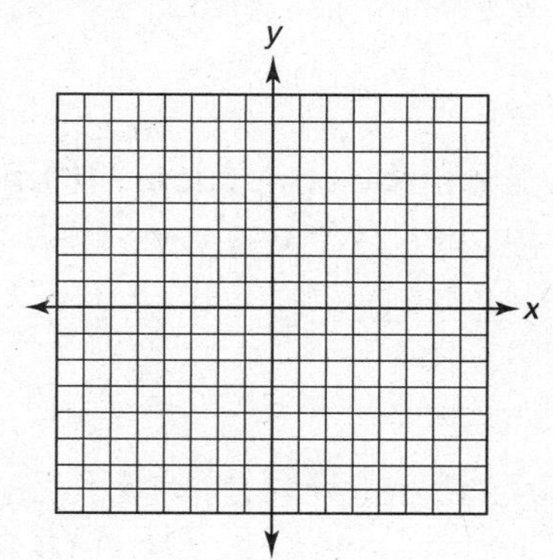

A. 34 sq. units
B. 72 sq. units
C. 56 sq. units
D. 90 sq. units

13. Find the perimeter of the triangle with vertices A (0, 0), B (0, 6), and C (–8, 0). (*Hint:* Plot the triangle on the coordinate plane first and remember, if necessary, you can use the Pythagorean theorem to find the length of the hypotenuse.)

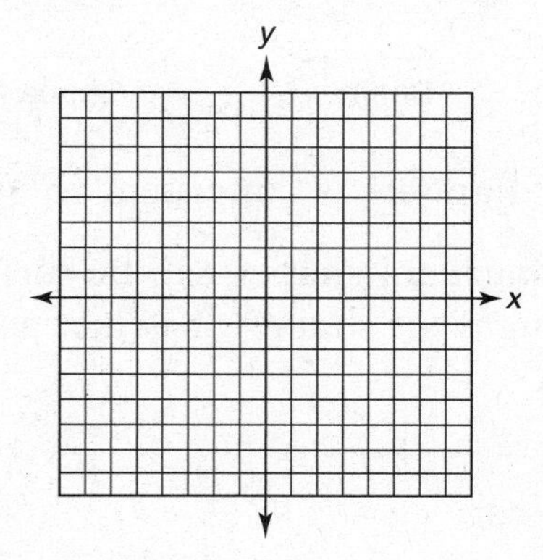

A. 24 units
B. 12 units
C. 48 units
D. 60 units

CONGRUENCY

When you take a geometry course in high school, you will learn a great deal about lines and shapes and how to determine if they are congruent. For now, we'll just review some main ideas and work with some simple shapes and lines.

What does congruent mean? *Congruent* really means the SAME SIZE and SHAPE.

Examples

A. Lines can be congruent. If line segments have the same length, then they are congruent. We use the symbol ≅ to replace the words "is congruent to."

A ———— B

D ———— E

If line segment AB = 15 ft, and line segment DE = 15 ft, then line segment AB ≅ line segment DE ($AB \cong DE$).

B. Angles can be congruent. If angles have the same measure, they are congruent.

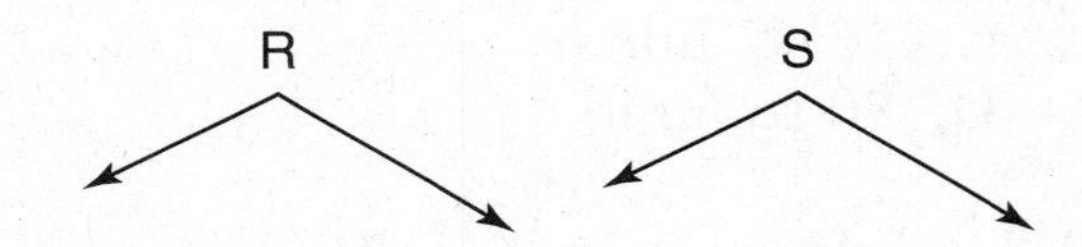

Angle R measures 120°.
Angle S also measures 120°.
Angle R is congruent to angle S.

C. Shapes can be congruent. For shapes to be congruent, their corresponding sides must be the same length, and their corresponding angles must be the same measure.

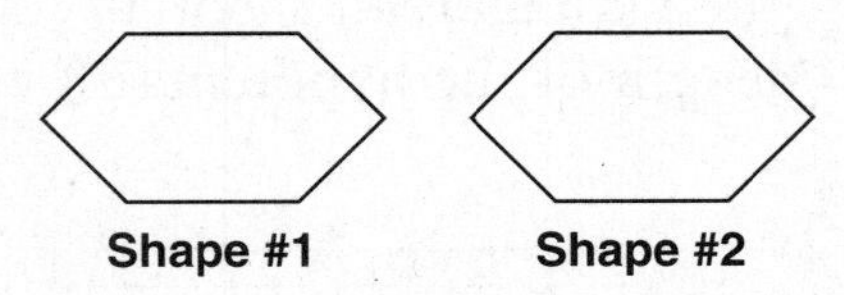

Shape #1 is congruent to Shape #2

Shape #3 ≅ Shape #4

D. Congruent shapes can be turned in different directions, but they are still congruent shapes. It is just a little more difficult to recognize them sometimes.

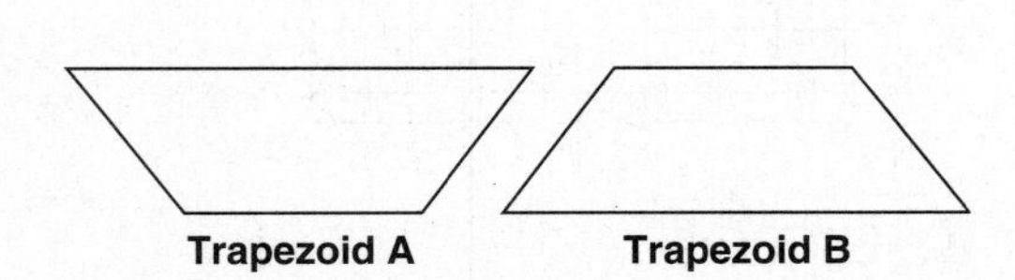

Trapezoid A ≅ Trapezoid B

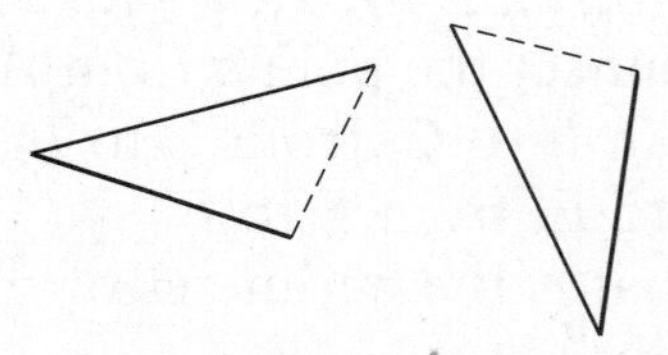

Triangle on the left ≅ Triangle on the right

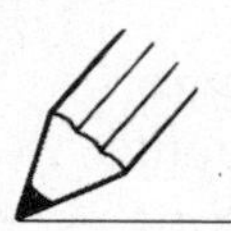

PRACTICE: Congruency

(For answers, see page 149.)

For this section it is a good idea to work on graph paper and use a ruler.

1. In triangle ABC, $m\angle A = 45°$, and $m\angle B = 90°$. In triangle PQR, $m\angle P = 45°$, and $m\angle Q = 90°$. Are these triangles congruent?

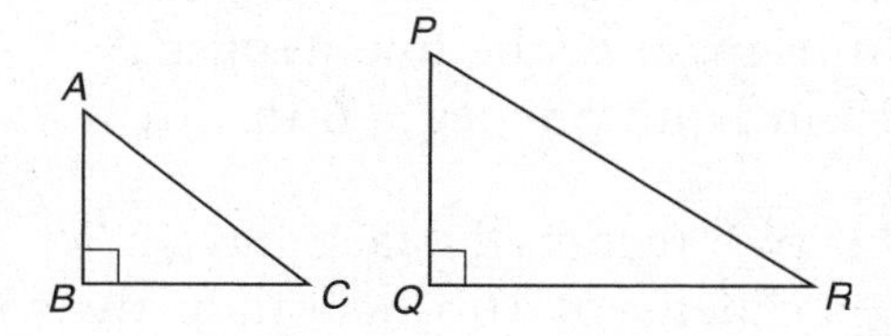

A. Yes
B. No
C. Not enough information

2. You are told that the corresponding angles of these two rectangles are *equal*, and the lengths of their sides are shown on the diagrams below. Are these rectangles congruent?

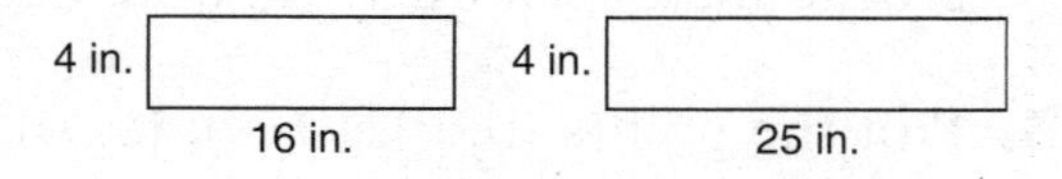

A. Yes, because their correspon-ding angles are equal.
B. No, because their correspon-ding sides are not equal.
C. Yes, because they look exactly the same.
D. Not enough information given.

3. Plot the following points and connect them to make two rectangles: A to B to C to D; then connect Q to R to S to T.

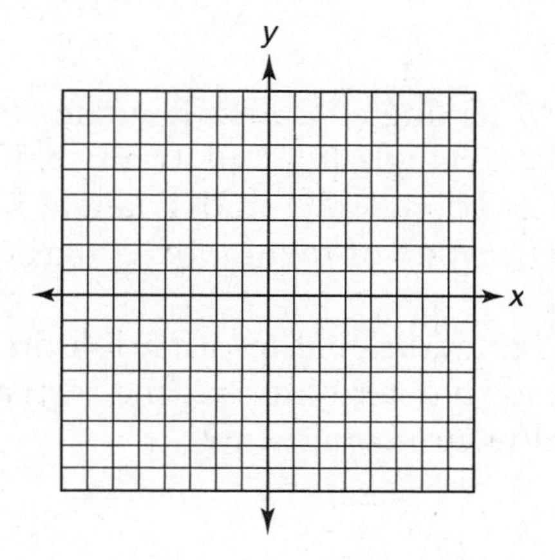

A (–4, 2), B (–4, 6), C (–1, 6), D (–1, 2)
Q (4, 3), R (4, 7), S (7, 7), T (7, 3)

Are these congruent rectangles? Show your work and explain your answer.

4. If two quadrilaterals have the same perimeter, are they always congruent? Show all work and draw diagrams to support your answer.

5. If two squares have the same perimeter, are they always congruent? Show all work and draw diagrams to support your answer.

6. Which two shapes *appear* to be congruent?

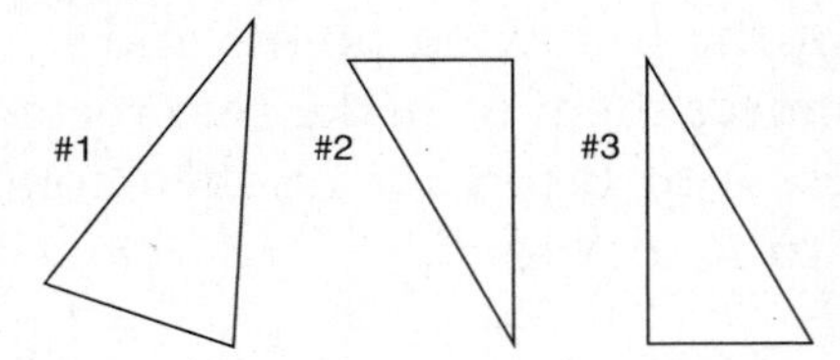

A. Triangle #2 and Triangle #3
B. Triangle # 1 and Triangle #2
C. Triangle # 1 and Triangle #3
D. None of them *look* congruent

7. What additional information do you need to determine that the rectangles below are congruent?

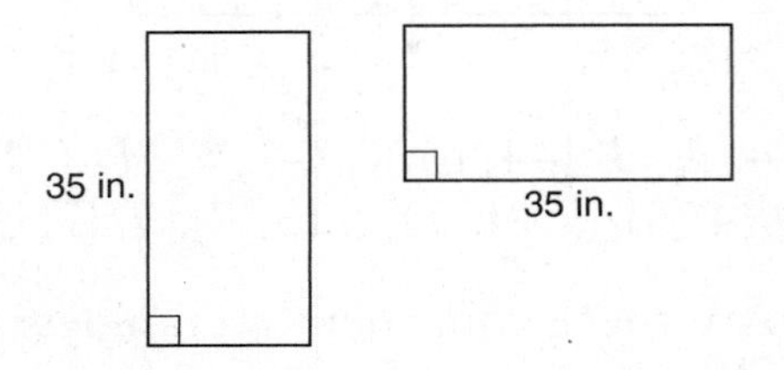

A. The measure of their angles
B. The length of the other long sides
C. The length of the short sides
D. No additional information is needed

8. Are these right triangles congruent? Show your work and explain your answer.

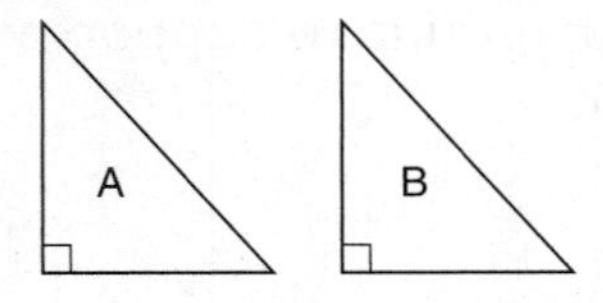

In triangle *A*, one leg = 3 in., and the other leg = 4 in.
In triangle *B*, the hypotenuse = 5 in., and one leg = 3 in.

9. Are these right triangles congruent? Show your work and explain your answer.

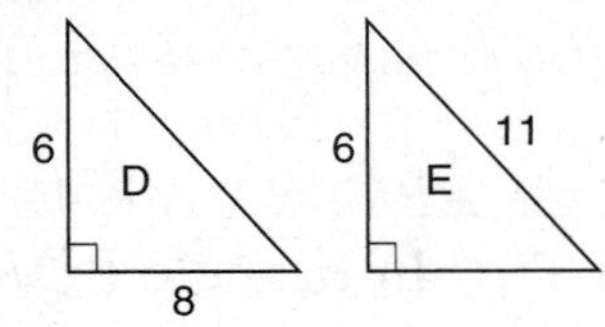

In triangle *D*, one leg = 6 in., and the other leg = 8 in.
In triangle *E*, the hypotenuse = 11 in., and one leg = 6 in.

10. If two rectangles each have a perimeter of 100, will they always be congruent rectangles? Give an example and explain your answer.

11. Are all circles congruent? Give an example and explain your answer.

12. Draw two congruent rectangles and explain why they are congruent.

13. Draw two squares that are *not* congruent and explain why they are not.

14. Plot the points (0, 0), (0, 4), (5, 0), and (5, 4), and connect them to make a rectangle. Now, plot the points (1, –1), (6, –1), and (6, –5). What should the other coordinate be if this is a rectangle *congruent* to the first one?

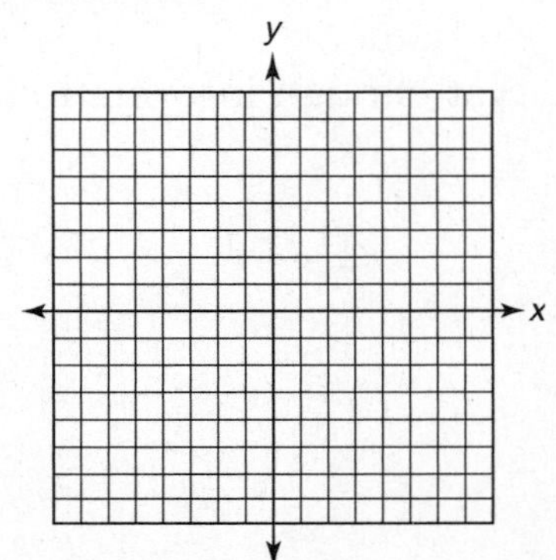

TRANSFORMING SHAPES

Transforming shapes means we are taking a plane figure and all of its points and moving them to another location according to specific rules. We might move a rectangle 3 units to the left, or move a triangle 5 units down. There are different ways to move a shape.

The new figure is usually congruent to (an exact copy of) the original. The exception is an image of a figure under dilation. Translations, reflections, rotations, and dilations are different types of transformations. Many computer games use transformations.

Examples

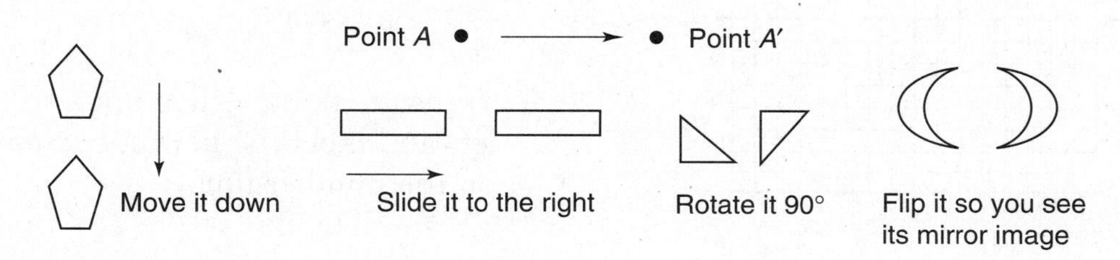

TRANSLATION

The result of a movement in one direction is a transformation called *translation*. A translation moves all points the same distance and in the same direction. Think of a train speeding along a track to the station. (When the front of the engine moves 10 yards ahead, the back of the engine moves 10 yards ahead, too.)

To move shapes a particular distance we need to first place them on a grid. See the figure to the right for a few reminders and vocabulary terms to remember.

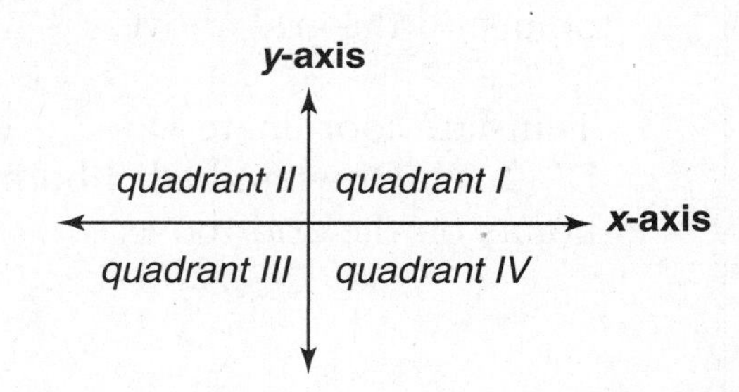

Translating Coordinate Points

Remember, a *translation* is the same as sliding a figure to a new location.

Example

Translate a coordinate (a point on the grid):
Coordinate A (5, 1) has been translated down 2 units to coordinate A′ (5, –1).

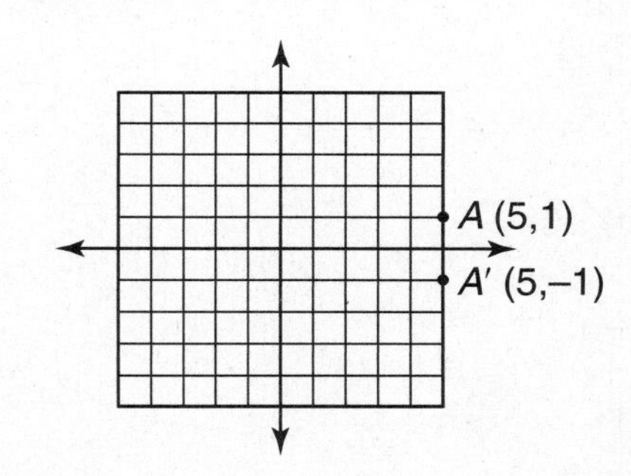

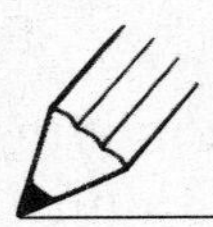

PRACTICE: Translating Points

(For answers, see page 150.)

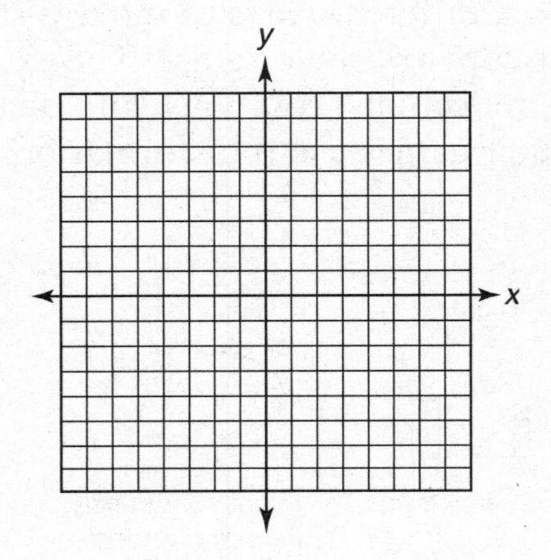

1. Translate a coordinate horizontally. Translate coordinate B (2, 4) to coordinate B' (–2, 4.) Draw and label both points on the grid above.

2. Translate coordinate C (1, 3) horizontally, then vertically to C' (–1, –3). Draw and label both points on the grid above.

3. Translate coordinate D (--2, –1) to D′ (2, 1). Draw and label both points on the grid above.

4. Translate point D four units to the right and label the new coordinate D' on the number line.

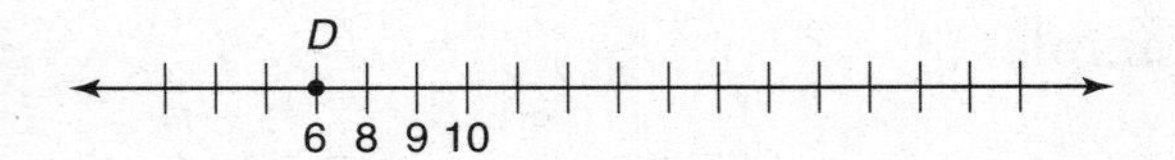

5. Translate point A five units to the left and label the new coordinate A' on the number line.

Translating Polygons

Examples

A. Translate a rectangle 6 units down.

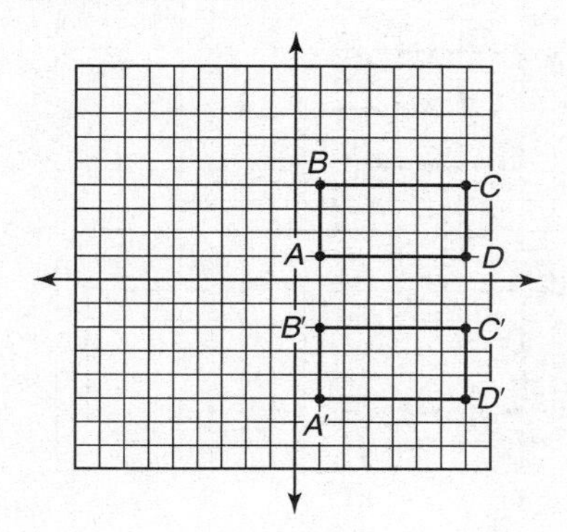

Rectangle *ABCD* has been translated down 6 units to rectangle *A′B′C′D′*.
From: *A* (1, 1), *B* (1, 5), *C* (8, 5), *D* (8, 1)
To: *A′* (1, –5), *B′* (1, –1), *C′* (8, –1), *D′* (8, –5)

Notice that there is no change in the *x*-values of the points. The difference between the *y*-values of the original point at its image is 6 for all four points.

B. Translate a triangle 11 units to the right.

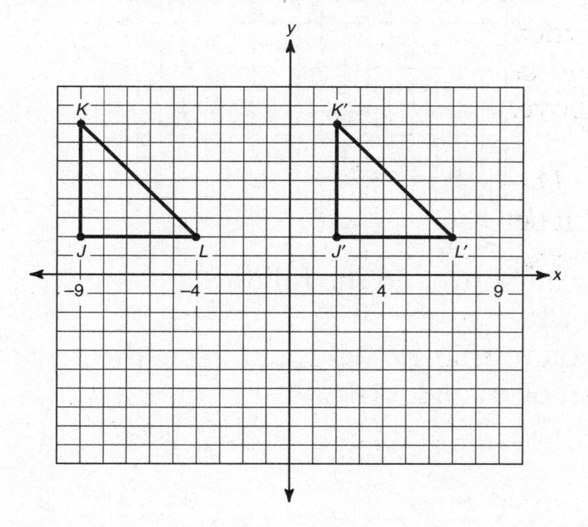

Triangle *JKL* has coordinates *J* (–9, 2), *L* (–4, 2), and *K* (–9, 8). It has been translated to *J′* (2, 2), *L′* (7, 2), and *K′* (2, 8), triangle *J′K′L′*. Notice that all points on the polygon moved (translated) 11 units to the right: *J* to *J′* (11 units to the right), *K* to *K′* (11 units to the right), and *L* to *L′* (11 units to the right). All points on the shape moved the same distance (11 units to the right). Translating a figure horizontally will show that there is no change in the *y*-axis coordinates.

C. Translate from quadrant II to quadrant III; then translate the same shape to quadrant IV. Here we'll take rectangle *PQRS* that begins in quadrant II and slide it down into quadrant III, and then we'll slide it right into quadrant IV.

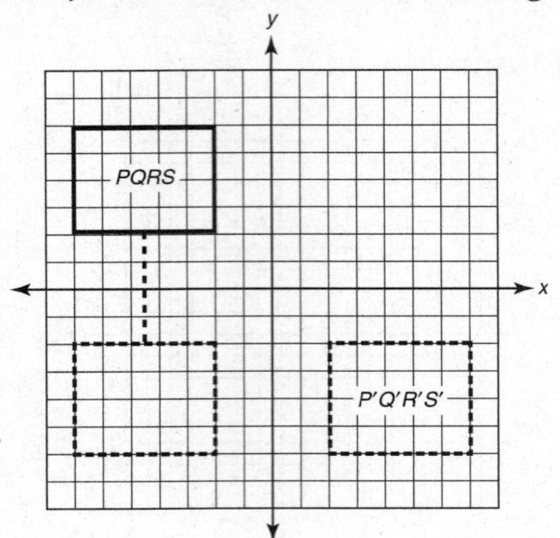

D. Triangle *ABC* has coordinates *A* (1, 4), *B* (1, 0), and *C* (4, 2).

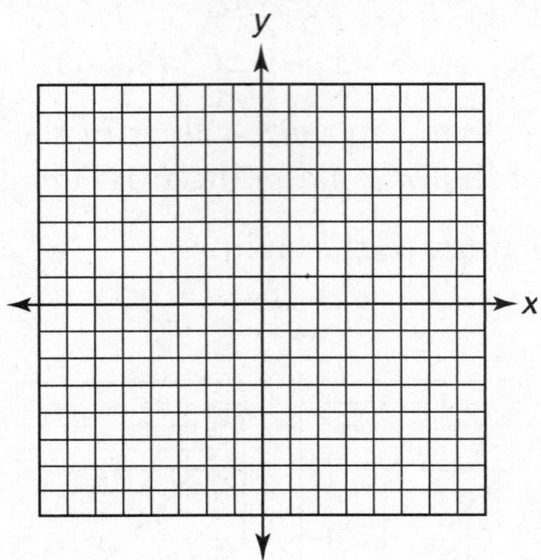

In the coordinate grid above, do the following:

1. Draw triangle *ABC*.
2. Draw a reflection of triangle *ABC*, and label the new triangle *A′B′C′*.
3. Draw a rotation of triangle *ABC*.

REFLECTION

Examples

A. To easily understand *reflection* think back to what you remember about symmetry. What do the following letters have in common?

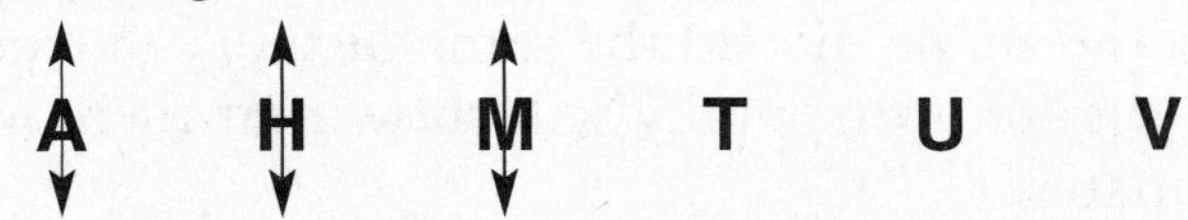

They all have a *vertical line of symmetry*. When you draw a vertical line through the center of each letter, you see that the shape on one side is a *mirror image* of the shape on the other side.

B. Now, think of letters in the alphabet that have a *horizontal line of symmetry.*

B **E** Can you think of any other letters with a horizontal line of symmetry?

C. Some letters have many lines of symmetry. How many lines of symmetry does the letter **X** or the letter **O** have? The letter **X** has two lines of symmetry, but the letter **O** has an infinite number of lines of symmetry.

D. Now think of some geometric shapes. How many lines of symmetry does a square have? It has four lines of symmetry—vertical, horizontal, and along the two diagonals. ⊠

In a *reflection*, the figure is reflected in a line on the plane. If you traced the figure on paper and cut it out, the new location of the figure would be "flipped" over the line of reflection.

E. See the original triangle (the triangle on the left) and the new reflected shape to the right. Here the line of reflection is the y-axis.

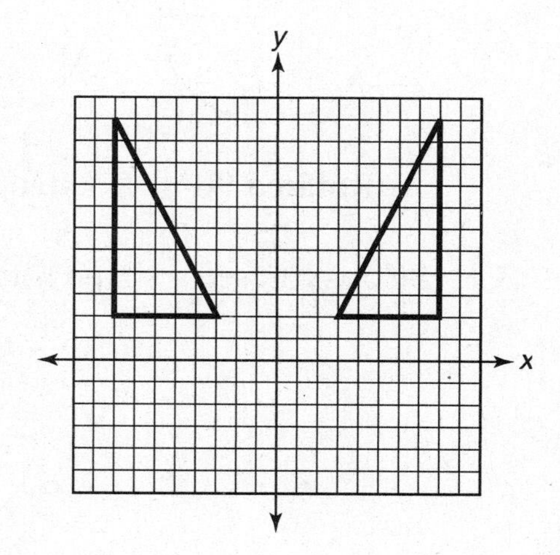

F. See how the trapezoid is reflected over the x-axis.

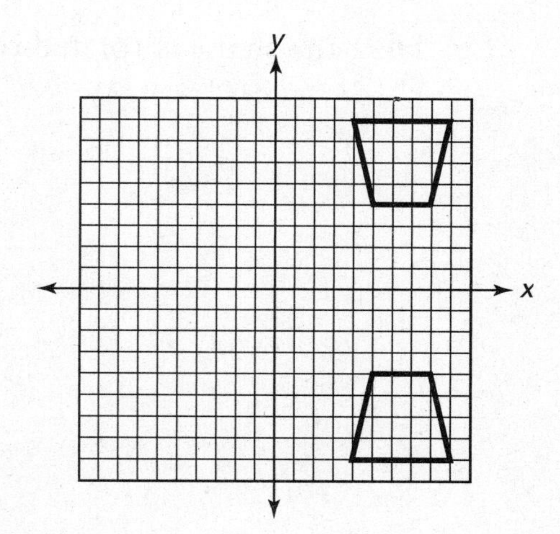

ROTATION

Examples

A. When a shape is *rotated* the figure is turned about a fixed point on the plane.

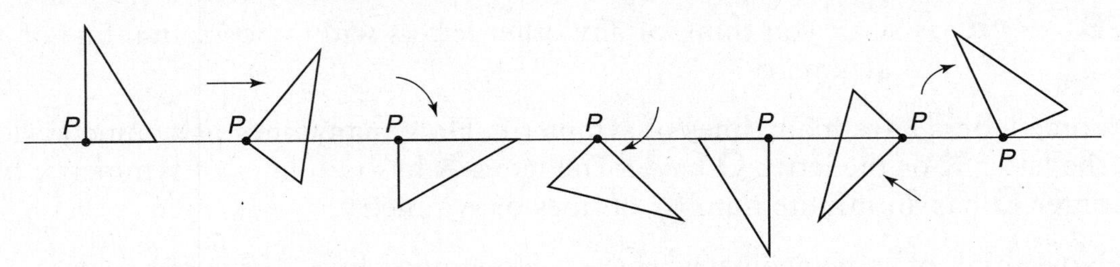

B. Think about the hour hand on a clock. The hand is pivoting from the center and rotates 360° throughout the day.

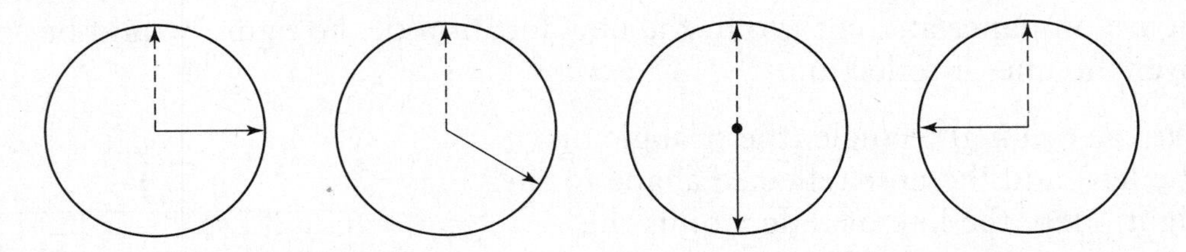

Rotated 90° **Rotated about 150°** **Rotated 180°** **Rotated 270°**

C. Below you see a shape rotated 180° around the fixed point *P*.

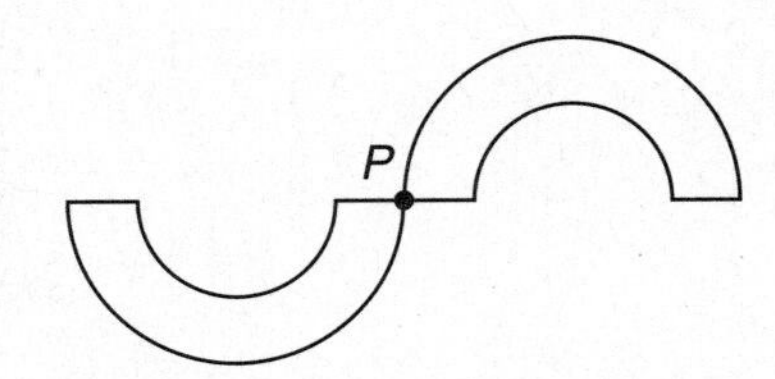

D. Here the shape is rotated only 90° around a fixed point *B*.

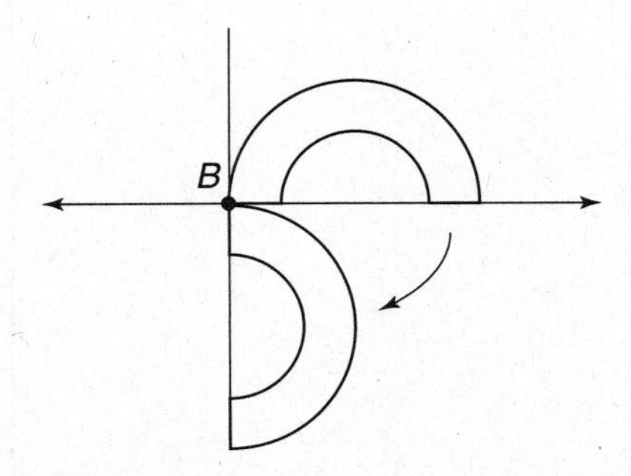

E. This shape is rotated 45° around a fixed point, from position *A* to *B*.

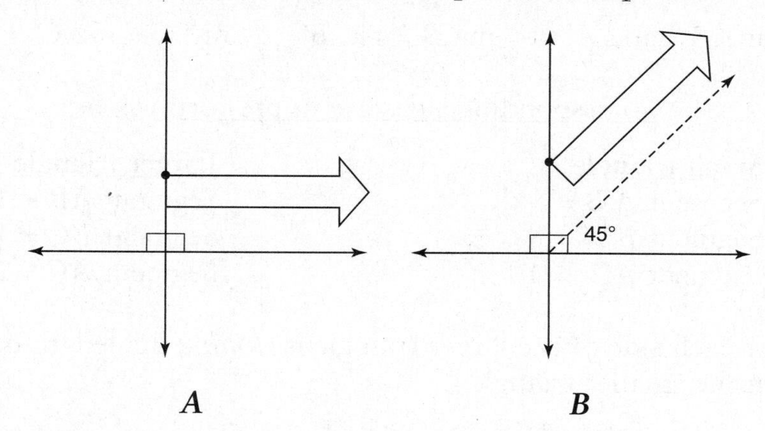

DILATION

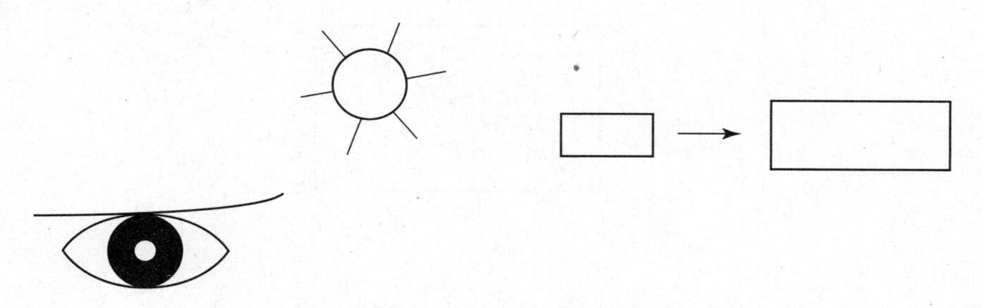

You probably have heard about how the pupil of your eye dilates (or enlarges) to adjust to the light. Well, a geometric shape can either shrink or stretch under *dilation*. The angle measurements of the original shape and the angle measurements of the new shape remain the same. The sides of the new shape are in the same proportion as the sides of the original shape. Angles remain the same. Dilating a shape creates a similar shape. (See the examples below.)

Examples

A. The larger triangle *ABC* is *similar* to the smaller triangle *A′B′C′*.

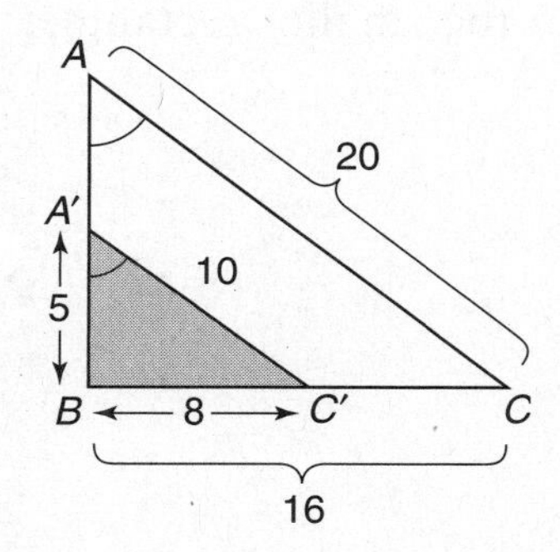

<u>Corresponding angles are equal</u>

$m\angle A = m\angle A'$ $m\angle B = m\angle B'$ $m\angle C = m\angle C'$

<u>Corresponding sides are in proportion</u>

Small triangle	**Larger triangle**
Segment $A'B = 5$	Segment $AB = 10$
Segment $BC' = 8$	Segment $BC = 16$
Segment $AC' = 10$	Segment $AC = 20$

Notice how each side of the larger triangle is *double* the length of its correspon-ding side in the smaller triangle.

B. The larger rectangle $DEFG$ is similar to the smaller rectangle $D'E'F'G'$.

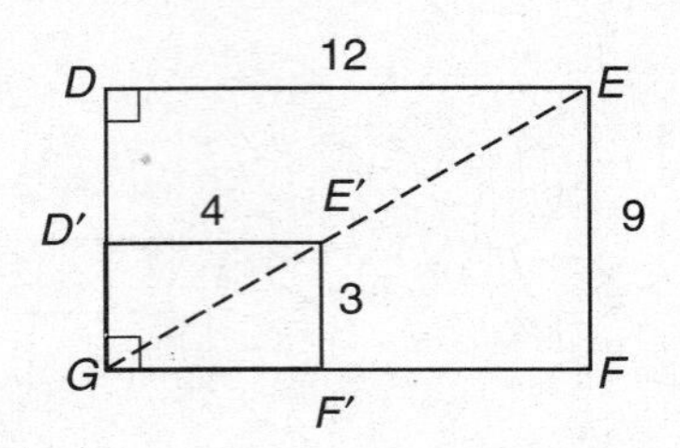

<u>Corresponding angles are equal</u>

$m\angle D = m\angle D'$ $m\angle E = m\angle E'$ $m\angle F = m\angle F'$ $m\angle G = m\angle G'$

<u>Corresponding sides are in proportion</u>

Small rectangle	**Larger rectangle**
Side $D'E' = 4$	Side $DE = 12$
Side $E'F' = 3$	Side $EF = 9$
Side $F'G = 4$	Side $FG = 12$
Side $GD' = 3$	Side $GD = 9$

Notice how each side of the larger rectangle is *three times larger* than the length of its corresponding side in the smaller rectangle.

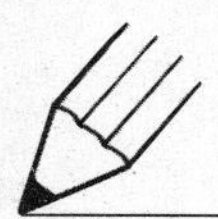

PRACTICE: Translating Polygons

(For answers, see page 150.)

1. Which pair of figures shows a 180° rotation?

A. 

B. 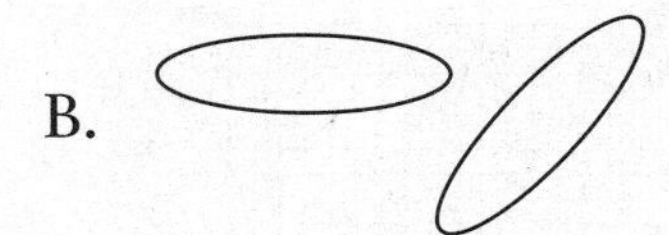

C. 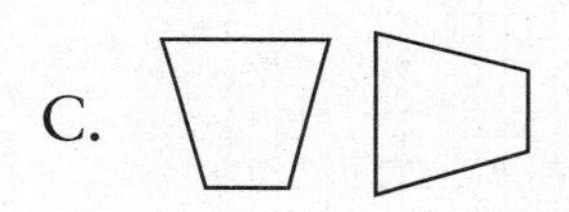

D. 

2. Which type of transformation is the following?

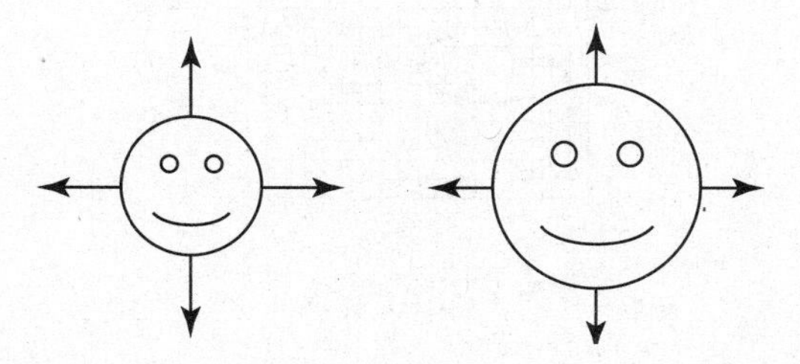

A. Translation
B. Reflection
C. Rotation
D. Dilation

Use the figure below for questions 3–5.

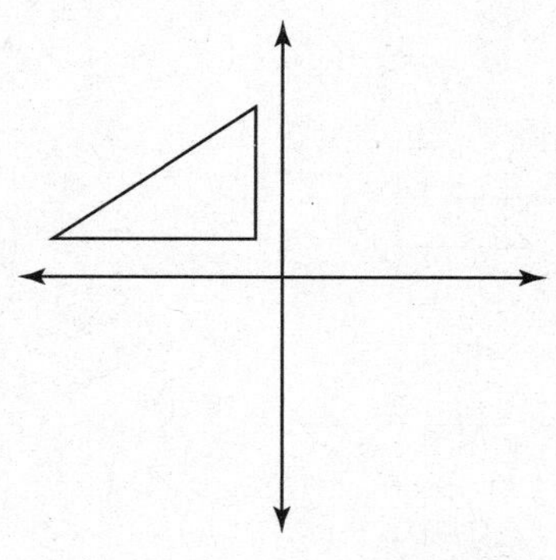

3. Sketch this shape so the x-axis is the line of reflection. Reflect triangle A over the x-axis.

4. If this shape was reflected over the x-axis and then over the y-axis what quadrant would it be in?

5. If we wanted to keep the figure above the same shape, but just slide it down 5 units, this transformation would be called a

A. translation
B. reflection
C. rotation
D. dilation

6. Which of the following has more than one line of symmetry?

A. 

B.

C.

D. 

Look at the four grids below and answer questions 7–10.

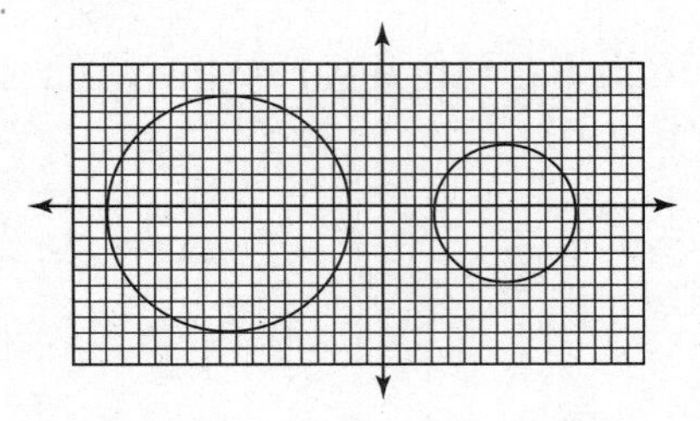

Figure 1

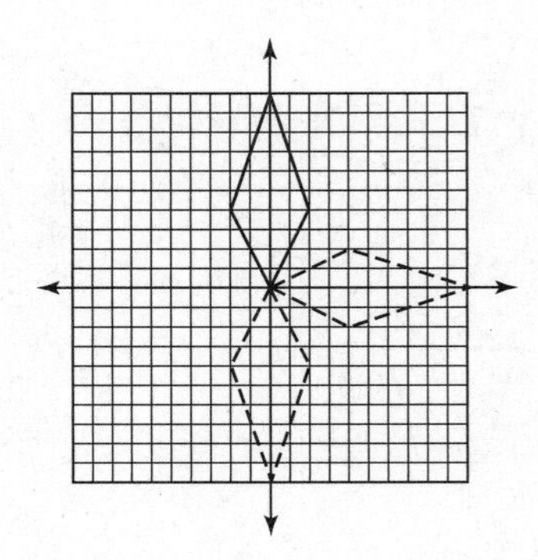

Figure 2

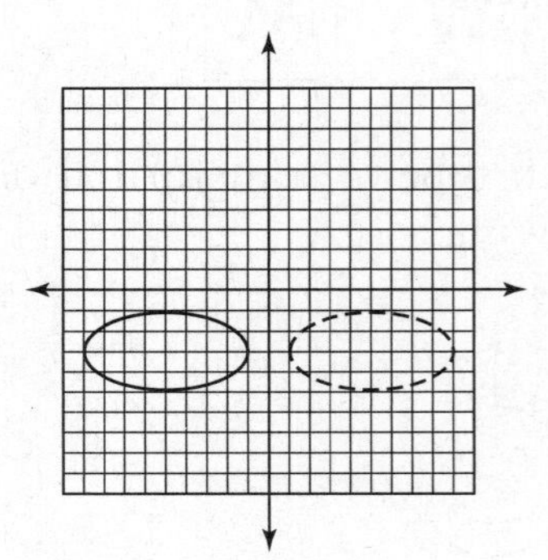

Figure 3

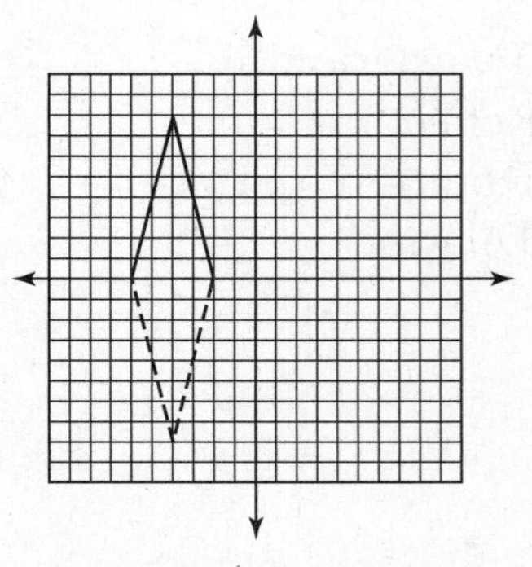

Figure 4

7. Which figure shows a shape that has undergone dilation?

   A. Figure 1
   B. Figure 2
   C. Figure 3
   D. Figure 4

8. Which figure shows a shape that has undergone translation?

   A. Figure 1
   B. Figure 2
   C. Figure 3
   D. Figure 4

9. Which figure shows a shape that has undergone rotation?

   A. Figure 1
   B. Figure 2
   C. Figure 3
   D. Figure 4

10. In Figure 2, what coordinates do all the shapes have in common?

11. What is the missing length of the base of the smaller triangle if you know these two triangles are similar?

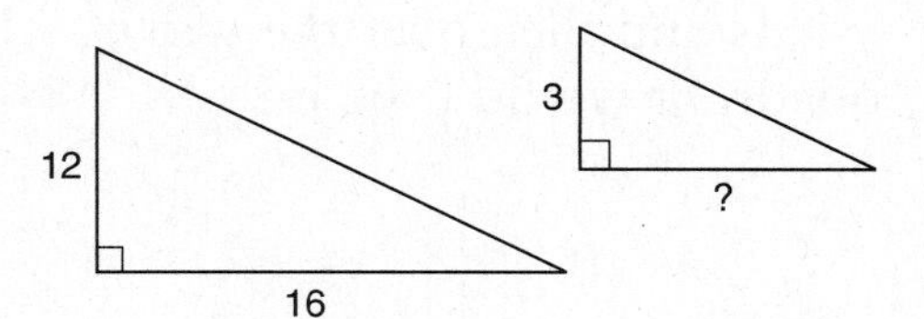

12. A landscape architect wanted to change the dimensions of a rectangular garden he had planned. He wants the new garden to be dilated so it is similar to the original one. Find the missing dimension of the new rectangular garden.

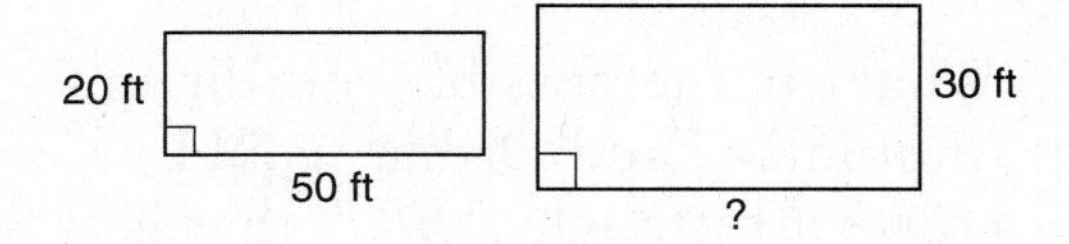

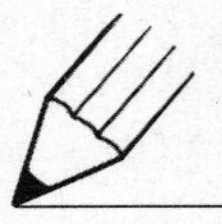

PRACTICE SCR NONCALCULATOR QUESTIONS

Each question is worth 1 point. No partial credit is given. (For answers, see page 151.)

1. If triangle ABC is translated over the x-axis and then over the y-axis, what quadrant would it be in?

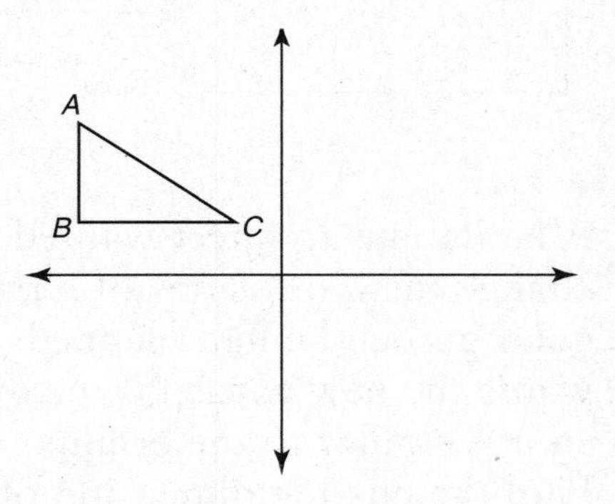

Answer: ___________

2. What are the possible coordinates of points C and D that would create a rectangle $ABCD$ that is 3 units tall?

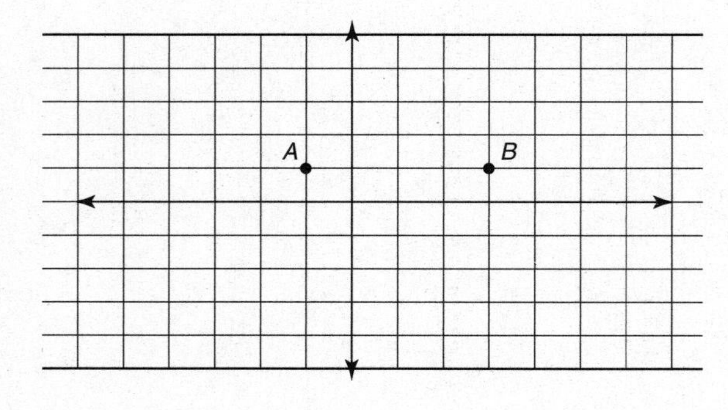

Answer: ___________

3. Which of the following polygons has a different area from the others?

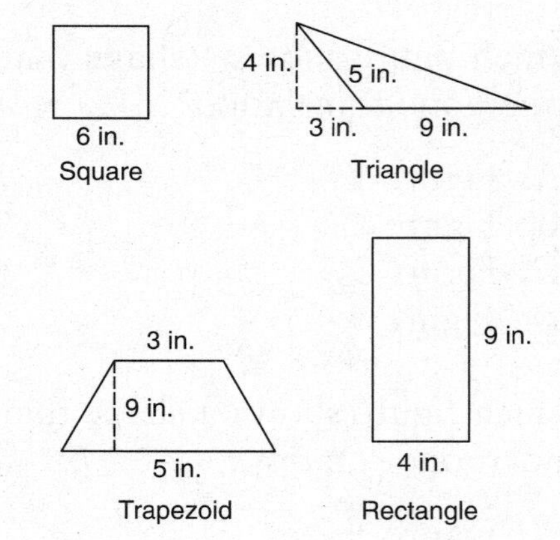

Answer: ___________

4. These polygons all have the same what?

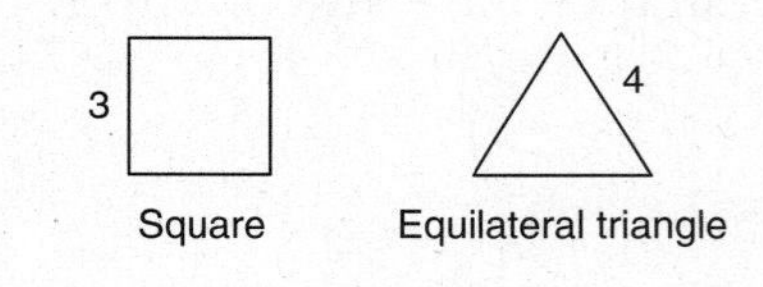

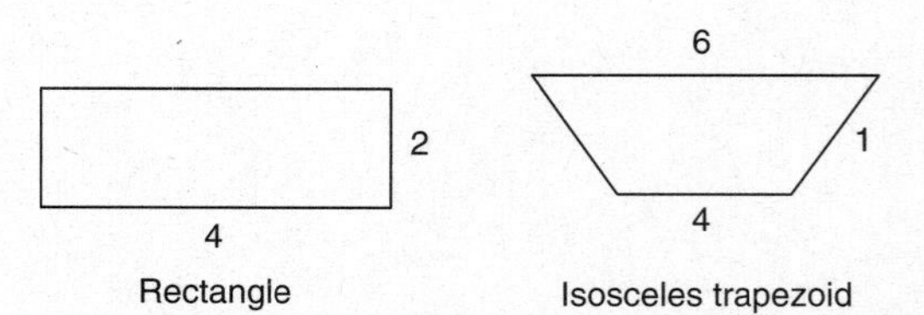

Answer: ___________

5. What is the value of x in the triangle shown?

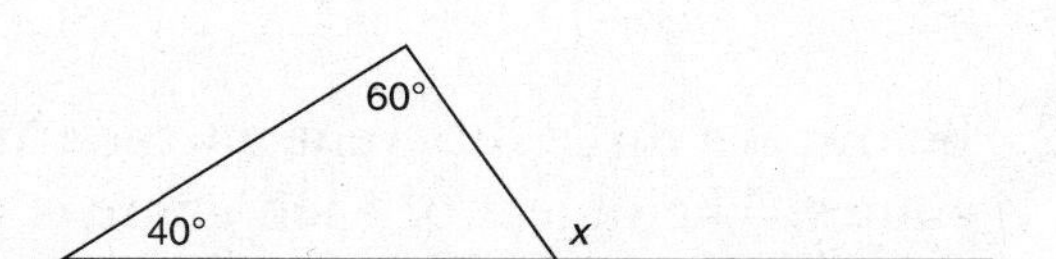

Answer: __________

6. Which two triangles are similar?

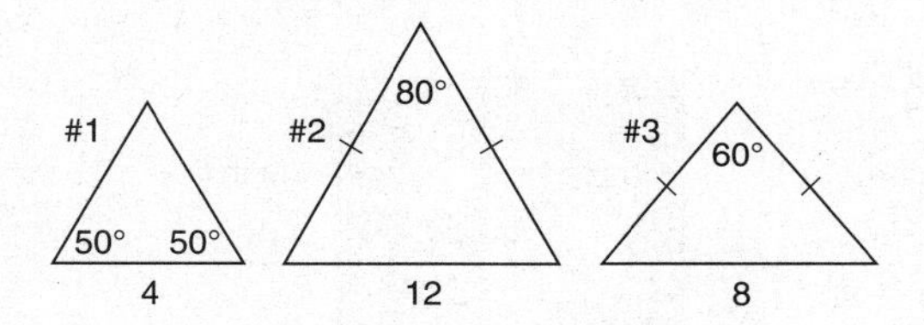

Answer: __________

7. Devan drove from his house 4 miles south and then 3 miles east to his friend Carl's house. What is the shortest distance from Devan's house to Carl's house?

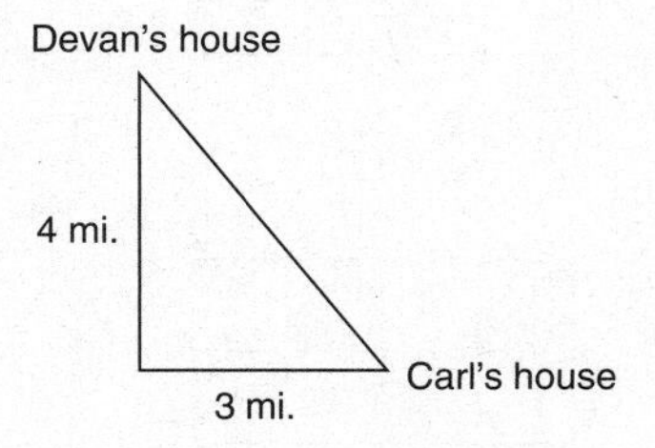

Answer: __________

8. Which has the smaller volume?

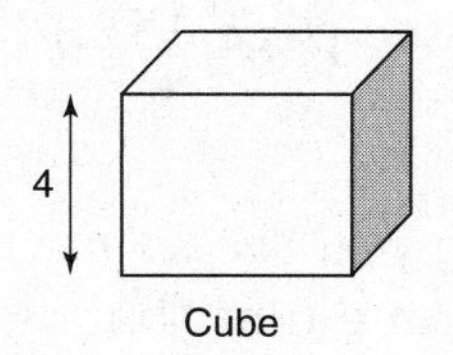

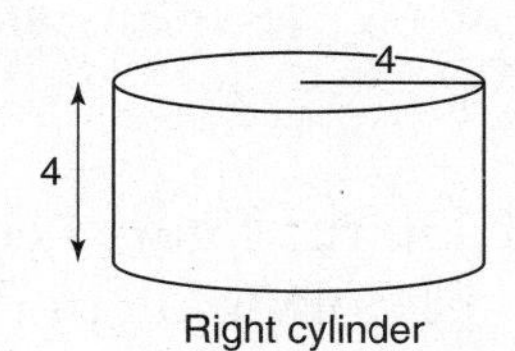

Answer: __________

9. What is the perimeter of a square that has an area of 25?

Answer: __________

10. What is the perimeter of the irregular figure drawn below? (All angles are right angles.)

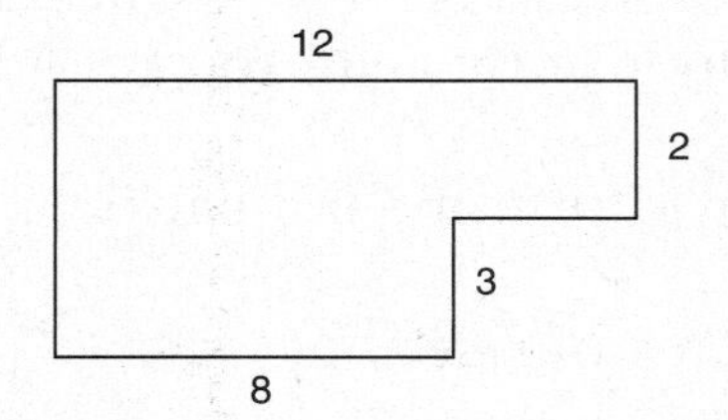

Answer: __________

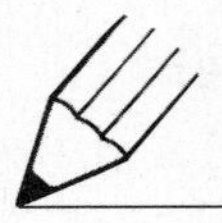

PRACTICE ECR QUESTIONS

(For answers, see page 152.)

1. In the figure below, you see a drawing of an A-frame tent that Nancy and Rod are taking on their camping trip.

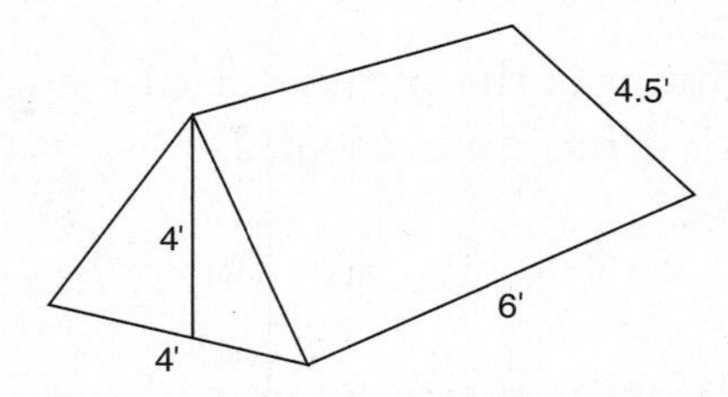

■ How much canvas was used to make this A-frame tent? Show all work. (Remember to include the floor of the tent, too.)

Area of front and back:

Area of sides:

Area of floor:

Total surface area:

2. Below are three different dartboard games. The object of each game is to throw a dart so it hits the shaded area of the dartboard. If you had your choice of any of the three boards below, which board would give you the best chance of getting the most points? Show all work and explain your answer.

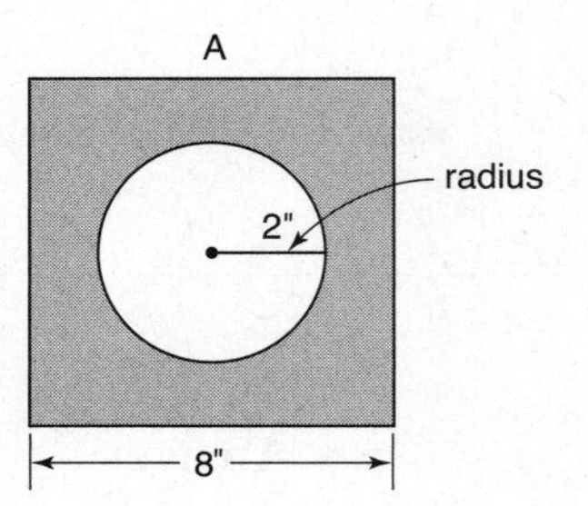

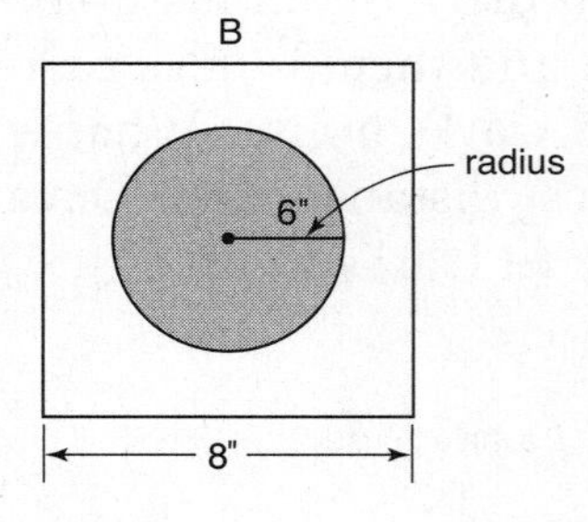

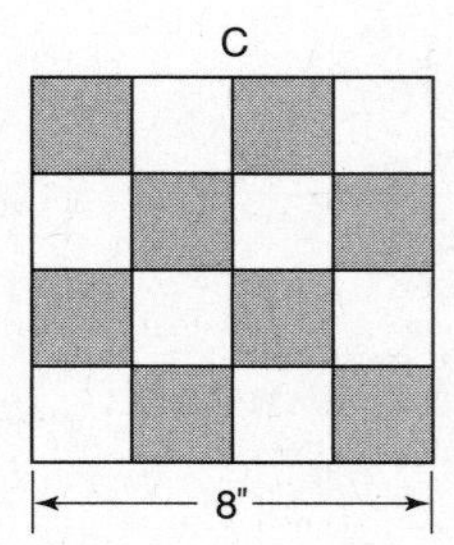

3. Holly wants to bake a large cake for her mom's birthday and needs to use the largest baking pan she can find. She has three choices:

   A rectangular pan that is 12″ long, 16″ wide, and 5″ high.

   A round pan that has a diameter of 16″ and is 5″ high.

   A square pan that measures 15″ on each side and is only 4″ high.

   - What is the maximum volume of each pan? Use 3.14 for pi.
   - Which would hold the most batter?
   - If she made a second cake and chose the smallest pan and filled it only 3″ deep, what would be the maximum volume of that batter?

4. Mr. and Mrs. B. are putting a new tile floor down in their entrance foyer. See figure below.

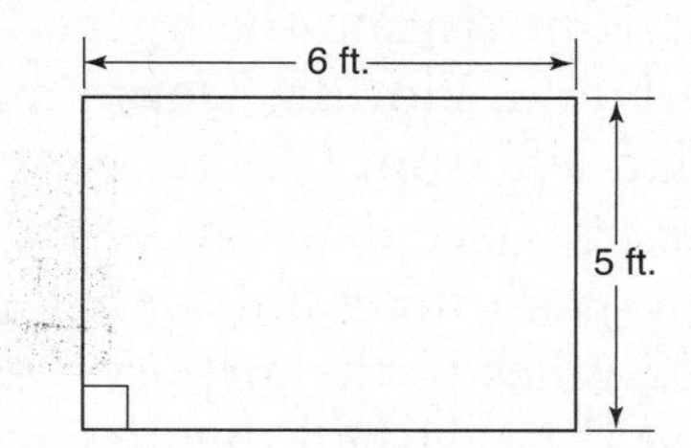

   - What is the maximum number of 6″ × 6″ square tiles that would fit on this floor?
   - The tiles are not sold separately. They are only sold in boxes of one dozen tiles for $21.60. What will it cost to buy the tiles to cover this floor? Be sure to include a 6% sales tax in your total cost.

5. Devan drove with his dad to pick up his mother and sister at the airport. His mother and sister were flying back from visiting their grandmother in Orlando, Florida. Devan and his dad left from Clarke, New Jersey, point "*C*" (see diagram below). They drove south for 9 miles, then east for 12 miles to the airport. Devan's dad said he wished the new road had been completed; it would have gone from point "C" directly to the airport.

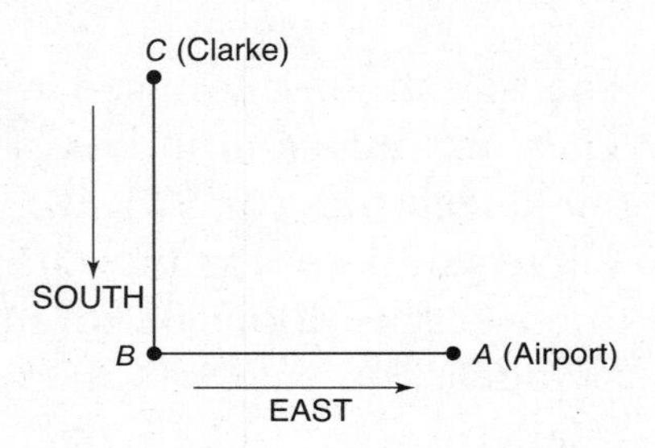

- How much shorter would the car ride have been if they could have driven from point "*C*" straight to point "*A*" instead of from *C* to *B* to *A*?
- The plane fare for Devan's mom and sister costs them each $325.00 round-trip. Considering the other expenses listed in the table below, what would have been the cost if his mother and sister had driven to Orlando?
- How much money would they have saved by driving instead of taking a plane?

Travel Information	Your work space to determine total cost
• 1,072 miles from Clark, New Jersey to Orlando, Florida. • 1,072 miles from Orlando back home to New Jersey Mom's car would have an average of 24 miles per gallon, and we'll estimate gasoline costs $2.40 per gallon.	Gasoline Cost:
• Tolls round-trip would be approximately $16.00.	Tolls:
• $68.50 hotel for one night while driving down to Florida • $59.00 hotel for one night on the drive home	Hotel Cost:
• Food purchased on trip from NJ to Florida ($15.00 + $28.00) • Food purchased on trip from Florida back to NJ ($12.00 + $6.00 + $18.00)	Food Cost:
	Total cost by car: ___________

SOLUTIONS TO PRACTICE QUESTIONS

POINTS, LINES, AND PLANES (page 83)

1. B 47°
2. B supplementary
3. 48°
4. 157°
5 30°
6. A complementary
7. C $m\angle 1 = 90°$
8. D $m\angle a = m\angle b$
9. C They are equal and vertical.
10. A 78° (Vertical angles are equal.)
11. C 102° (supplementary)
12. D Vertical angles are equal.
13. B $m\angle 1 = 46°$
14. B 140°
15. B
16. A
17. B
18. B
19. A
20. D
21. 1. B
    2. I
    3. D
    4. A
    5. C
    6. H
    7. E
    8. F

AREA OF FLAT SHAPES (page 92)

1. 84 square inches
2. 59.45 or 59 square units
3. 30 square feet
4. 60 square inches
5. a. B 7.5
   b. D 8.6
   c. D 113 square feet
   d. D 255 square feet
   e. B 113 square feet
6. 19.6 square centimeters

7. $r = 2$ Circumference $= 2\pi r = 2\pi 2 = 4\pi$
Area $= \pi r^2 = \pi(2)(2) = 4\pi$
8. D $r^2 \sim 81$; therefore, $r \sim 9$
9. 1×72 2×36 3×24 4×18 6×12
10. 1×36 2×18 3×6 4×9 6×6

AREA OF SHADED REGION (page 97)

1. 30.96
2. D 20–21 square feet
3. 20 square inches
4. 22 square units
5. C $25 \times 25 = 625$ (area not 100)
6. C 102 square units
7. B $24 + {\sim}6.28 = 30.28$; use 30 sq. cm
8. A (using 3.14 for π) 497.44 sq. ft
9. B $54 \times 44 = 2{,}376$ sq. ft
10. D $(9 \times 4)/2 = 36/2 = 18$
11. 300 square inches; $A_{rectangle} = 30 \times 20 = 600$; $A_{triangle} = \frac{1}{2}(30)(20) = 300$

SURFACE AREA AND VOLUME (page 104)

1. C $3 \times 3 = 9$; 9×6 sides $= 54$ square units
2. a. 160 sq. units = Total surface area
b. 96 cubic units = Volume
c. Various solutions:
if use 3, 4, 8 for sides
Volume $= 3 \times 4 \times 8 = 96$ cu. units
Surface area = 136 sq. units
if use 12, 4, 2 for sides
Volume $= 12 \times 4 \times 2 = 96$ cu. units
Surface area = 256 sq. units
3. B 3 in.
4. B 4 cm
5. C 9 sq. ft
6. B 8 feet
7. D 125 cu. in.

PERIMETER (page 108)

1. D 62 ft
2. C $8 \times 5 = 40$ (pentagon)
3. a. $42 = 12 + 3 + 8 + 6 + 4 + 9$
b. 60 sq. units $12 \times 3 + 6 \times 4 = 36 + 26 = 60$ and
$9 \times 4 + 8 \times 3 = 36 + 24 = 60$
4. B $9 \times 3 = 27$ inches
5. B $10 + 8 + 8 = 26$ cm

6. 13 — $36 + 6 + 8 + 7 = 57$; $70 - 57 = 13$ is the length of side AB
7. ~40 ft — $C = \pi d$; $3.14(12.6) = 39.56$ feet or ~ 40 ft
8. ~50 — If you use 3.14: $(3.14)(16) = 50.24$
   If you use π: $(\pi)(d) = (\pi)(16) \sim 50.265$
9. a. more like a square
   b. 25×25
   c. Answers may vary: A *long thin rectangle:* could measure 48 by 2

      $48 + 48 + 2 + 2 = 100$ perimeter
      $(48)(2) = 96$ sq. ft (a small area)

      A *square rectangle:* could measure 25 by 25

      $25 + 25 + 25 + 25 = 100$ perimeter
      $(25)(25) = 625$ sq. ft (a much larger area)
10. C 12 ft
11. B 27 in.
12. C 48 cm
13. 1. Equilateral triangle = C. All sides are the same length
    2. Isosceles triangle = A. Two sides are the same length
    3. Perimeter = E. The distance around a shape
    4. Circumference = B. The distance around a circle
    5. A regular shape = D. Any shape where each side measures the same length

TRIANGLES (page 112)

1. $40° + 50° = 90°$, $180° - 90° = 90°$
2. 120° $180° - 60°$
3. 70° $180° - 90°$ (right angle) $= 90°$; $90° - 20° = 70°$
4. 55° Vertical angles are equal.
5. 95° $55° + 30° = 85°$; $180° - 85° = 95°$
6. isosceles (2 sides $\cong$)
7. equilateral (all sides $\cong$)
8. scalene (no sides $\cong$)
9. acute (all angles less than 90°)
10. obtuse (one angle = 102°)
11. right ($50° + 40° = 90°$)
12. $\angle y$ $\angle y$ is opposite the longest side.
13. Triangle A is larger. Triangle A $(5 \times 5)/2 = 25/2 = 12.5$ sq. units
    Triangle B $(4 \times 5)/2 = 20/2 = 10$ sq. units
14. 80°, vertical
15. 100°, 180°
16. 50°, 180° $\angle 2 = 100°$, $\angle 7 = \angle 2 = 100°$, $\angle 4 = 30°$
17. 130° $\angle 6 + \angle 5 = 180°$; $\angle 6 = 50°$, $\angle 5 = 130°$
18. 4, 4
19. 60° $\angle 8$ and $\angle 10$ are vertical angles and are equal.
20. 120° Supplementary angles = 180°

21. $\angle A$, $\angle B$, and $\angle D$ Each angle in a rectangle = 90°.
22. 36° (Right angles) 90° – 54° = 36°

OTHER POLYGONS (page 115)

3 and 4. Answers will vary:

8-sided polygon	6 triangles
6-sided polygon	4 triangles
5-sided polygon	3 triangles
4-sided polygon	2 triangles

5. 180°
6. Answers will vary: (number of triangles × 180°)
   8 sided = 6 × 180 = 1080°
   6 sided = 4 × 180 = 720°
   5 sided = 3 × 180 = 540°
   4 sided = 2 × 180 = 360°
7. D a composite shape
8. B 5
9. B A regular hexagon
10. A Area square = $s^2 = 6 \times 6 = 36$ sq. ft
11. B $P = 2(l + w) = 2\ (3 + 6.2) = 18.4$
12. C 84 sq. units Explanations may vary:
    One example: (6 × 8) = 48; (12 × 3) = 36; 48 + 36 = 84
    Other example: 2(3 × 2) = 12; (9)(8) = 72; 12 + 72 = 84
13. A
14. Answers will vary.

RIGHT TRIANGLES (page 120)

1. C 15 cm long — Use $a^2 + b^2 = c^2$: $12^2 + 9^2 = c^2$; $144 + 81 = c^2$, $225 = c^2$, $15 = c$
2. D 7.8 inches long — Use $a^2 + b^2 = c^2$: $6^2 + 5^2 = c^2$; $36 + 25 = c^2$, $61= c^2$, $7.8 = c$
3. A 15 cm — Use $a^2 + b^2 = c^2$: $8^2 + b^2 = 17^2$; $64 + b^2 = 289$, $225 = b^2$, $15 = b$
4. B 26 feet tall — Use $a^2 + b^2 = c^2$: $24^2 + 10^2 = c^2$; $576 + 100 = c^2$, $676 = c^2$, $26 = c$
5. ~10.23 yards — Use $40^2 + 12^2 = \text{Hypotenuse}^2$: $1{,}600^2 + 144^2 = \text{Hypotenuse}^2$; $1{,}744 = \text{Hypotenuse}^2$, 41.76 ~ Hypotenuse
   Walking around the 2 sides = 40 + 12 = 52 yards;
   Walking along the diagonal line (hypotenuse) = 41.76 yards
   52 – 41.76 is approximately a 10.23-yard shorter walk
6. ~7 meters — $5^2 + 5^2 = c^2$: $25 + 25 = c^2$; $50 = c^2$, approximately $7 = c$
7. ~13.7 ft — $x^2 + 6^2 = 15^2$: $x^2 + 36 = 225$; $x^2 = 225 - 36$, $x^2 = 189$, $x \sim 13.7$ ft
8. C ~11.3 cm.

COORDINATE GEOMETRY (page 124)

1. 6
2. 5
3. 10 From –4 to 6 is 10 units; or think $|-4| + |6| = 4 + 6 = 10$.
4. 9 from –3 to 6 is 9 units; or think $|-3| + |6| = 3 + 6 = 9$
5. *W*
6. *C*
7. (5, –6)
8. (–2, 2)
9. **D** 24
10. **A** Area of rectangle = (10)(9) = 90 sq. units
11. 20 units — $AB = 6$, $BC = 4$, $CD = 3$, $DE = 2$, $EF = 3$, $FA = 2$
    Add all sides: 6 + 4 + 3 + 2 + 3 + 2 = 20 units is perimeter
    18 sq. units — Divide the shape into rectangles and find the area of each rectangle and add them together. One solution is: (3)(4) = 12 and (2)(3) = 6; 12 + 6 = 18.
12. **B** Area = $8 \times 9 = 72$
13. **A** $BC = 6$, $CA = 8$, $AB = 10$ (hypotenuse)
    Use $a^2 + b^2 = c^2$: $6^2 + 8^2 = 100^2$; 36 + 64 = 100;
    Perimeter (add all sides) = 6 + 8 + 10 = 24

CONGRUENCY (page 127)

1. **B** No, their corresponding angles are =, but we don't have information about their sides. To be congruent their corresponding sides must be the same length.
2. **B** No, because their corresponding sides are not equal.
3. Yes When plotted, you have two rectangles that are ≅. $AB = 4$ and $QR = 4$, $BC = 6$ and $RS = 6$, and so on. They have all 90° angles, and the corresponding sides are the same length.
4. No Answers will vary. The following two shapes have the same perimeter, but they are not congruent: A rectangle could have dimensions of 12, 10, 12, 10 ($P = 44$); a trapezoid could have dimensions 11, 14, 11, 8 ($P = 44$).
5. Yes
6. **A** Triangle #2 and Triangle #3
7. **C** The length of the short sides
8. Yes
9. No
10. No
11. No They could be different sizes.
12. Answers will vary.
13. Answers will vary.
14. (1, –5)

TRANSLATING POINTS (page 130)

1.

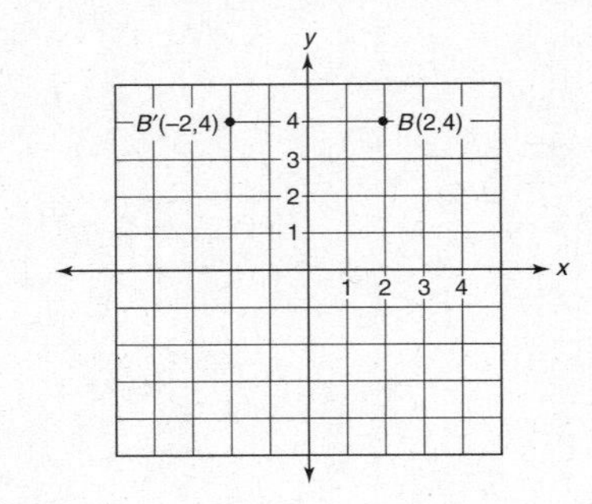

2.

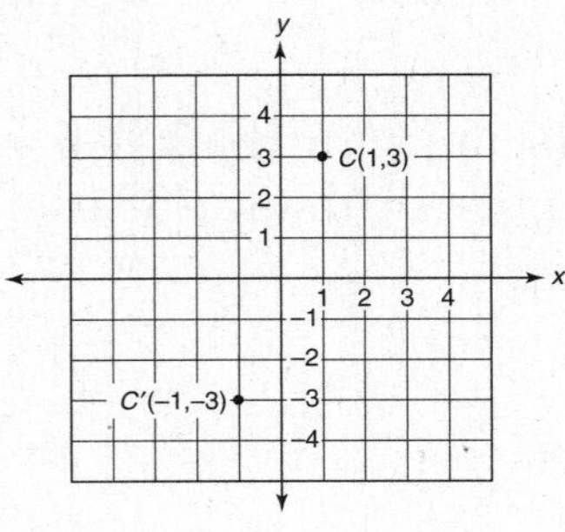

3.

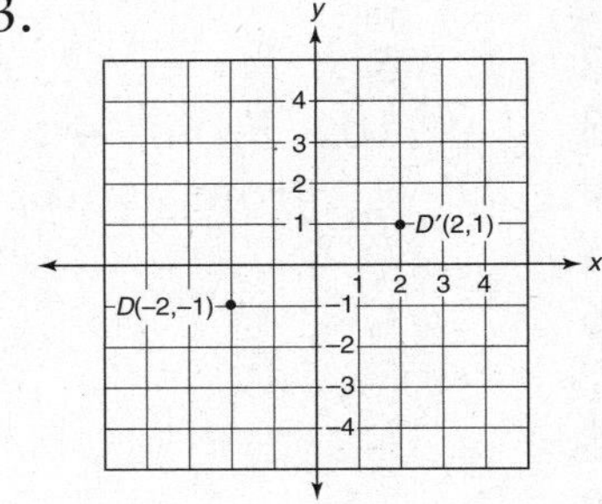

4. Show D' moved to the right at coordinate 10.
5. Show A' moved to the left at coordinate –2.

TRANSLATING POLYGONS (page 137)

1. D W M
2. D Dilation (The same shape is enlarged.)
3.

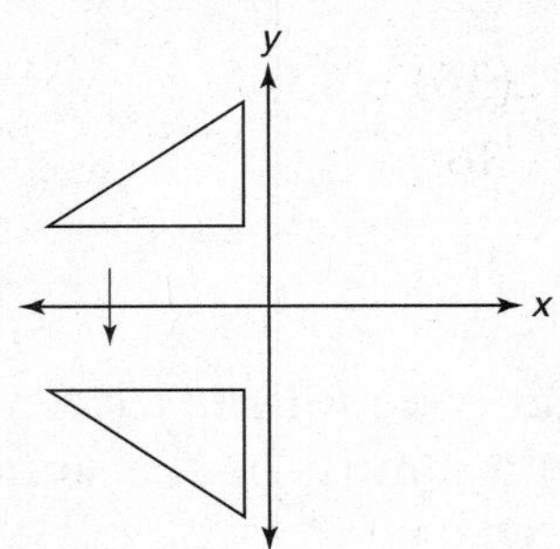

4.

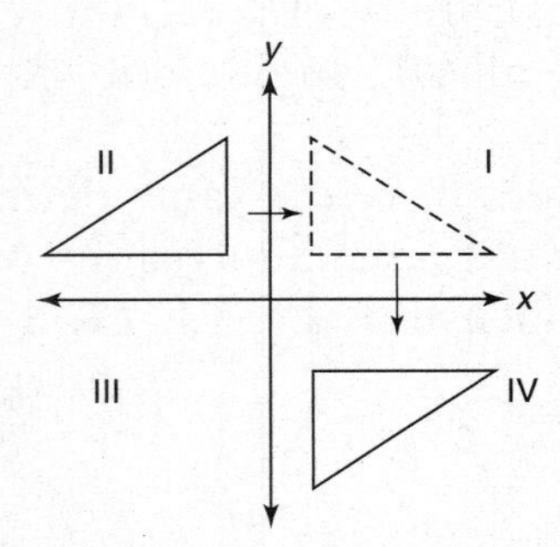

IV Bottom right, quadrant IV

5. A translation (just a slide)
6. A
7. A Figure 1
8. C Figure 3
9. B Figure 2
10. (0, 0)
11. 4 $12/16 = 3/x$, $(3)(16) = (12)x$, $48 = 12x$, $4 = x$
12. 30×75 $20/50 = 30/x$, $(30)(50) = (20)x$, $1{,}500 = 20x$, $75 = x$

SCR NONCALCULATOR QUESTIONS (page 140)

1. Quadrant IV

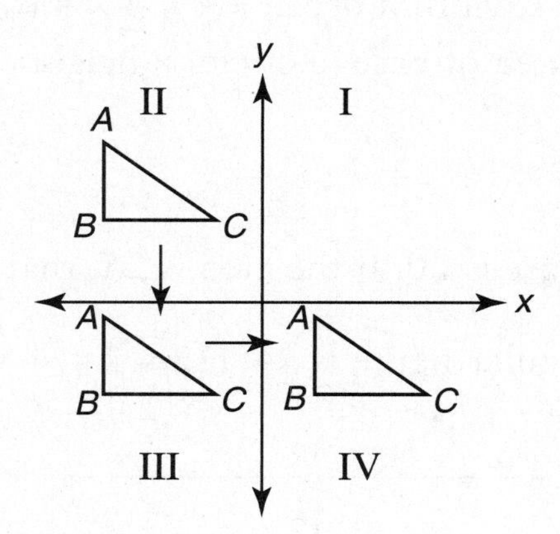

2. There are two possible correct answers: (–1, 4) and (3, 4) or (–1, –2) and (3, –2).

3. The area of the obtuse traingle is different.
   Area of square: $6^2 = 6 \times 6 = 36$
   Area of obtuse triangle: ½(base)(height) = ½(9)(4) = ½(36) = 18
   Area of trapezoid: $[\frac{1}{2}(3 + 5)] \times 9 = \frac{1}{2}(8) \times 9 = 4 \times 9 = 36$
   Area of tall rectangle: $4 \times 9 = 36$

4. The polygons all have the same perimeter, 12.

5. $x = 100°$

6. Triangles #1 and #2 are similar.

7. The shortest distance from Devan's house to Carl's house is 5 miles. You might recognize this as a special right triangle, a 3–4–5 right triangle. You could also use the Pythagorean theorem:

$$A^2 + b^2 = c^2$$
$$3^2 + 4^2 = c^2$$
$$9 + 16 = c^2$$
$$25 = c^2$$
$$5 = c$$

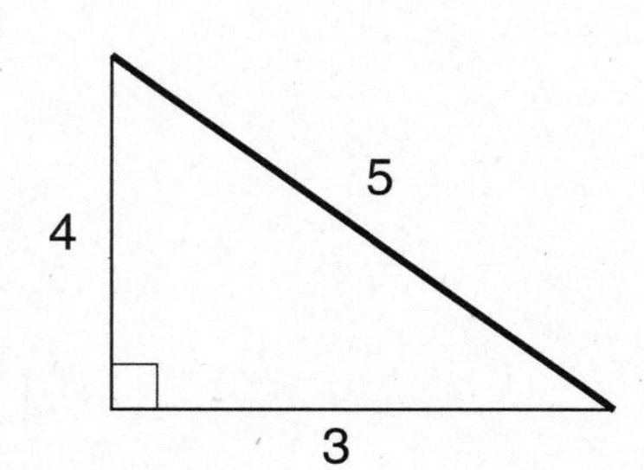

8. The volume of the cube is smaller.
   Volume of cube = length × width × depth = 4 × 4 × 4 = 16 × 4
   Volume of the cyliner = area of base (a circle) × height
   $\pi r^2 \times 4$
   (3.14) × 4 × 4 × 4 = 3.14 × 16 × 4

9. The perimeter of the square is 20. If the area is 25, that means each side = 5.

10. The perimeter of th eirregular figure is 34. (12 + 2 + 4 + 3 + 8 + 5)

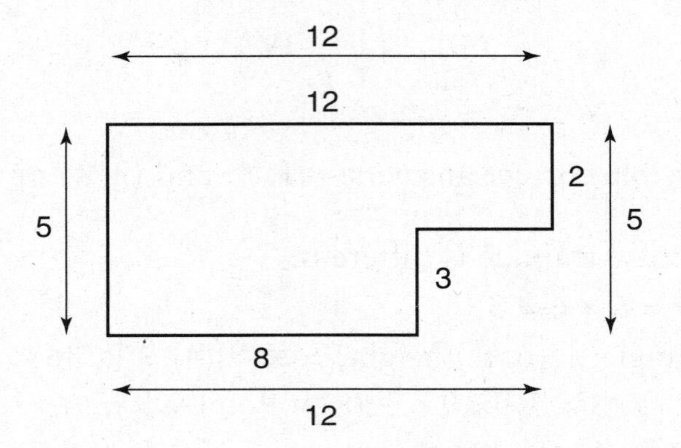

ECR QUESTIONS (page 142)

1. 94 sq. feet

   Area of front and back = area of 2 triangles = $(2)\frac{bh}{2} = (2)\frac{4(4)}{2}$ = 16 sq. ft.

   Area of 2 sides (rectangles) = 2 (length)(width) = 2(6)(4.5) = 54 sq. ft.
   Area of floor (also a rectangle) = (length)(width) = (4)(6) = 24 sq. ft.
   Total surface area = 16 + 54 + 24 = 94 sq. ft.

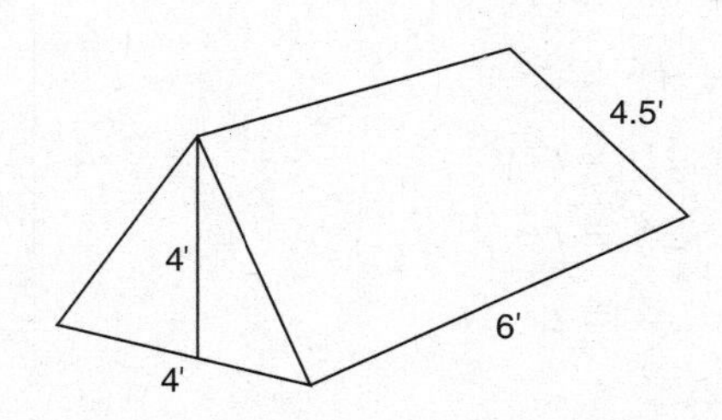

2. **Board B**

   All the boards are the same size and shape. The total area of each dartboard is the area of a square = (side)(side) = (8) (8) = 64 square inches. The dartboard with the largest shaded area would be the best.
   Board A: Area of unshaded region (circle) = (πr^2)(h) = (π)(4) = 12.566 sq. in.
   Board B: Area of shaded region (square – circle) = 64 – 12.566 = 35.72 sq. in.
   Board C: Area of shaded region is $\frac{1}{2}$ total area of entire board = $\frac{64}{2}$ = 32 sq. in.

   The correct answer is Board B, since its shaded region has the largest area.

3. ■ 960 cubic inches = the maximum volume of the rectangular pan: (12)(16)(5) = 960.

   1004.8 cubic inches = the approximate maximum volume of the round pan: [Volume = (area of the base)(height) = (πr^2)(h) = (3.14)(8)(8)(5) = 1004.8.]
   900 cubic inches = the maximum volume of the square pan: (15)(15)(4) = 900.

   ■ The round pan would hold the most batter (1004.8 cubic inches)

   ■ The square pan filled only 3″ high would hold 675 cubic inches of batter.
   Volume = (side)(side)(depth) = (15)(15)(3) = 675 cubic inches.

4. ■ 120 tiles that are 6 square inches would be needed to cover this floor.

   Change the dimensions of the floor from feet to inches: 5 ft × 6 ft = (5)(12 in/ft) × (6)(12in./ft.) = 60″ × 72″ = 4,320 sq. in. = area of floor.

   Find the area of one tile: (6″)(6″) = 36 sq. in. = area of one tile.

   4,320 (area of floor) ÷ 36 (area of one tile) = 120 tiles.

   ■ The total cost to purchase the tiles would be $228.96 including sales tax.
   120 sq. ft. (area of floor) ÷ 12 tiles per box = 10 boxes would be needed.
   (10 boxes)($21.60 per box) = $216.00 + 6% sales tax.
   ($216.00)(1.06) = $228.96, the total cost including the sales tax

5. **6 miles shorter**

Since you have a right triangle you can use the Pythagorean formula:

$a^2 + b^2 = c^2$
122 + 92 = c^2
144 + 81 = c^2

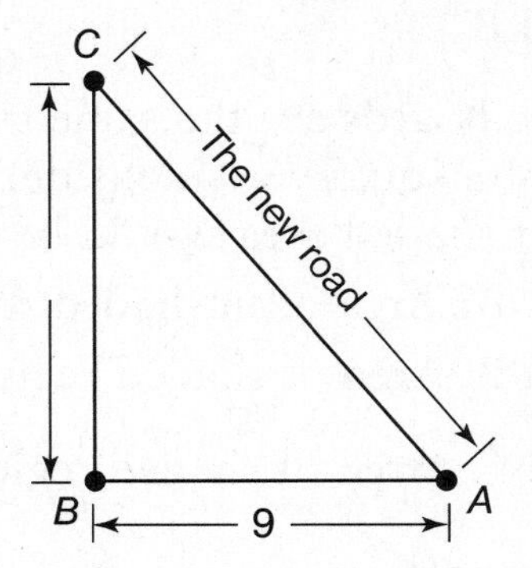

$225 = c^2$; $\sqrt{225} = \sqrt{c^2}$; 15 miles = c, the distance from "C" to "A"

The longer distance (from "C" to "B" to "A") is 12 + 9 or 21 miles.
The shorter distance would be 21 – 15 (from "C" to "A") = 6 miles shorter.

Travel Information	Your work space to determine total cost
• 1,072 miles from Clark, New Jersey to Orlando, Florida. • 1,072 miles from Orlando back home to New Jersey Mom's car would have an average of 24 miles per gallon, and we'll estimate gasoline costs \$2.40 per gallon.	Gasoline Cost: 1,072 + 1,072 = 2,144 total miles $\frac{2,144}{2}$ miles per gallon = 89.33 gallons (89.33 gallons)(\$2.40/gallon) = **\$214.40**
• Tolls round-trip would be approximately \$16.00.	Tolls: **\$16.00**
• \$68.50 hotel for one night while driving down to Florida • \$59.00 hotel for one night on the drive home	Hotel Cost: 68.50 + 59.00= **\$127.50**
• Food purchased on trip from NJ to Florida (\$15.00 + \$28.00) • Food purchased on trip from Florida back to NJ (\$12.00 + \$6.00 + \$18.00)	Food Cost: 15 + 28 = \$43.00 12 + 6 + 18 = \$36.00 Total food cost = **\$79.00**
	Total cost by car: \$436.90
Total amount saved by driving instead of flying	Plane travel \$650.00 Car travel –436.90 \$213.10 **They would have saved \$213.10 by driving**

Name: ______________________ Date: ______________

Cluster II Test

35 minutes
(Use the *NJ ASK 8 Mathematics Reference Sheet* on page 265.)

MULTIPLE-CHOICE QUESTIONS

DIRECTIONS FOR QUESTIONS 1 THROUGH 12: Each of the questions or incomplete statements below is followed by four suggested answers. Select the one that is the best in each case, and fill in the corresponding lettered circle. Be sure the circle is filled in completely so you cannot see the letter. Unless you are told to do so in the question, do NOT include sales tax in your answer to questions involving purchases.

1. To find the area of a circle with diameter of 12, what buttons should I press on my calculator?

A. [3] [.] [1] [4] [×] [6] [=]
B. [3] [.] [1] [4] [×] [12] [=]
C. [3] [.] [1] [4] [×] [1] [4] [4] [=]
D. [3] [.] [1] [4] [×] [6] [×] [6] [=]

2. Use the diagram of the square below. What is the area of the shaded region?

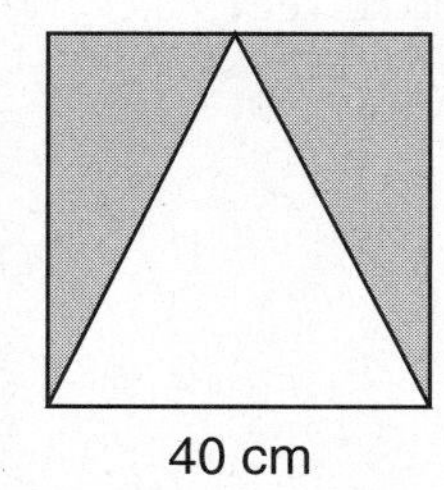

A. 80 sq. cm
B. 400 sq. cm
C. 800 sq. cm
D. 1,000 sq. cm

Ⓐ Ⓑ Ⓒ Ⓓ

3. The following two right triangles are congruent. Angle *A* measures 60°. What is the measure of angle *F*?

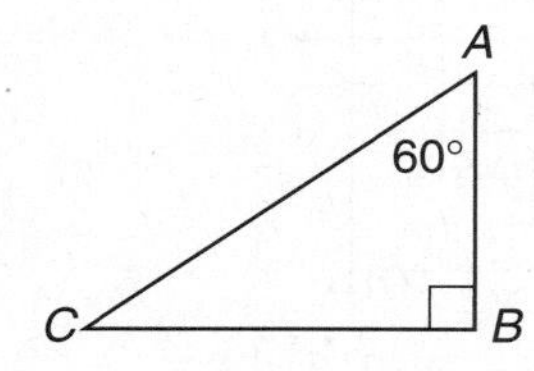

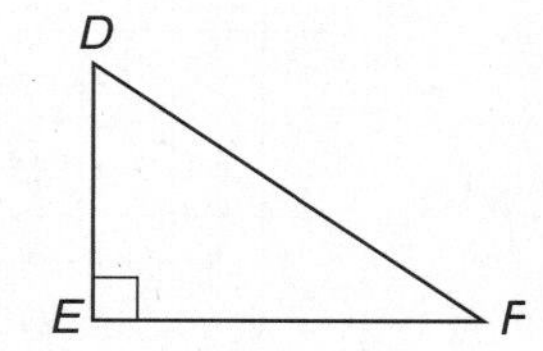

A. 20°
B. 30°
C. 40°
D. 60°

4. The perimeter of this figure is 50 yards and all angles are right angles. What is the measure of line segment *FG*?

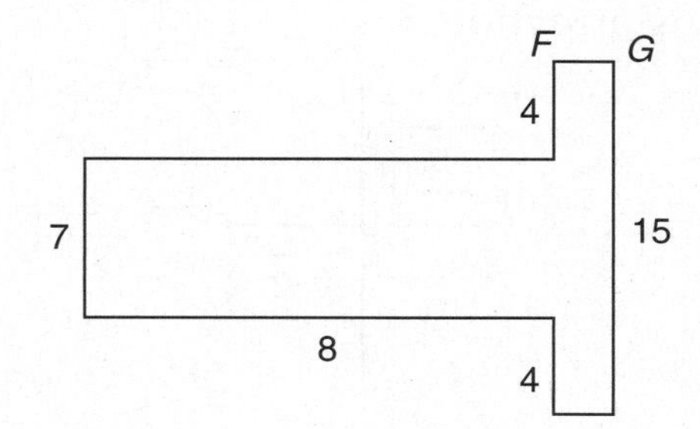

A. 2 yd
B. 4 yd
C. 7 yd
D. 8 yd

GO ON TO THE NEXT PAGE

5. Point Q has the coordinates (–2, 4). What are the coordinates of its image point if it is translated 3 units to the right and then reflected over the x-axis?

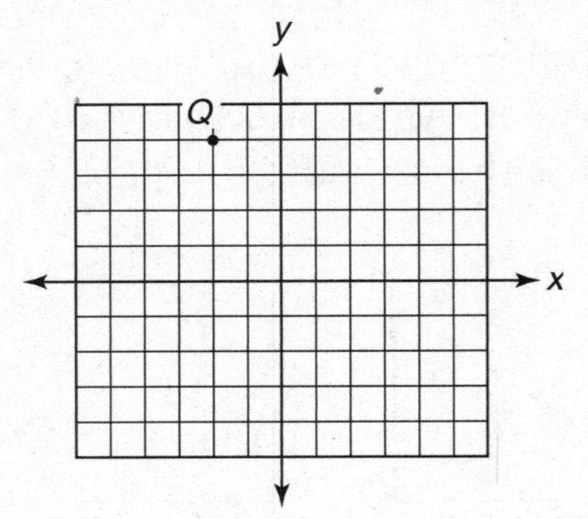

A. (3, 4)
B. (–3, 4)
C. (1, –4)
D. (1, 4)

Ⓐ Ⓑ Ⓒ Ⓓ

6. Figure B' is the result of a sequence of transformations of Figure B. Which of the following does *not* describe a correct possible sequence of transformations?

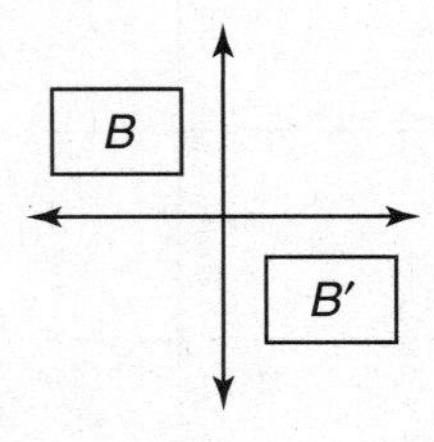

A. A translation of figure B over the x-axis and then a translation over the y-axis?
B. A reflection over the y-axis and a second reflection over the x-axis?
C. A translation to quadrant I and then a translation over the x-axis?
D. A translation over two axes and then a dilation.

Ⓐ Ⓑ Ⓒ Ⓓ

7. What is the surface area of the box drawn below? (What is the area we would need if we were to cover this box with self-stick wrapping paper without overlapping?)

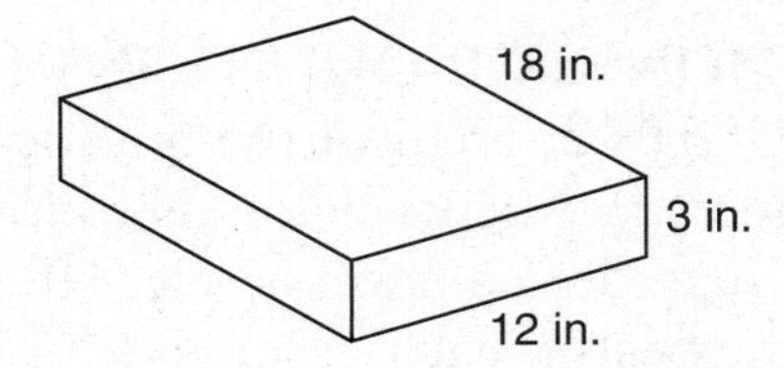

A. 648 square inches
B. 612 square inches
C. 306 square inches
D. 252 square inches

8. The perimeter of a square flower garden is 112 feet. What is its area?

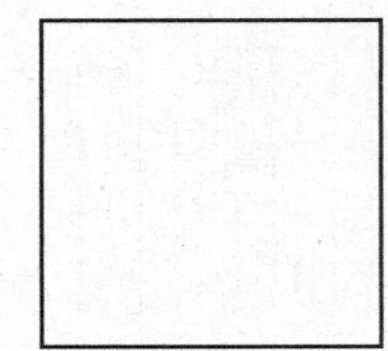

A. 784 square feet
B. 56 square feet
C. 112 square feet
D. 224 square feet

GO ON TO THE NEXT PAGE

9. When diagonals of a rectangle intersect, they create vertical angles. Using the diagram below, what is the measure of angle 2?

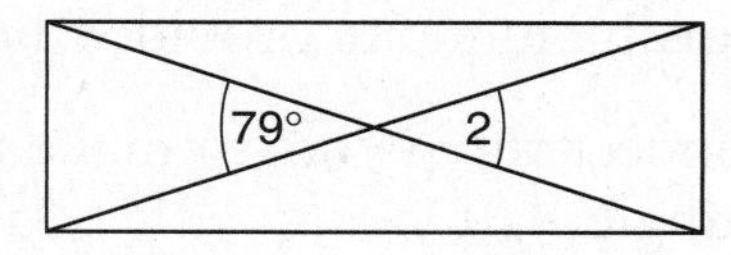

A. 105°
B. 79°
C. 15°
D. 25°

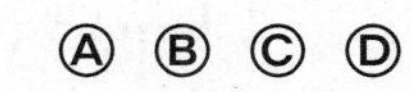

10. What is the sum of the interior angles of the pentagon drawn below?

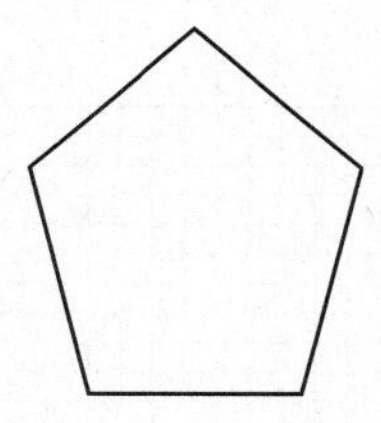

A. 540 degrees
B. 270 degrees
C. 720 degrees
D. 360 degrees

11. Nicole is playing with wooden cubes. She just built a box-shaped structure that is 7 cubes long, 5 cubes wide, and 4 cubes high. How many wooden cubes did Nicole use in all?

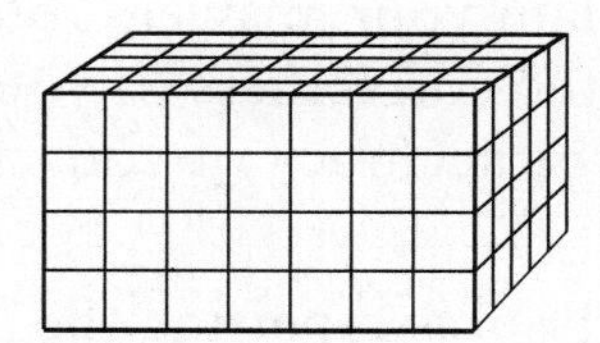

A. 70 cubes
B. 228 cubes
C. 140 cubes
D. 221 cubes

12. Which figure has an area of 64 square meters?

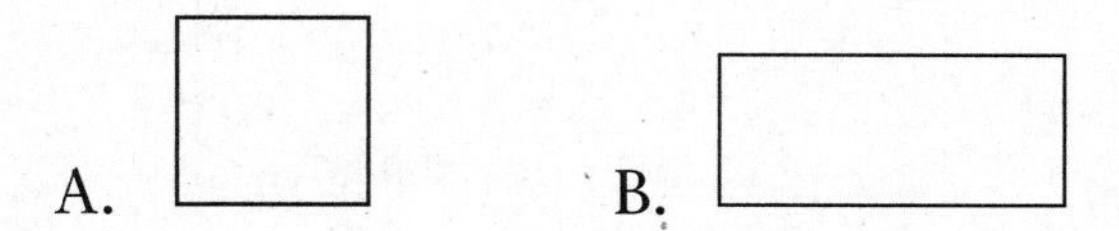

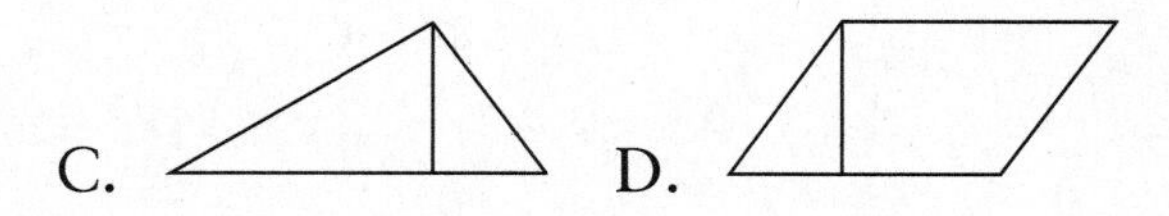

A. A square with one side measuring 8 meters.
B. A rectangle that is 20 meters wide and 12 meters high.
C. A triangle that has a base of 16 meters and is 4 meters high.
D. A parallelogram that has a base of 16 meters and is 2 meters high.

GO ON TO THE NEXT PAGE

EXTENDED CONSTRUCTED RESPONSE QUESTIONS

DIRECTIONS FOR QUESTIONS 13 AND 14: Respond fully to the ECR questions that follow. Show your work and clearly explain your answer. You will be graded on the correctness of your method as well as the accuracy of your answer.

13. The school is repairing the flooring in some of the older science rooms. Find the number of square feet of flooring that will be needed for this irregularly shaped room. Use the diagram below, and show how you found the area of the floor. Show all your work.

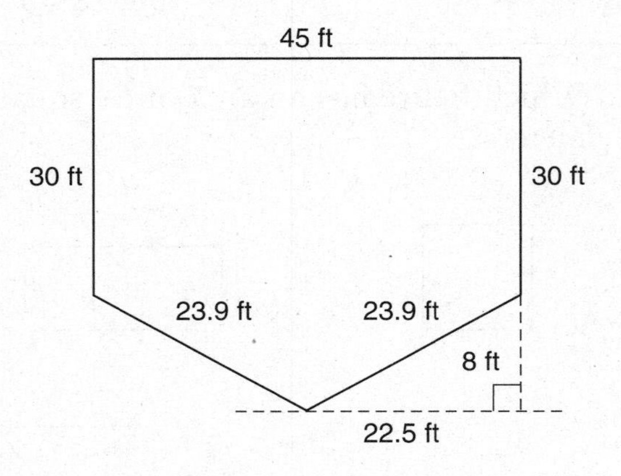

14. A triangle ABC is made by connecting the points A (0, 0), B (5, 0), and C (5, 6).
    1. Plot and label the points on the coordinate plane provided below.
    2. Connect the points to make the triangle ABC.
    3. Classify the triangle as right, isosceles, equilateral, or obtuse.
    4. Use the Pythagorean theorem (Pythagorean formula) to find the length of the side AC.
    5. Find the perimeter of triangle ABC.
    6. Find the area of triangle ABC.

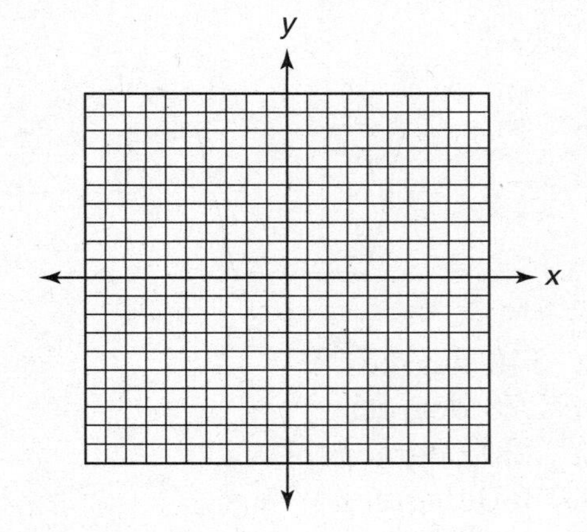

SOLUTIONS TO CLUSTER II TEST

MULTIPLE-CHOICE QUESTIONS

1. D	2. C	3. B	4. A	5. C	6. D
7. B	8. A	9. B	10. A	11. C	12. A

EXTENDED CONSTRUCTED RESPONSE QUESTIONS

QUESTION 13

- The irregular shape should be divided into geometric shapes. There are two possible solutions drawn below. The **bold** lines show how you might divide this shape.
- Next, you should demonstrate how you found the area of the various shapes. (You need to show your work.)

Area rectangle = 45 × 30 = 1,350
Area triangle = (45)(8)/2 = 180
Rectangle + triangle = Total area
1,350 + 180 = 1,530 sq. ft

Area large rectangle = 45 × 38 = 1,710
Area one triangle = (22.5)(8)/2 = 90
Area large rectangle – 2 triangles = Total area
1,710 – 2(90) = 1,710 – 180 = 1,530 sq. ft

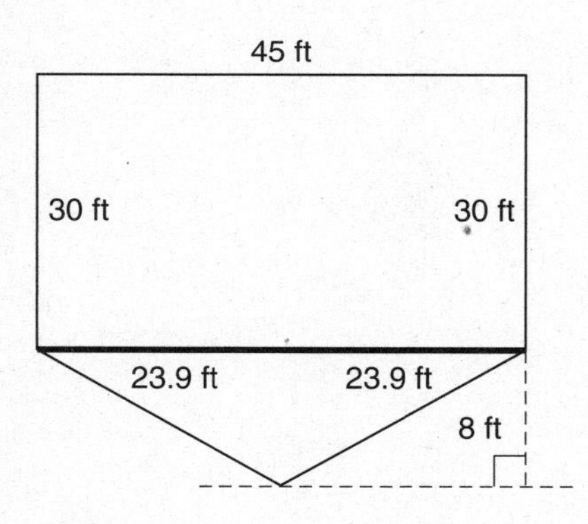

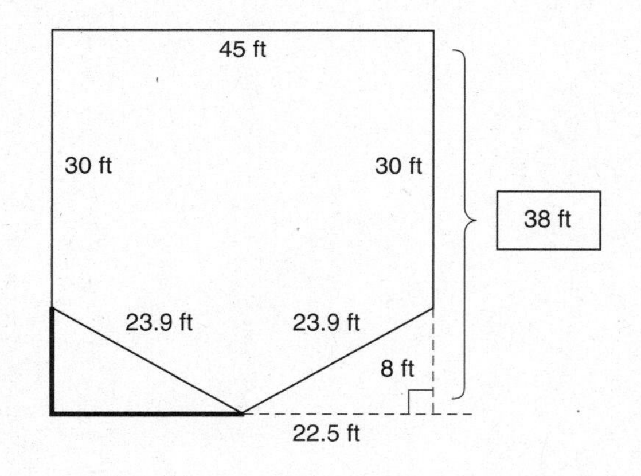

QUESTION 14

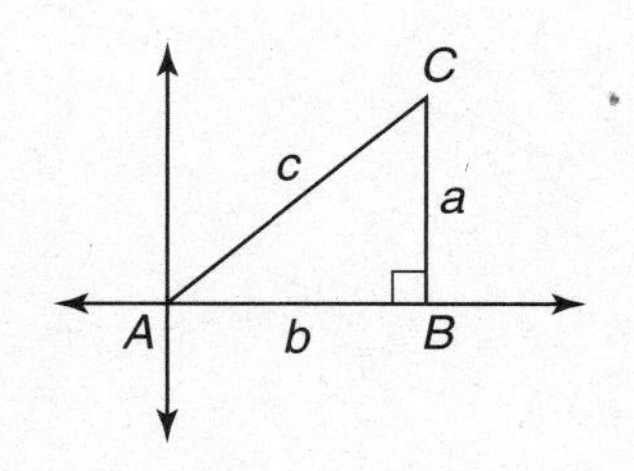

1. See the diagram.
2. See the diagram.
3. This is a right triangle.
4. $a^2 + b^2 = c^2$
   $5^2 + 6^2 = c^2$
   $25 + 36 = c^2$
   $61 = c^2$
   $7.8 = c$ (= Hypotenuse)
5. 6 + 5 + 7.8 = 18.8 units = Perimeter of triangle
6. $\frac{(5)(6)}{2} = \frac{30}{2}$ = 15 sq. units = Area of triangle

Cluster III: Data Analysis, Probability, Statistics, and Discrete Mathematics

WHAT DO ASK 8 DATA ANALYSIS AND PROBABILITY AND STATISTICS QUESTIONS LOOK LIKE?

MULTIPLE-CHOICE QUESTIONS (MC)

Example 1: Probability
Using a six-sided cube numbered from 1 to 6, what is the probability of getting a 4?

A. 4%
B. 17%
C. 25%
D. 50%

Example 1: Strategies and Solutions
Correct choice is B.

$$\frac{\text{Favorable outcomes}}{\text{Total possible outcomes}} = \frac{1}{6}$$

$$\frac{\text{Only one \#4}}{\text{6 different numbers}} = \frac{1}{6} = 0.1666$$

or approximately 0.17 or 17%

Note: Each side of this number cube contains one number: 1, 2, 3, 4, 5, or 6.

Example 2: Data Analysis and Statistics
The following are scores Jill received on her math quizzes this marking period: 80, 82, 80, 83, 75. If she gets an 80 on her next quiz, which of the following is true?

A. The mean will change.
B. The median will change.
C. The mode will change.
D. All will remain the same.

Example 2: Strategies and Solutions
Correct choice is D.
First, organize original data in order from lowest to highest: 75, 80, 80, 82, 83

- Mean = 80 $\frac{75+80+80+82+83}{5} = \frac{400}{5} = 80$
- Median = 80 (the number in the middle when they are all arranged in numerical order)
- Mode = 80 (the number that appears most often)

If her next test grade is an 80, all will remain the same.

SHORT CONSTRUCTED RESPONSE QUESTION (SCR)

Example 3: Data Analysis
(No calculator permitted.)
Use the stem-and-leaf plot below.

4	1	2	8	8	6	
3	0	6	1	0		
2	5	4	3	3	8	3
1	0	2	6	9	0	

What is the mode?

Answer: ________

Example 3: Data Analysis, Strategies and Solutions
First put the numbers in numerical order.

4	1	2	6	8	8	
3	0	0	1	6		
2	3	3	3	4	5	8
1	0	0	2	6	9	

Now it is easy to see that the number 23 appears the most often. It appears three times. That is the mode.

Answer: 23

EXTENDED CONSTRUCTED RESPONSE QUESTION (ECR)

Example 4: Use the box-and-whisker plot below.

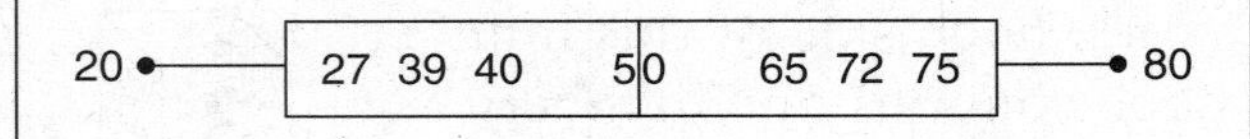

- What is the mean average?
- Explain why the vertical line in the box is not the same as the mean average.

Example 4: Strategies and Solutions

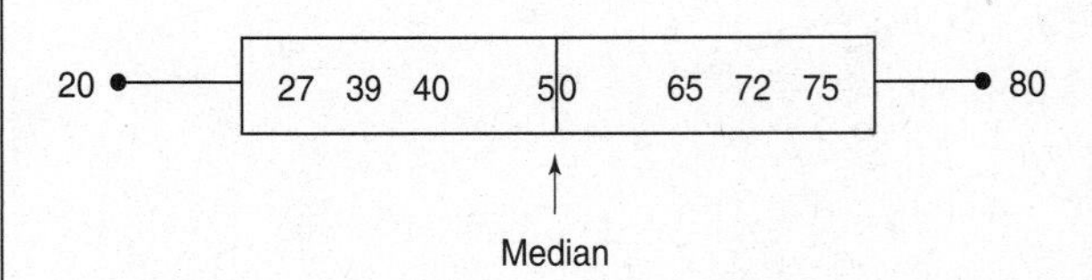

- The mean average is 52. Add all the data: $80 + 75 + 72 + 65 + 50 + 40 + 39 + 27 + 20 = 468$, then divide by $9 = \frac{468}{9} = 52$.
- The vertical line in the box represents the median (the number in the middle). The median of this data is 50.

FACTS YOU SHOULD KNOW

Before beginning this chapter, review the following facts that you should know:

- Number of days in a week: 7
- Number of days in a year: 365
- Number of seconds in a minute: 60
- Number of weekdays (Monday–Friday): 5
- Number of days in a leap year: 366
- Number of minutes in an hour: 60
- Number of cards in a regular deck of playing cards: 52
- A deck of cards is numbered 1–10 plus picture cards (Jacks, Queens, and Kings)
- Number of suits in a deck of cards: 4 (spades, clubs, hearts, and diamonds)
- Number of colors in a deck of cards: 2 (spades and clubs are black; hearts and diamonds are red)
- Each suit contains 13 different cards (1, 2, 3, 4, 5, 6, 7, 8, 9, 10, Jack, Queen, King)

PROBABILITY AND STATISTICS

Probability is used to make predictions. Probabilities can be written in many forms such as fractions, decimals, or percents.

- If an event will never happen (an elephant flies on its own), the probability of that happening is 0.
- If the event will definitely happen (it will rain in the rainforest this year), then the probability of that happening is 1.
- Sometimes percent is used, for example, if the weather forecaster says there is a 70% chance of snow tomorrow, then there is a 70% probability that it will snow tomorrow.

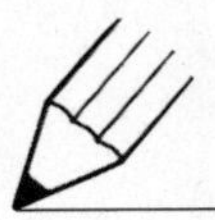

PRACTICE: Probability

(For answers, see page 205.)

1. What is the probability that you will pick one card that is a number 12 out of a regular deck of playing cards?

   A. 0
   B. 0.50
   C. 1
   D. not enough information given

2. What is the probability that you will pick an even number out of the numbers 2, 4, 6, 8, 10?

   A. 0
   B. 0.50
   C. 1
   D. not enough information given

3. What is the probability that it will snow in Alaska this year?

   A. 0
   B. 0.50
   C. 1
   D. not enough information given

4. What are your chances of spinning a number less than 2 using the following spinner?

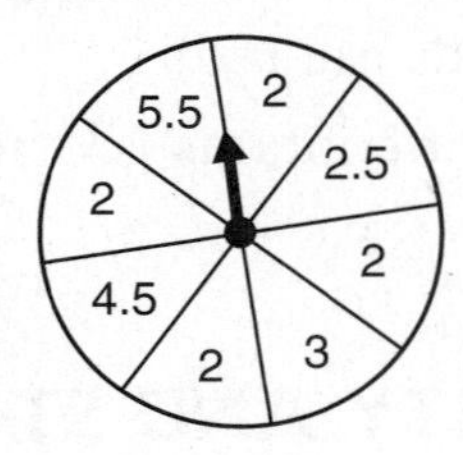

   A. 0
   B. 0.50
   C. 1
   D. not enough information given

EXPERIMENTAL PROBABILITY

An *experiment* is an activity where the results are observed. Each time you repeat the activity, it is called a *trial*, and the result of each trial is called the *outcome*.

- Flip a coin 5 times (5 trials); the outcome is tails, heads, heads, tails, heads (2T, 3H)
- Spin a spinner 4 times (4 trials); the outcome is red, blue, red, yellow (2R, 1B, 1Y)

The *sample space* is the set of all possible outcomes of an experiment.

Experiment	Sample Space (all the different possible outcomes)
Flip a penny	Heads, tails (These are the only possibilities.)
Pick a number from 1 to 10	1, 2, 3, 4, 5, 6, 7, 8, 9, 10
Roll a die	1, 2, 3, 4, 5, 6
The spinner lands on a season	Summer, fall, winter, spring
Pick a letter of the alphabet	The 26 letters in the alphabet (A, B, C, . . .)

Experimental probability can be found by using

$$\frac{\text{The number of times an event occurs}}{\text{Total number of trials}} \quad \text{or} \quad \frac{\text{The number of favorable outcomes}}{\text{Total number of outcomes}}$$

When we talk about the probability of an event occurring, we can give the answer as a percent.

Examples

A. If you role a number cube 30 times and 4 of those times you role a 2, then based on that experiment, the probability of rolling a 2 is $\frac{4}{30} = 0.1\overline{333} \sim 13\%$.

B. If a spinner lands on green 5 times out of 20 spins, the *experimental probability* of landing on green is $\frac{5}{20} = \frac{1}{4} = 0.25 = 25\%$.

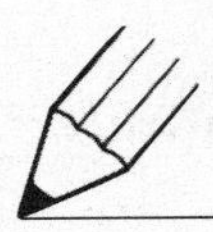

PRACTICE: Experimental Probability

(For answers, see page 205.)

1. If Janet picks 3 red jellybeans out of the 10 that she picked from the jar, what is the experimental probability of picking a red jellybean from that jar?

   A. 3%
   B. 30%
   C. 300%
   D. 70%

2. If John takes a handful of mixed nuts from a can and picks 6 salted peanuts out of the 12 he picked, what is the experimental probability of picking a salted peanut?

   A. 10%
   B. 20%
   C. 40%
   D. 50%

3. Jack has a Halloween bag filled with small pieces of candy. If he takes a handful and gets 4 chocolate kisses out of the 8 pieces he picked, what is the experimental probability of picking a chocolate kiss from that bag?

   A. 5%
   B. 50%
   C. 24%
   D. 40%

4. Approximately what is the probability that Mrs. Davis' birthday will be on a Monday?

   A. 8.33%
   B. 14.29%
   C. 20%
   D. 3.33%

5. Ronnie is correct when she says that there is about an 8% chance that anyone's birthday is

   A. in the spring
   B. on a Wednesday
   C. in May
   D. an even number

6. If you spin the spinner below, what is the probability that it will not land on an odd number?

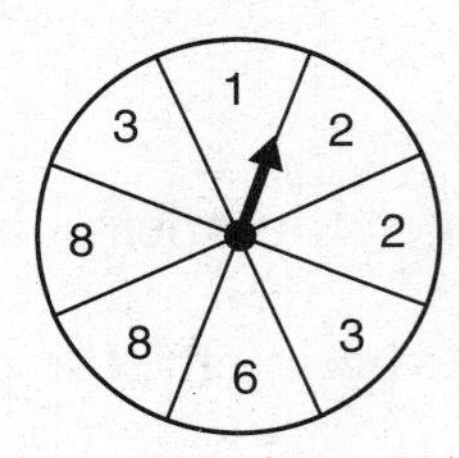

   A. $\frac{5}{8}$ or 62.5%

   B. $\frac{4}{8}$ or 50%

   C. $\frac{3}{8}$ or $37\frac{1}{2}$%

   D. $\frac{2}{8}$ or 25%

FINDING PROBABILITIES OF OUTCOMES IN A SAMPLE SPACE

Examples

A. What is the probability of landing on yellow using the spinner and table below?

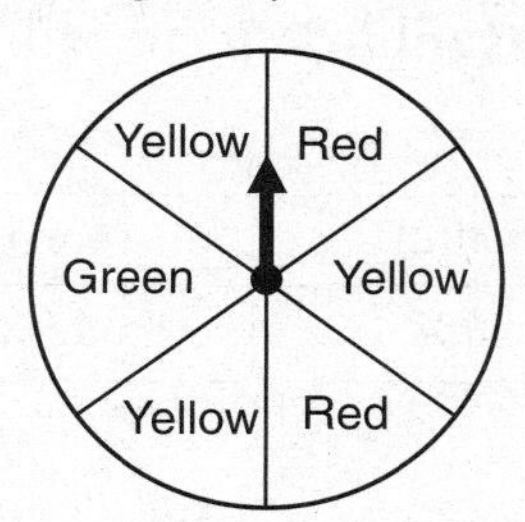

Outcome	Red	Yellow	Green
Probability	2	3	1
Sample Space	Red, yellow, green (6 spaces)	Red, yellow, green (6 spaces)	Red, yellow, green (6 spaces)

The probability of the spinner landing on yellow is $\frac{3}{6}$ or $\frac{1}{2}$ or $P(\text{yellow}) = \frac{1}{2}$.

The probability of the spinner landing on red is $\frac{2}{6}$ or $\frac{1}{3}$ or $P(\text{red}) = \frac{2}{3}$.

The probability of landing on green is $\frac{1}{6}$. Check to be sure the sum of all the possibilities equals 1. (Make common denominators and add.) $\frac{1}{2}+\frac{2}{6}+\frac{1}{6} =$

$\frac{3}{6}+\frac{2}{6}+\frac{1}{6} = \frac{6}{6} = 1$ ✔ Correct!

B. If you have a regular deck of cards, what is the probability of choosing a red 5?

Since there are only 2 red 5's (a 5 of hearts and a 5 of diamonds) and since there are 52 cards in a regular deck of cards, we can create the fraction of $\frac{\text{Favorable outcomes}}{\text{Total possible outcomes}}$.

This is $\frac{2}{52} = 0.03846$ or ~ 3.8% chance of picking a red 5.

FINDING PROBABILITIES OF EVENTS

Examples

A. When you are finding the probability of *one or more* events you *add* the probabilities. What is the probability that Rahul *or* John will win the race?

Runner	Rahul	Kevin	Joe	John
Probability of winning	30%	15%	20%	35%

$$P(\text{Rahul or John}) = 30\% + 35\% = 65\%$$

There is a 65% chance that Rahul or John will win the race.

B. When you find the probability that *one and another* event will happen, then you find the probability of each one and *multiply* them.
What is the probability that it will rain on Monday and Wednesday? Remember, there is less of a chance that it will rain on both days than on one day.

Day of week	Monday	Tuesday	Wednesday	Thursday	Friday
Probability of rain	10%	15%	20%	25%	70%

$$P(\text{Mon and Wed}) = (10\%)(20\%) = (0.10)(0.20) = 0.02 = 2\%$$

There is only a 2% probability that it will rain on Monday and Wednesday.
Remember: You are multiplying decimal numbers less than 1, so your product will be a smaller number. Be careful multiplying small decimal numbers!

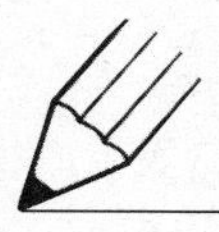

PRACTICE: Probability of Events

(For answers, see page 205.)

1. The sports announcer said the New Jersey Devils have a 60% chance of winning its next game at the Meadowlands. Complete the chart (chances of winning, chances of losing).

Outcome	The Devils win	The Devils lose
Probability	?	?

A. 0.06, 0.04 B. 0.60, 0.40 C. $\frac{4}{10}, \frac{6}{10}$ D. 40%, 60%

2. The table below charts the probability of the Trenton Thunder minor league baseball team getting a certain number of runs. What is the probability they will get 1 or 2 runs?

Runs	0 runs	1 run	2 runs	3 runs	4 runs
Probability of getting a run	0.10	0.30	0.25	0.15	0.05

A. 15% B. 55% C. 65% D. 0.075%

3. Kyle and Zak are having similar success in basketball this year. According to the latest school basketball statistics, there is a 50% probability that Zak or Kyle will make a basket during the last quarter of the game.

   a. From the choices shown, Kyle's chances of making a basket are most likely

   A. 25% B. 50% C. 75% D. 100%

   b. Estimate the probability that Zak and Kyle will both make a basket?

   A. 6–7% B. 25–30% C. 50% D. 75–80%

4. The latest weather forecast gave a 50% chance of rain in New York City tomorrow, and a 20% chance of rain the next day. What is the probability it will rain on both days?

A. 70% B. 50% C. 30% D. 10%

5. Jaime's brother is in his last year at college. He needs four more classes to complete his schedule.

Number of classes	1	2	3	4
Probability of classes being available	0.05	0.50	0.40	0.70

a. Approximately, what is the probability that he will get into class 1 **or** class 2?

A. less than 10% B. about 25% C. 50% D. 55%

b. What is the probability that he will get into class 3 **and** class 4?

A. 28% B. 30% C. 55% D. 110%

THEORETICAL PROBABILITY

Understanding probability and odds is helpful in making intelligent, informed choices.

Theoretical Probability	Odds in Favor	Odds Against
$\dfrac{\text{Number of favorable outcomes}}{\text{Number of possible outcomes}}$	$a : b$ a = number of favorable outcomes b = number of unfavorable outcomes	$b : a$ a = number of favorable outcomes b = number of unfavorable outcomes
One number 4 is on a die Six numbers are on a die Probability of rolling a 4 is $\frac{1}{6}$ or sometimes written 1 : 6.	$\dfrac{\text{Number favorable outcomes}}{\text{Number possible outcomes less Number favorable outcomes}}$	$\dfrac{\text{Number possible outcomes less Number favorable outcomes}}{\text{Number favorable outcomes}}$
	If the probability of an event is $\frac{1}{5}$, then the odds *in favor* of the event are $\frac{1}{5-1} = \frac{1}{4}$ or 1 : 4.	If the probability of an event is $\frac{1}{5}$, then the odds *against* the event are $\frac{5-1}{1} = \frac{4}{1}$ or 4 : 1.
	The odds are one to four in favor of winning this match.	The odds are four to one against winning this match.

DATA COLLECTION AND ANALYSIS

MEAN, MEDIAN, MODE, AND RANGE

The *mean* is the average of a set of data. 74 is the average of the following set of data: 60, 80, 70, 90, 70. You find the *mean* by adding the data and dividing by the number of numbers.

$$60 + 80 + 70 + 90 + 70 = 370; \frac{370}{5} = 74$$

The *median* is the number in the middle after a set of data is organized from lowest to highest.

60, 70, **70**, 80, 90

The *mode* is the number that appears most frequently. (*Remember:* mode = most often.) In this case, it is 70.

In a set of data, the spread of the numbers given from the lowest to the highest is the *range*.

$$90 - 60 = \mathbf{30}$$

There are 30 whole numbers between 90 and 60.

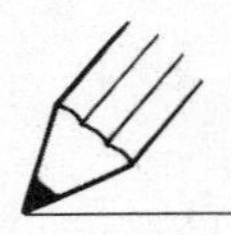

PRACTICE: Mean, Median, Mode, and Range

(For answers, see page 206.)

1. Find the mean, median, and mode of the data set: 15, 7, 9, 12, 21, 11, 13, 12, 8

a. The mean A. 11.2 B. 12 C. 13 D. 14

b. The median A. 11 B. 12 C. 12.5 D. 13

c. The mode A. 12 B. 21 C. 13 D. none

2. Find the *range* of the data set: 40, 33, 46, 50, 36, 51, 47, 35, 51, 47, 52, 53

A. 12 B. 13 C. 19 D. 20

3. The following are the amounts Jane saved each week in her new savings account: $50, $75, $60, $100, $60, $70, $90.

a. What is her approximate average savings each week (the mean average)?

A. $57.85 B. $69.39 C. $70.71 D. $72.10

b. What is the mode of this data set?

A. $50 B. $60 C. $70 D. $90

c. How much should she save the following week to raise her mean average by $10?

A. $152.10 B. $82.10 C. $92.10 D. $89.10

4. The following are the batting averages of the local school's best hitters: 310, 285, 300, 270, 300, 275. If a new hitter joins the team with a batting average of 290, what would change?

A. the mean, the mode, and the median
B. the mean and the median
C. the median only (mean remains at 290, mode stays at 300)
D. nothing would change

5. The teacher asks you to create a graph of the class' grades on a recent test. The data set is 79, 66, 82, 98, 86, 75, 85, 80, 92, 78, 82, 85, 77, 63, 88, 100, 88, 92, 82, 80, 77, 98. What is the range you would use to set up the axis to represent these grades?

A. 47 B. 37 C. 50 D. 100

THE FUNDAMENTAL COUNTING PRINCIPLE

In order to understand how many different ways a large number of items can be arranged, it is very helpful to understand the Fundamental Counting Principle:

> If there are m ways to select a first item and n ways to select a second item, then there are $(m)(n)$ ways to select both items.
>
> This is the same as what you did earlier to find probability of it raining on Monday *and* on Friday. You multiplied the two numbers.

Examples

A. How many different ways can you order an ice cream if the choices are:

Type of cone	Number of scoops	Ice cream flavor	Topping
Sugar or waffle	One or two	Chocolate, vanilla, coffee, or pistachio	Sprinkles, syrup, or none
Here you have 2 choices.	Here you have 2 choices.	Here you have 4 choices.	Here you have 3 choices.

(2) (2) (4) (3) = 48 different possible choices

B. A computer randomly generates a 5-character serial number that must contain 3 letters followed by 2 single-digit numbers. An example might be JAC42 or BBX52. How many possible serial numbers can be generated?

First Letter	Second Letter	Third Letter	First Digit	Second Digit
26 choices (A–Z)	26 choices (A–Z)	26 choices (A–Z)	10 choices (0–9)	10 choices (0–9)

(26) (26) (26) (10) (10) = 1,757,600 different possible ways to write a 3-letter followed by 2-digits as the serial number

C. What are the chances that the serial number WXT47 will be generated?

$$P(\text{WXT47}) = \frac{\text{Number of favorable outcomes}}{\text{Number of total possible outcomes}}$$

$$= \frac{1}{(26)(26)(26)(10)(10)} = \frac{1}{1{,}757{,}600} = 0.0000005$$

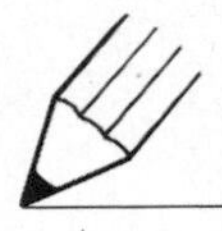

PRACTICE: The Fundamental Counting Principle

(For answers, see page 206.)

1. How many different ways can you order a meat and cheese sandwich given the following choices?

Bread	Meat	Cheese	Spread
Rye, white, or roll	Ham or bologna	Swiss or American	Mayonnaise, ketchup, mustard, or none

A. 11 B. 18 C. 24 D. 48

2. If the sandwiches in the above chart were prepackaged and you just randomly selected one, what is the probability that you would get a ham and Swiss cheese sandwich on a roll with mustard?

A. 50% B. 40% C. 10% D. 2%

3. How many different ways can you order a pizza given the following choices?

Large Size	Medium Size	Small Size	1 Slice
Cheese, pepperoni, or broccoli	Cheese, pepperoni, or broccoli	Cheese or pepperoni	Cheese

A. 9 B. 18 C. 19 D. 27

4. In New Jersey, many license plates have 3 letters, followed by 2 digits, then by 1 letter. One example would be NWP65B. Find the total number of possible license plates.

Letter	Letter	Letter	Number	Number	Letter
26	26	26	10	10	26

A. 114 B. 11,400 C. 45,697,600 D. 10,400

THE FUNDAMENTAL COUNTING PRINCIPLE AND TREE DIAGRAMS

Another technique for solving similar problems is to draw a tree diagram. Naturally, drawing a tree diagram is appropriate when the quantities are small.

Examples

A. How many different outfits can be made from the following? Use a tree diagram to show this.

Footwear	Pants	Shirts
Boots, sandals, or sneakers	Shorts or jeans	Red, blue, or white

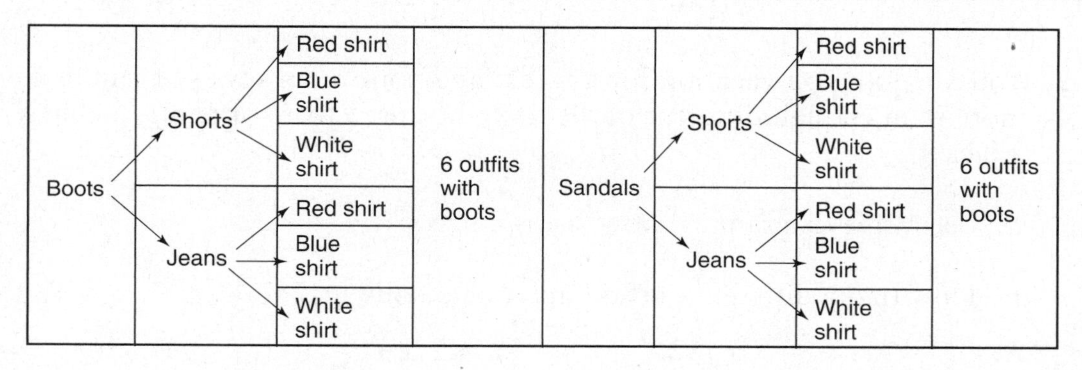

Looking at the tree diagrams above, you can see that one outfit would be *boots, shorts and a red shirt;* another outfit would be *boots, shorts, and a blue shirt.* In a similar way, you could create six outfits with sneakers. Therefore you can create a total of 6 × 3 (6 with boots, 6 with sandals, and 6 with sneakers) or 18 outfits.

B. If you flip three different fair coins, how many of the outcomes will have exactly two tails? Notice that when each coin is tossed there are two possibilities, either heads (H) or tails (T).

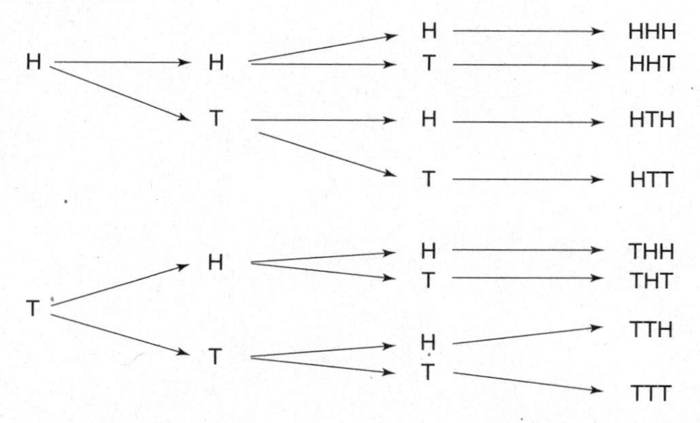

a. What is the probability of getting exactly two heads? $P(\text{exactly 2 tails}) = \frac{3}{8}$.

b. What is the probability of getting three heads? $P(\text{only 3 heads}) = \frac{1}{8}$.

PRACTICE: The Fundamental Counting Principle and Tree Diagrams

(For answers, see page 207.)

1. The school dietician is ordering supplies and needs to find out how many different variations of meals she can prepare if each meal is to have meat, potatoes, and one vegetable. She can use: turkey or roast beef, mashed potatoes or baked potatoes, and peas or corn or carrots. Use a tree diagram to demonstrate your solution.
   How many different meals can she plan with the food choices listed above?

   A. 9 B. 12 C. 24 D. 36

2. You are going on vacation for a week, and your mom says you can make more than enough different outfits if you bring 2 pairs of pants, 3 shirts, and 2 jackets.

   a. Use a tree diagram to show she is correct.

   b. How many different outfits can you actually make?

   A. 9 B. 12 C. 24 D. 36

3. If you flip 2 coins at the same time, what are your chances of getting 2 heads? Draw a tree diagram to show this.

   A. 10% B. 20% C. 25% D. 50%

COMBINATIONS AND PERMUTATIONS

Look at the chart below to gain an understanding of the differences between *combinations of data* and *permutations of data.* There are shortcut ways to determine combinations and permutations. You can use *factorials* to do this quickly. First, we'll review definitions and then learn about factorials.

Combinations	Permutations
Are a way of selecting things from a group when the *order does not matter.*	Are ways of selecting things from a group when the *order does matter.*
Select any two toppings for your pizza. You can, for example, select one with cheese and peppers or one with cheese and mushrooms.	*Vote for the president and vice president of your 8th grade class from the list of six different candidates.* One student becomes president and a different student becomes vice president.
A *combination* is a group of items chosen from a larger group of items. The difference between one combination and another is the items, not the order of the items. For example, the difference is whether you select ham and Swiss cheese, or baloney and American cheese, and not which one goes on the sandwich first.	A *permutation* is an arrangement of a group of items in a particular order; the items are arranged in a certain way. A particular item must be first, another second, and so on. In the preceding example, it makes a difference which person is selected first (as president) and which is selected second (as vice president).

Now, let's review factorials. Here, we'll evaluate expressions containing *factorials.* The symbol used as an exclamation point is the same one used as the factorial symbol in mathematics!

- $0! = 1$ (This is read as "zero factorial." Zero factorial is always equal to 1.)
- $7!$ (seven factorial) means $(7)(6)(5)(4)(3)(2)(1) = 5{,}040$
- $\dfrac{5!(\text{five factorial})}{3!(\text{three factorial})} = \dfrac{(5)(4)(3)(2)(1)}{(3)(2)(1)} = \dfrac{(5)(4)\cancel{(3)(2)(1)}}{\cancel{(3)(2)(1)}} = (5)(4) = 20$
- $\dfrac{8!}{(9-4)!} = \dfrac{8!}{5!} = \dfrac{(8)(7)(6)\cancel{(5)(4)(3)(2)(1)}}{\cancel{(5)(4)(3)(2)(1)}} = (8)(7)(6) = 336$
- $\dfrac{4!}{2!(8-5)!} = \dfrac{4!}{2!(3!)} = \dfrac{(4)\cancel{(3)(2)(1)}}{(2)(1)\cancel{(3)(2)(1)}} = \dfrac{4}{2} = 2$ (*Remember:* Work inside parentheses first.)
- $\dfrac{4!}{(9-9)!} = \dfrac{4!}{0!} = \dfrac{4!}{1} = \dfrac{(4)(3)(2)(1)}{1} = \dfrac{24}{1} = 24$ (*Remember:* $0! = 1$.)

Example

There are five folders on a shelf; each folder is a different color (red, yellow, blue, green, or orange). How many different ways can you arrange these folders on the shelf?

There are five choices for the red folder: it can be in the first, second, third, fourth, or fifth position on the shelf. The shortcut way in math to determine this is to use *factorials*. $5! = (5)(4)(3)(2)(1) = 120$ or 5! (five factorial) different ways to arrange these folders on the shelf.

Many times the data being considered is more complex than just five folders; therefore, it is essential to learn the two different kinds of groupings and the related formulas to use.

Permutations

The number of permutations of n things taken r at a time is

$${}_{\mathrm{v}}P_{\rho} = \frac{n!}{(n-r)!}$$

Examples

A. There are six racehorses in a small local race held in Hunterdon County, New Jersey.

a. Find the number of orders in which all six horses can finish.

$${}_6P_6 = \frac{6!}{(6-6)!} = \frac{6!}{0!} = \frac{6!}{1} = (6)(5)(4)(3)(2) = 720 \text{ permutations}$$

This means there are 720 different orders that the six horses can finish.

b. Find the number of different ways the six horses can finish first and second. Here the n things being considered are the 6 horses taken r (or 2) at a time. This reduces the possibilities.

$${}_6P_2 = \frac{6!}{(6-2)!} = \frac{6!}{4!} = \frac{(6)(5)(4)(3)(2)(1)}{(4)(3)(2)(1)} = \frac{(6)(5)\cancel{(4)(3)(2)(1)}}{\cancel{(4)(3)(2)(1)}} = 30 \text{ permutations}$$

This means there are 30 ways that six horses can finish in first and second place.

B. If you can use each number 1, 2, and 3 only once, there are 6 permutations of these three numbers.

$${}_3P_3 = \frac{3!}{(3-3)!} = \frac{3!}{0!} = \frac{(3)(2)(1)}{1} = (3)(2)(1) = 6$$

Since this is a rather small set of data you can see all the permutations:

1, 2, 3 1, 3, 2 2, 1, 3 2, 3, 1 3, 1, 2 3, 2, 1 = (3)(2)(1) or 3! = 6

PRACTICE: Combinations, Permutations, and Factorials

(For answers, see page 208.)

1. If no letter is used more than once, how many permutations are there of the letters DEFG?

   A. 8
   B. 16
   C. 24
   D. 25

2. In how many ways can six people line up at the bus stop?

   A. 12
   B. 18
   C. 36
   D. 720

You did examples earlier where the order of the items did *not* matter. These are called *combinations*. An example would be the variations of ingredients in a sandwich. The sandwich could contain meat, cheese, and tomato or tomato, cheese, and meat; it doesn't matter what order they are in. Here, we'll use *factorials* to solve combinations with larger numbers quickly and we'll learn how to determine all possible combinations when you select groups of two or more items where the order does not matter.

Combinations

The number of combinations of n things taken r at a time is

$${}_{\mathrm{v}}C_{\mathrm{p}} = \frac{n!}{r!(n-r)!}$$

Examples

A. Rosa's pizza store is offering a special sale: two toppings instead of one for the same price. The following six choices of toppings are available: mushroom, sausage, meatball, cheese, onion, vegetables. Find the number of different two-topping pizzas that can be ordered.

Look at the equation above and compare it to our solution below. The six represents the *total choices* and the two tells us we are taking things *two at a time*. (The order does not matter, so pepperoni and cheese counts the same as cheese and pepperoni; this counts as one possibility.)

$${}_6C_2 = \frac{6!}{2!(6-2)!} = \frac{6!}{2!4!} = \frac{(6)(5)\cancel{(4)}\cancel{(3)}\cancel{(2)}\cancel{(1)}}{(2)(1)\cancel{(4)}\cancel{(3)}\cancel{(2)}\cancel{(1)}} = \frac{30}{2} = 15$$ different combinations of pizza with two toppings

You can see this if we let A = one topping, B = another topping, and so on.

- 1 AB, 1 AC, 1 AD, 1 AE, 1 AF = 5 combinations
- 1 BC, 1 BD, 1 BE, 1 BF = 4 combinations
- 1 CD, 1 CE, 1 CF = 3 combinations
- 1 DE, 1 DF = 2 combinations
- 1 EF = 1 combination

B. At Mario's Sandwich Shoppe in addition to meat you can add **three** items to the sandwich at no additional cost. The five items you may select from are peppers, onions, lettuce, tomato, cheese. How many three-item combinations are there?

$${}_5C_3 = \frac{5!}{3!(5-3)!} = \frac{5!}{3!2!} = \frac{(5)(4)\cancel{(3)}\cancel{(2)}\cancel{(1)}}{\cancel{(3)}\cancel{(2)}\cancel{(1)}(2)(1)} = \frac{20}{2}$$ = 10 different combinations of three different items on a sandwich

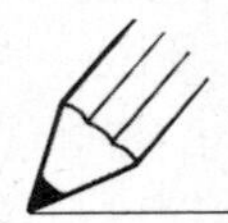

PRACTICE: Combinations and Factorials

(For answers, see page 208.)

1. At Gold's Deli you can order your sandwich with any two of the following sides: pickles, coleslaw, macaroni salad, potato salad, hot peppers, sweet peppers.

   How many 2-item combinations are there?

   A. 30
   B. 15
   C. 10
   D. 6

2. The Office Supply Store at the mall has a special sale for students. If you buy two large binders, you get two of the following items free: pen, pencil set, ruler, compass, protractor, black marker, yellow highlighter.

   How many free two-item combinations are there?

   A. 30
   B. 35
   C. 12
   D. 21

STATISTICS

This section of the chapter deals with different ways to collect data, to work with large sets of numbers, and to organize, display, and analyze the data.

FREQUENCY TABLES AND INTERVALS

Here is an example of a *frequency table*. In this case, the column on the left gives you a *range* of scores on a quiz (60 points is actually the maximum score), and the column on the right tells you how many students scored within that range (the *frequency* of that data being used).

Frequency Table

Range of quiz scores (interval)	Number of students (frequency)
1–9	0
10–19	0
20–29	1
30–39	6
40–49	15
50–59	5
60–69	1

Use this table to answer the following questions:

- How many students scored in the 20 to 39 point range?

  (20 to 29) = 1 student + (30 to 39) = 6 students = 7 students

- What percentage of students scored above 39 points?

  $$\frac{\text{21 students scored above 39 points}}{\text{28 students in all}} = \frac{3}{4} = 0.75 = 75\%$$

- Can you tell the *median* score? If not, explain why not?

  No, you do not know the exact individual scores so you cannot find the middle score.

- Is it possible to find the *mean average* score? If not, explain why not.

  No, you do not know the exact individual scores.

- What is the *range* of scores on this quiz?

  You cannot find the exact range. The scores might be between 20 and 69, then the range would be 69–20 or 49. However, they also could be between 29 and 60, and then the range would be only 31.

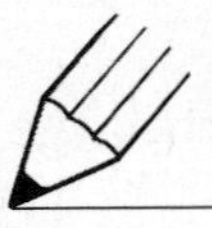

PRACTICE: Frequency Tables

(For answers, see page 208.)

1. Ten cars were monitored as they drove down a hill in a 50-mile-per-hour zone. They were clocked at the following miles per hour:

   44, 58, 50, 62, 54, 48,
   55, 60, 52, 48 .

   Set up a frequency table using the above data.

2. Find the median and range of this data. Explain how you arrived at your answers.

STEM-AND-LEAF PLOTS

Using a *stem-and-leaf plot* is one way to display a great deal of information where all of the information is seen. For example, using a stem-and-leaf plot, the numbers 246, 243, 251, and 258 can be represented as follows:

24	6 3
25	1 8
Stem	Leaves

The numbers 204 and 206 are represented as

Stem	Leaves
20	4 6

Example

The following is a display of the various scores received by students in Mrs. Barrow's class on an ASK 8 Review Test. These scores represent the percentage correct.

75	88	90	80	95	85
85	92	72	95	75	98
92	78	74	68	82	82
95	80	85	78	86	88

Make a stem-and-leaf plot for this data.

First, write the *stems* from least to greatest to the left of the vertical line. Then write each *leaf* to the right of its stem.

6	8									
7	5	2	5	8	4	8				
8	8	0	5	5	2	2	0	5	6	8
9	0	5	2	5	8	2	5			

Now, rearrange the *leaves* for each stem in order from least to greatest. Title the stem-and-leaf plot.

Scores on NJ ASK 8 Review Test

6	8									
7	2	4	5	5	8	8				
8	0	0	2	2	5	5	5	6	8	8
9	0	2	2	5	5	5	8			

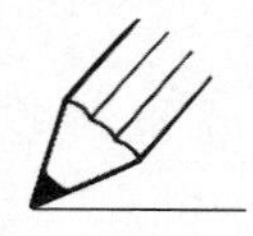

PRACTICE: Stem-and-Leaf Plotting

(For answers, see page 209.)

1. What is the mode of the test scores shown on the stem-and-leaf plot below.

Grades on Chapter Review Test

5	9							
6	8	8						
7	0	2	5	5	8	8		
8	2	5	5	8	8	8		
9	0	0	0	0	2	2	5	8

A. 0 B. 8 C. 88 D. 90

2. What is the median of the grades on the above Chapter Review Test?

A. 90 B. 85 C. 88 D. 75

3. How many students had grades higher than 85 on the above test?

A. 11 B. 14 C. 8 D. 3

HISTOGRAMS

Data displayed in a frequency table can also be displayed as a histogram. A *histogram* is simply a bar graph that is used to show frequencies. Notice that in this kind of bar graph there are no spaces between the bars.

First we'll review a *bar graph* then we'll look at a *histogram*.

PRACTICE: Bar Graphs

(For answers, see page 209.)

1. Bar graphs are used for data that can be grouped into specific categories. Use the bar graph below. How many more students said they enjoyed reading biographies than novels?

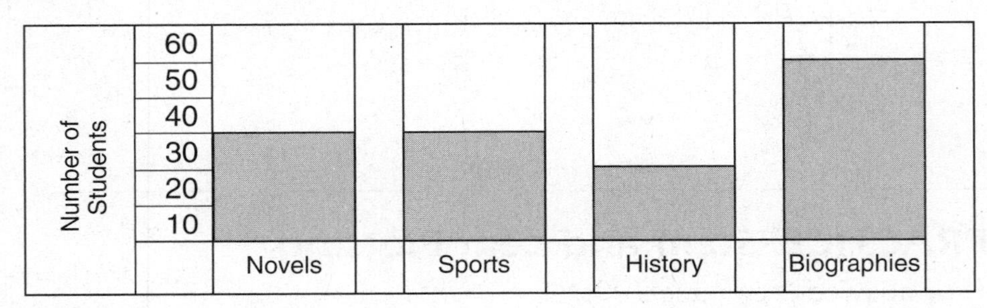

A. 10 students B. 20 students C. 30 students D. 40 students

Example

Now let's look at a *histogram* with similar information. Here the data is grouped in intervals. You cannot tell exactly how many students chose each category.

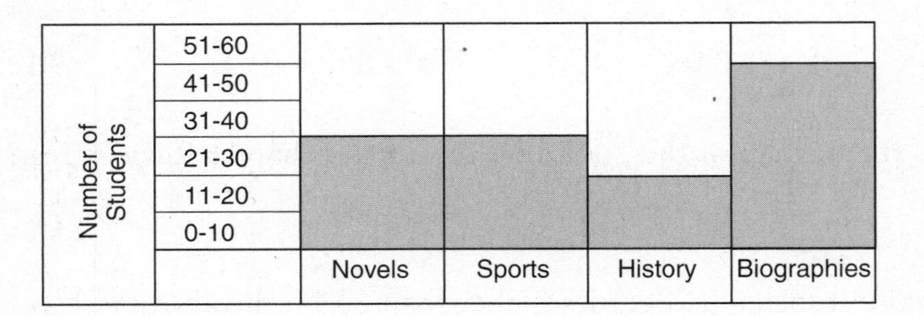

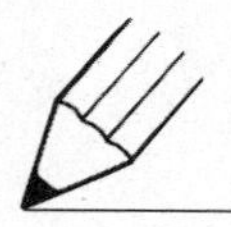

PRACTICE: Histograms

(For answers, see page 209.)

1. Use the histogram shown to answer the following questions.

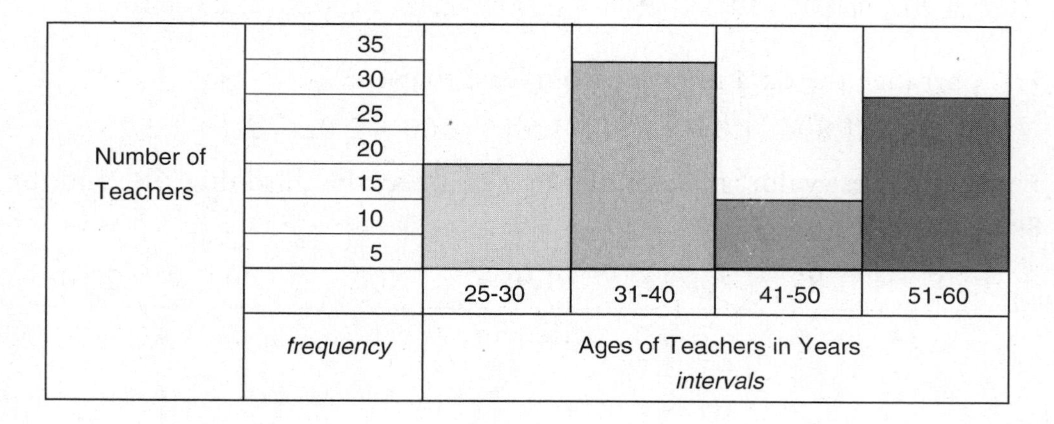

a. How many teachers are between 41 and 50 years old?

b. How many teachers are younger than 41?

c. How many teachers are included in this chart all together?

d. What percentage of teachers are older than 50?

e. What percentage of teachers are older than 22?

BOX-AND-WHISKER PLOTS

Sometimes it is easier to "see" data if it is divided into quarters called *quartiles*. This is what is done in a *box-and-whisker plot*. Look at the following data arranged in order from lowest to highest and study the different terms. The *median* divides the set of data into two parts. The median is the number in the middle. The *first quartile* is also a *median*; it is the number in the middle when you consider all the numbers to the left of the median of all the data. The *third quartile* is the number in the middle when you consider all the numbers to the right of the actual median of all the data.

Example

We'll review how to make a box-and-whisker plot for the following data.

Gasoline Prices over the past year (price is per gallon)								
1.90	1.80	1.95	1.74	2.00	2.10	2.02	2.23	1.85

- First, arrange the data in order from least to greatest.

  1.74 1.80 1.85 1.90 1.95 2.00 2.02 2.10 2.23

- Find the lowest value, median, the first quartile, the third quartile, and the greatest value.

1.74	1.80	1.85	1.90	**1.95**	2.00	2.02	2.10	**2.23**
Lowest value (lower extreme)	↑			Median		↑		Greatest value (highest extreme)
	1.80	1.85				2.02	2.10	
	1.825					2.06		

First quartile
(the median of the values from the least to the overall median, but not including the overall median) In this case, it is the average of the two middle numbers in the first quartile.

$$\frac{1.80 + 1.85}{2} = \frac{3.65}{2} = 1.825$$

Third quartile
(the median of the values between the median and the greatest, including the greatest)

$$\frac{2.02 + 2.10}{2} = \frac{4.12}{2} = 2.06$$

A box-and-whisker plot simply shows this information in a box, with the *lower* and *upper extremes* at the ends of the "whiskers." This is what a box-and-whisker plot looks like.

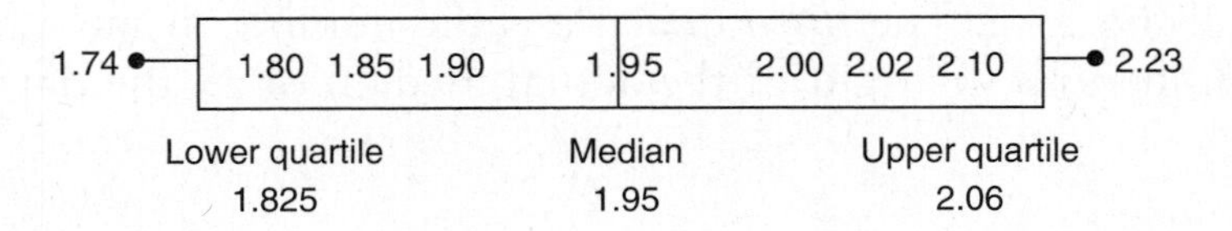

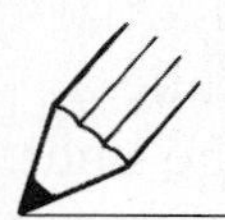

PRACTICE: Box-and-Whisker Plots

(For answers, see page 209.)

Use the box-and-whisker plot below for questions 1 to 4.

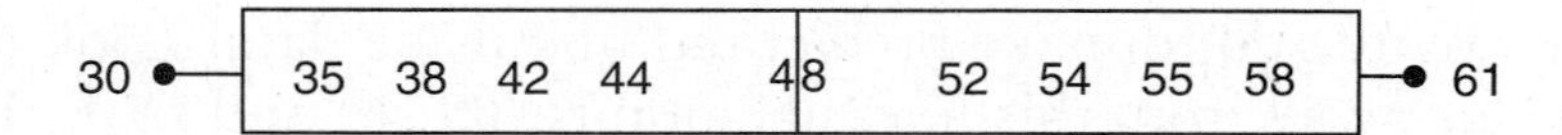

1. What is the least value? What is the greatest value?

2. What is the median value?

3. What are the first quartile and the third quartile values?

4. What is the range of the data?

SCATTER PLOTS (SCATTERGRAMS)

A scatter plot looks like a graph on a grid with many points that are not connected.

Examples

A. First, we'll "read" the points to understand the basic information.

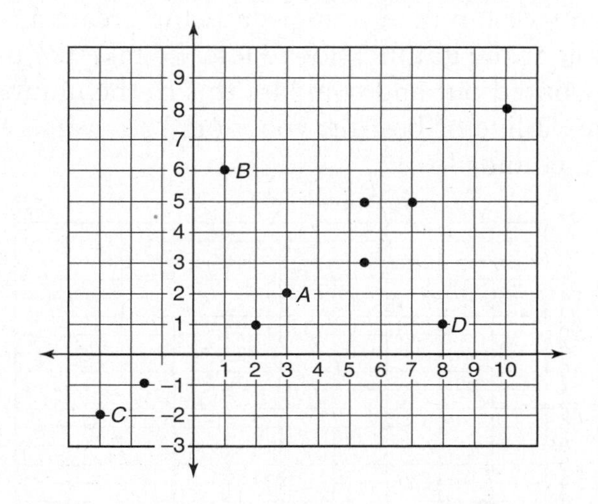

Point A: (3, 2) Point B: (1, 6) Point C: (−3, −2) Point D: (8, 1)

The points in a *scatter plot* (sometimes called a *scattergram*) do not lie on a particular line. But, many points seem to lie near a line. Sometimes you can take a thin piece of spaghetti and place it on top of a scattergram to see where the line seems to cover the most number of points. This line is called a *trend line.* Do *not* connect the points to make a trend line. There is no data between the points; the points in a *scatter plot* do not match a particular linear equation. The trend line is used to estimate and to make predictions about the data. Look at the data that seems to be far away from this line, like points $B(1, 6)$ and D $(8, 1)$ on page 187. We call these points *outliers.* Some scatter plots show a *positive correlation* among the data graphed, and some show a *negative correlation.* Study the next two graphs to see examples of each.

B. Negative correlation A local gym took a survey to compare the number of hours a group of women worked out at the gym compared to their weight. We would expect to see that the more hours worked out each week would show the person weighed less. If you were to draw a line-of-best-fit, you would see a line with a *negative slope* (more hours, less weight).

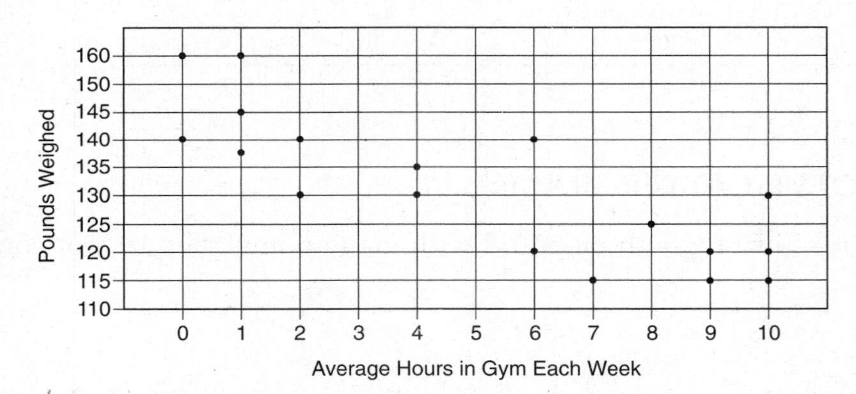

C. Positive correlation We can take similar data and create a scatter plot with a *positive slope.* Look at the graph below. Here we take the number of hours a group of women worked out and compare this to the number of pounds lost. If you were to draw a line-of-best-fit, you would see a line with a *positive slope* (more hours, more pounds lost).

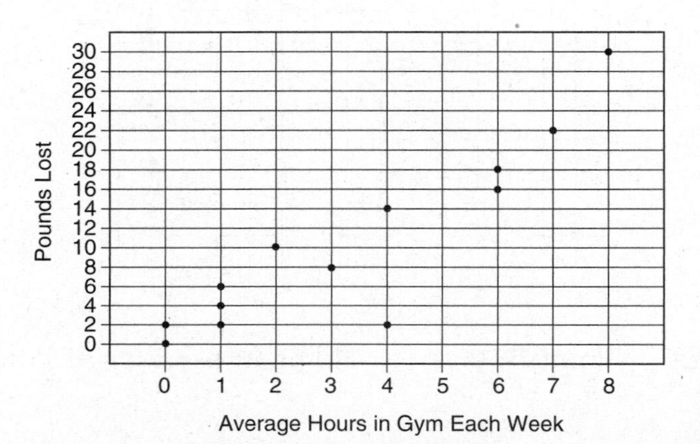

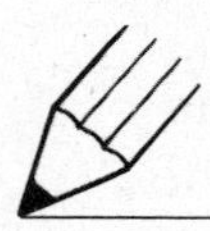

PRACTICE: Scatter Plots

(For answers, see page 209.)

Use the scatter plot below to answer questions 1–2.

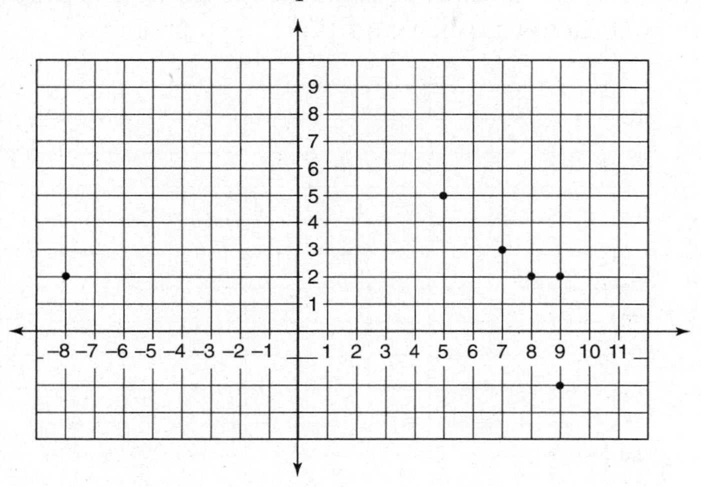

1. Which of the following points seems to be an outlier?

   A. (8, –2) B. (–8, 2) C. (8, 2) D. (9, –2)

2. Does the point with coordinates (5, 5) seem to be on the trend line?

   A. yes B. no
   C. sometimes D. not enough information given

3. If this scatter plot represents Margie's deposits into a new savings account, how much has she saved so far? (*Hint*: Notice what happens at week # 8.)

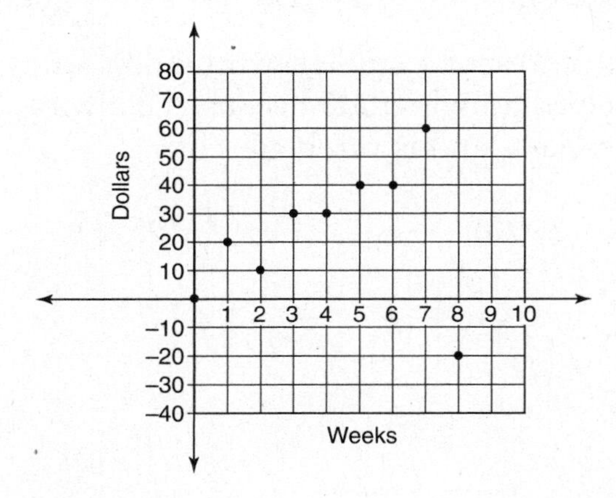

   A. $ 260 B. $ 240 C. $210 D. $ 200

4. Mrs. Teller, an 8th grade math teacher, was concerned that her students had not been doing well on their quizzes and tests lately. On yesterday's test, she asked the students to write at the top of their tests the amount of time they studied. She then created a scatter plot to see if there was a positive correlation between the time students studied and their testing scores.

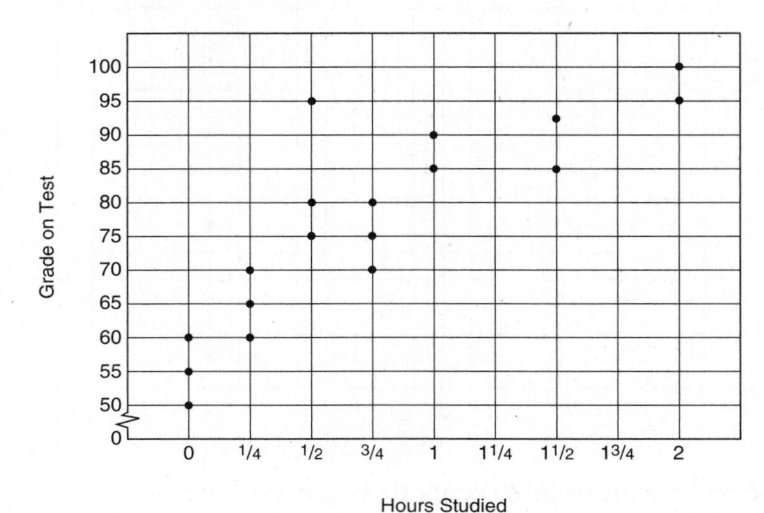

a. Does this graph show a positive or negative correlation?

A. positive B. negative
C. neither D. not enough information shown

b. If you draw a line-of-best-fit are there any *outliers*?

A. yes (0 hours and 55%) B. yes ($\frac{1}{2}$ hr and 95%)
C. yes (2 hours and 95%) D. no

c. Which point might represent a student who didn't study very long the night before but still received a high grade because he always does his homework, takes good notes in class, and is rarely absent?

A. (2, 95) B. (2, 90) C. ($\frac{1}{2}$, 95) D. ($1\frac{1}{2}$, 90)

5. This scatter plot below shows the relationship between the number of hours a family takes for its Sunday car rides compared to the price of gasoline for their car. Does this show a positive or negative correlation?

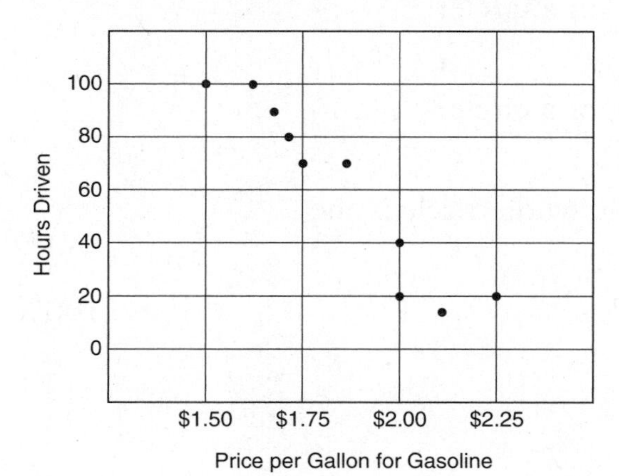

A. positive
B. negative
C. neither
D. not enough information shown

CIRCLE GRAPHS

A *circle graph* compares parts of the circle to the whole circle.

Sometimes a circle graph is divided into degrees. There are 360° in a whole circle.

Examples

A. How many degrees are in $\frac{1}{2}$ of a circle? $\frac{360}{2} = 180°$

B. How many degrees are in $\frac{1}{5}$ of a circle? $\frac{360}{5} = 72°$

C. Sometimes a circle is divided into uneven sections. How many degrees are in the empty section?

$180° + 30° + 110° = 320°$

$360° - 320° = 40°$

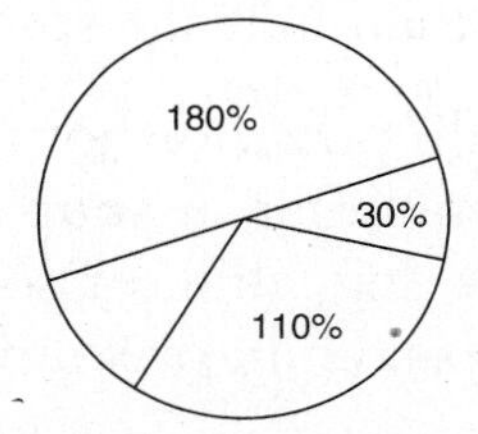

Other times a circle graph is divided into percents. The whole circle equals 100%.

Examples

A. What percent is $\frac{1}{2}$ of a circle? $\frac{100\%}{2} = 50\%$

B. What percent is $\frac{1}{5}$ of a circle? $\frac{100\%}{5} = 20\%$

C. What percent of the whole circle is the empty section?

$60\% + 10\% = 70\%$

$100\% - 70\% = 30\%$

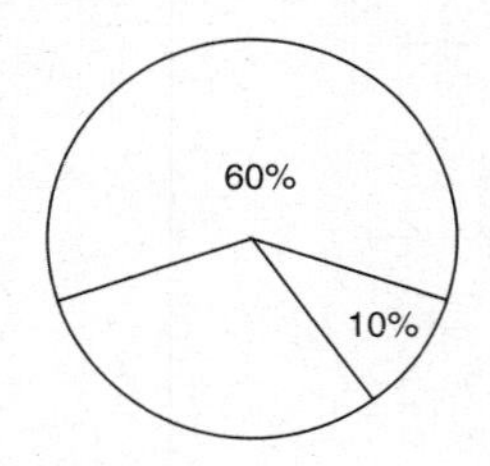

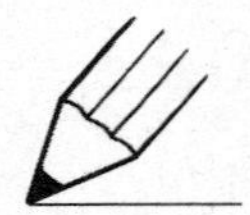

PRACTICE: Circle Graphs

(For answers, see page 210.)

Use the circle graph below for questions 1–4.

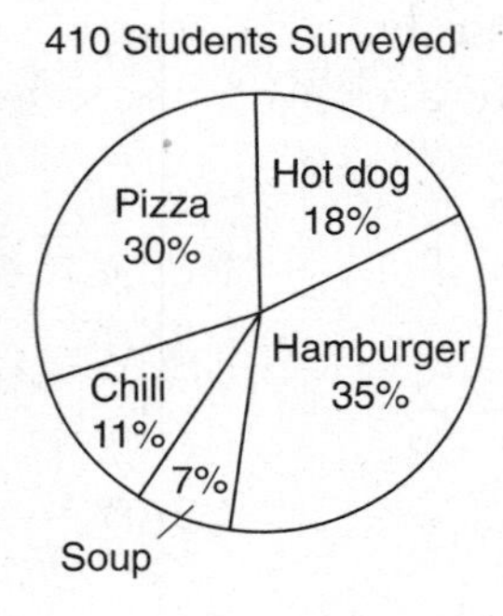

1. What percent of the students surveyed selected hamburger or pizza as their favorite?

A. 30% B. 35%
C. 65% D. 5%

2. Approximately how many students chose chili as their favorite?

A. 11 B. 41
C. 4.5 D. 45

3. Which equation shows how many more students chose hamburger than soup?

A. $(410)(0.35) - (410)(0.07)$
B. $(410)(0.35 - 0.07)$
C. $(410 - 0.35) + (410 - 0.07)$
D. $\frac{410}{0.35} - \frac{410}{0.07}$

4. More than half the students chose

A. chili, pizza, or hot dog
B. hamburger or soup
C. hot dog, chili, or soup
D. pizza, chili, or soup

Use the circle graph below for questions 5–7.

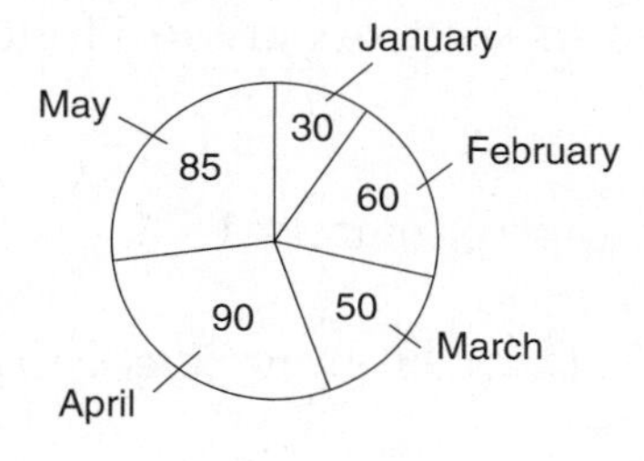

5. What percent of students surveyed had birthdays in February?

A. 19% B. 20%
C. 40% D. 60%

6. Approximately what percent of students surveyed have birthdays either in April or May?

A. 175% B. 180%
C. 56% D. 58%

7. What percent of student birthdays are in the fall?

A. 0% B. 50%
C. 25% D. 100%

VENN DIAGRAMS

Example

The Guidance Department of a local high school used computer data to analyze the number of 9th grade students in honors math and science classes. They collected the data and arranged it in the table below. But, this showed too many numbers and was confusing.

	Honors			College-Prep Level					Basic Level			RR	
Math	29	27	**56**	26	26	25	25	**102**	21	20	**41**	3	**3**
Science	24	24	**48**	25	26	25	26	**102**	21	20	**41**	3	**3**

So they arranged the same data in a *Venn diagram.*

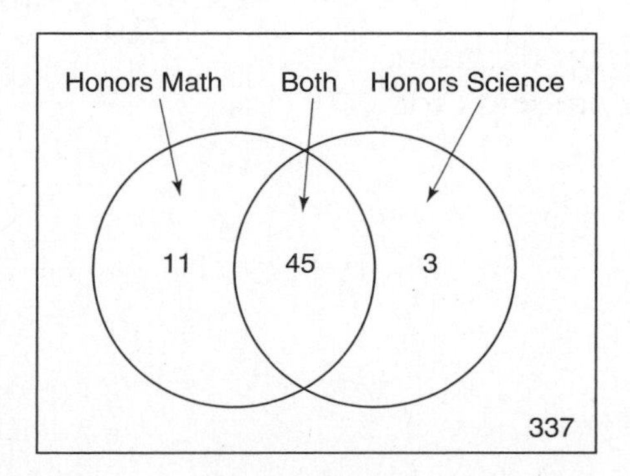

In this Venn diagram, you can see that the center section (where the two intersecting circles overlap) tells us that 45 students are taking *both* Honors Math and Honors Science. In the other sections, you see that 11 students are in Honors Math *only* (56 – 45), and 3 are in Honors Science *only* (48 – 45), and 337 are *not* in Honors Math or Honors Science.

- If there is a total of 396 students, what percent of students are in Honors Math *and* Honors Science all together?

$$\frac{45+11+3}{396} = \frac{59}{396} = 0.148989 \text{ or approximately } 15\%$$

- What percent of students are *not* in Honors Math or Honors Science?

$$\frac{337}{396} = 0.8510 \text{ or approximately } 85\%$$

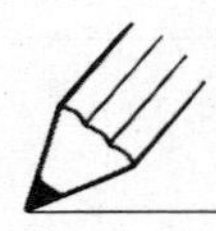

PRACTICE: Venn Diagrams

(For answers, see page 210.)

1. Use the Venn diagram below to answer the following questions. Valley High School has a total of 396 students in grades 10 and 11.

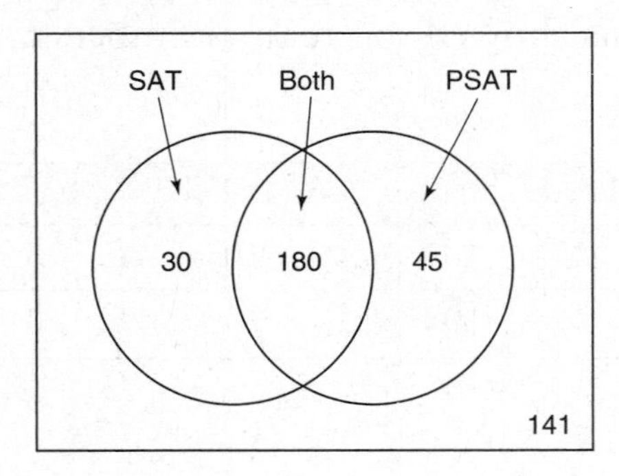

a. According to the Venn diagram, how many of these students took the PSAT?

A. 180
B. 45
C. 180 + 45
D. 18

b. What percent of the students who took these tests, took both the PSAT and the SAT?

A. $\frac{210}{255}$ = about 82%

B. $\frac{180}{255}$ = about 71%

C. $\frac{30+45}{255} = \frac{75}{255}$ = about 29%

D. $\frac{180}{210}$ = about 86%

c. What percent of *all* students in grades 10 and 11 took either the PSAT, the SAT, or both?

A. $\frac{255}{273}$ = about 93%

B. $\frac{255}{396}$ = about 64%

C. $\frac{180}{255}$ = about 72%

D. $\frac{75}{255}$ = about 29%

2. Use the Venn diagram below to answer the following questions.

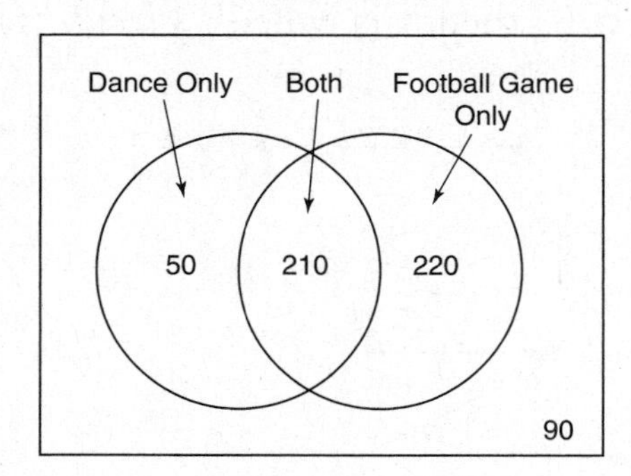

a. How many students did not attend either the Friday night dance or the Saturday afternoon football game?

A. 90
B. 30
C. 70
D. 160

b. How many students were surveyed in all?

A. 360
B. 480
C. 570
D. 350

c. Which expression would give the percent of all students who attended at least one activity?

A. $\frac{50+210+220}{570}$

B. $\frac{50+210+210+220}{570}$

C. $\frac{50+220}{570}$

D. $\frac{430+260}{570}$

NETWORKS

A *network* is a figure that has points (*vertices*) and lines (called *edges*). The lines connect the points to create a network. Here are some examples.

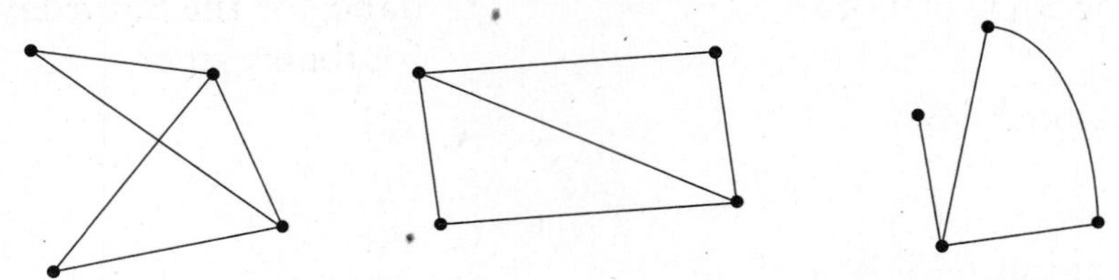

A particular route along these lines that has a beginning (at one vertex) and an ending (at another vertex) is called a *path* (from home to school).

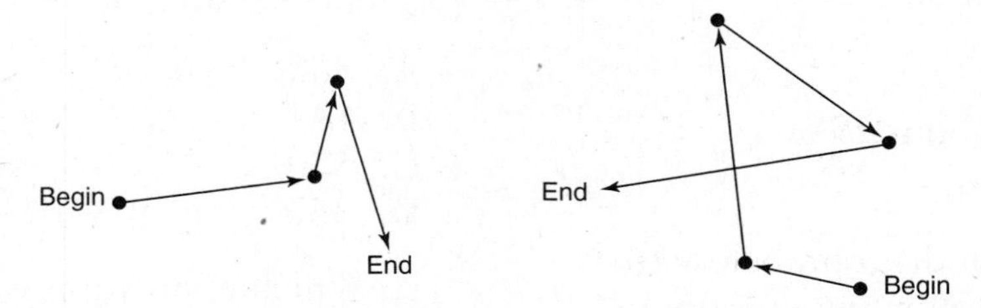

If a particular path begins and ends at the same vertex, then it is called a *circuit* (from home to the store and back home again).

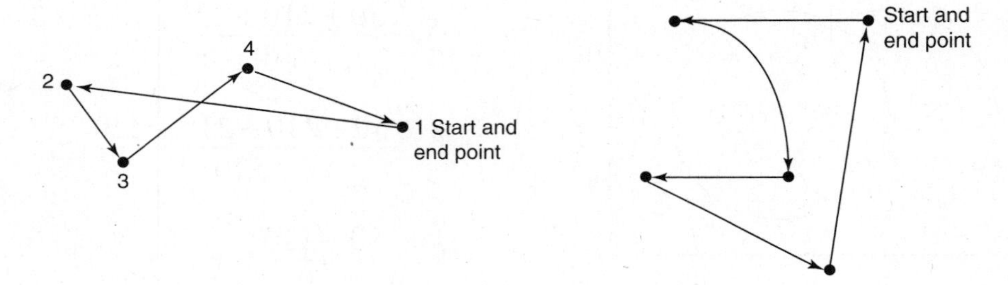

Example

Real-life situations can often be modeled using a network. Look at the network below to discover which students are in the same classes during the day at school.

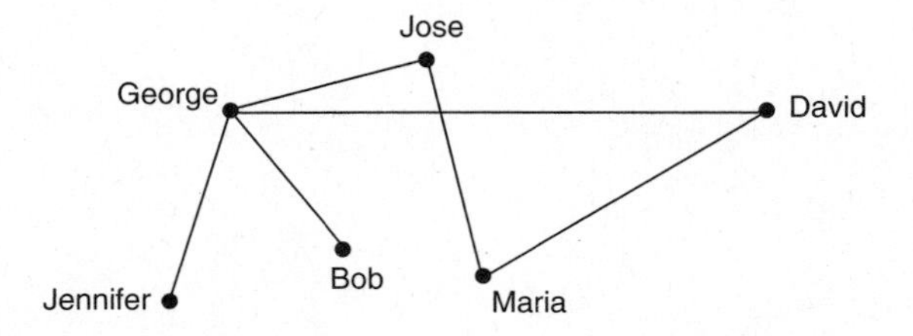

Here you see that George is in a class with Jose, Bob, Jennifer, and David, but Bob is not in a class with Jose, Jennifer, or David; he is only in a class with George.

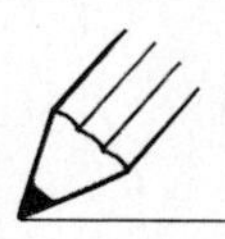

PRACTICE: Networks

(For answers, see page 210.)

1. Which route seems to be the shortest if you want to go from a school located at point A to a school located at point B? (*Hint*: Let your knowledge of geometry help you here.)

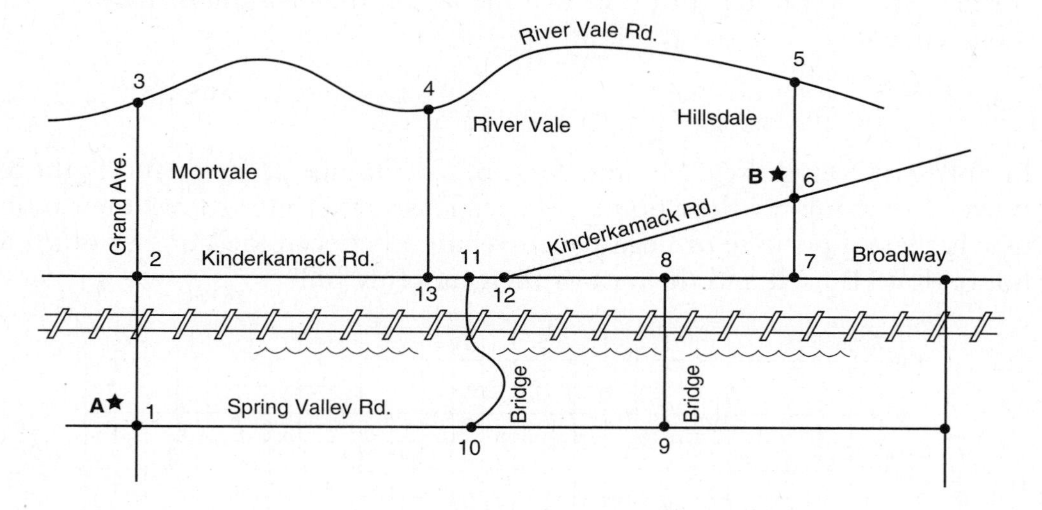

A. 1, 9, 8, 7, 6
B. 1, 10, 12, 6
C. 1, 2, 3, 4, 5, 6
D. 1, 2, 12, 6

2. Which network below best matches a salesman's delivery route? He makes deliveries to six of his stores, three are in one town and three are in another?

A. 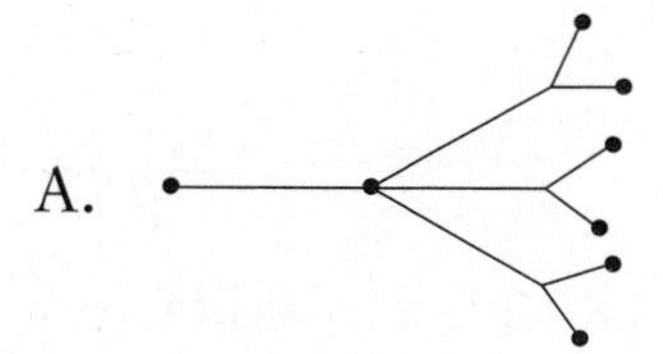

B. 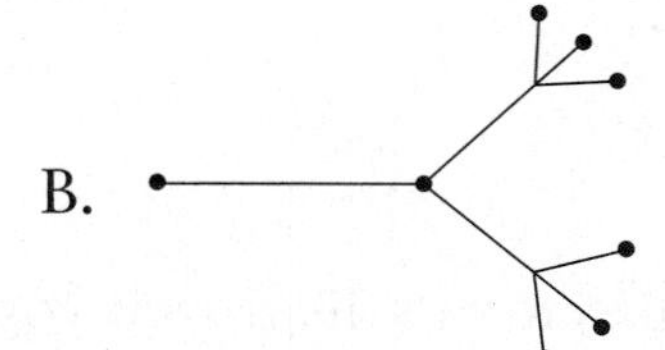

C. 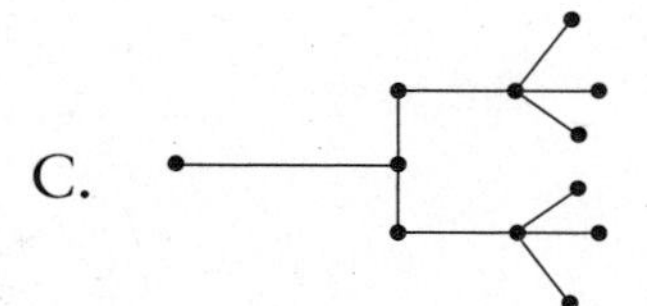

D. 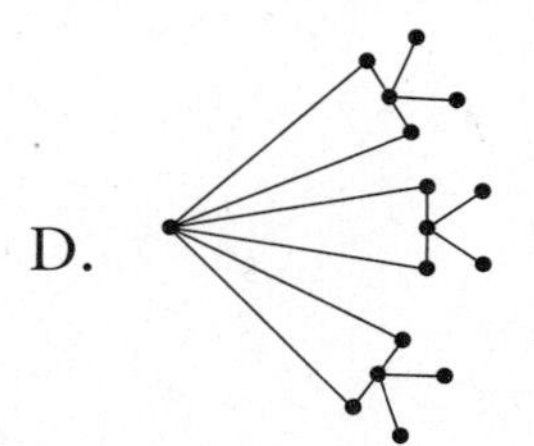

PRACTICE SCR NONCALCULATOR QUESTIONS

Each question is worth 1 point. No partial credit is given.
(For answers, see page 211.)

1. Jessica's four tests scores are: 80 75 80 85.

   If she scores 80 on her fifth test, will the *mean*, the *median*, or the *mode* change?

   Answer: ___________

2. In northern New Jersey, Mr. and Mrs. B. use Orange & Rockland Light & Power Company for their home electric. This scatter plot shows their monthly use. Is there a positive or negative correlation between the number of kilowatt hours (KWH) used and the cost of their monthly bill?

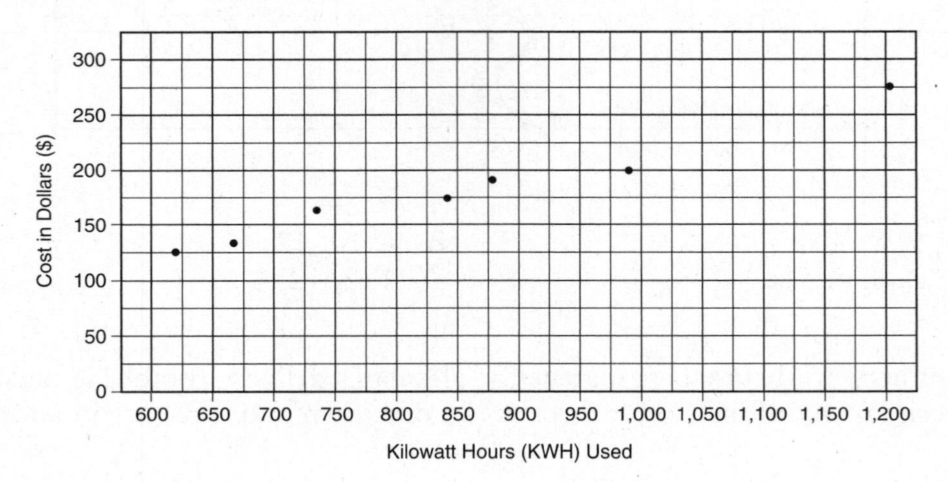

   Answer: ___________

3. The school cafeteria sells frozen yogurt. You can get chocolate, vanilla, or swirl. You can choose chocolate sprinkles, multicolored sprinkles, or no sprinkles. How many different combinations of frozen yogurt are available?

   Answer: ___________

4. In a deck of playing cards, there are 13 red hearts, 13 red diamonds, 13 black clubs, and 13 black spades. What is the probability of getting a *red 5* from this deck of cards?

Answer: ___________

5. Use the circle graph below. Of the 200 students surveyed, how many students take an art class this year?

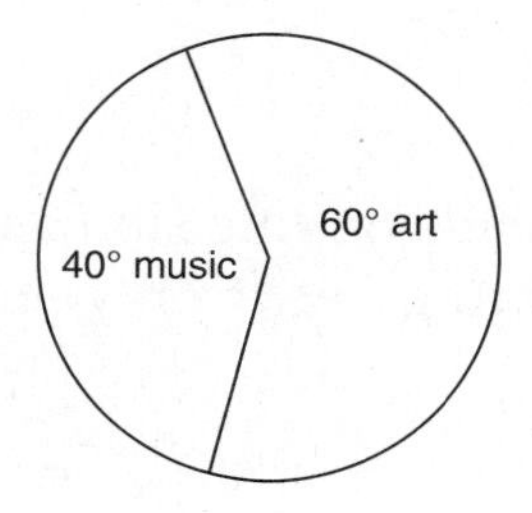

Answer: ___________

6. Dave mowed lawns five days last week. What was the average numbers of lawns he mowed?

Day of the week	Saturday	Monday	Tuesday	Thursday	Friday
Number of lawns mowed	4	2	1	1	2

Answer: ___________

7. Miquel worked part-time during the past two months. Using the table below, what was his *median* salary?

Week #	1	2	3	4	5	6	7
Salary $	$150	$200	$125	$150	$225	$150	$100

Answer: ___________

8. What is the probability of a dart hitting the shaded area of these two squares?

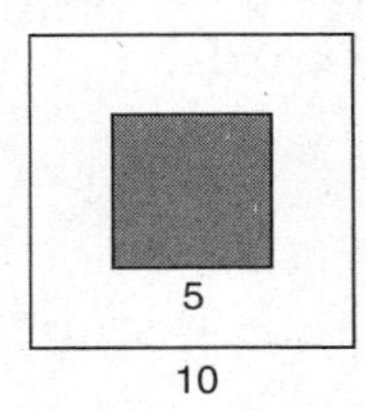

Answer: ___________

9. When you flip a fair coin (heads on one side and tails on the other), what is the probability of getting *heads* on your 10th flip?

Answer: ___________

10. Using the histogram drawn below, about how many students chose fruit for dessert on Wednesday and Friday?

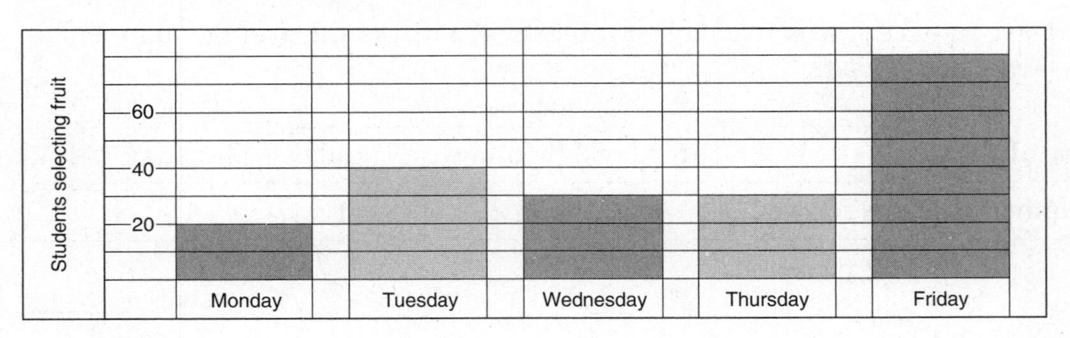

Answer: ___________

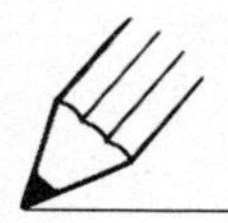

PRACTICE ECR QUESTIONS

(For answers, see page 212.)

1. Suppose two different varieties of tomatoes are growing in pots. The regular tomato plant is already 12 inches tall and is growing at a rate of $\frac{3}{2}$ inches per week. The cherry tomato plant is 6 inches tall and is growing at a rate of 2 inches per week. Complete the chart below to show how their heights change each week. Show all work.

	Regular Tomato Plant		Cherry Tomato Plants	
	Show work!	Height of Plant (inches)	Show work!	Height of Plants (inches)
Now	12		6	
Week 1	$12 + \frac{3}{2} =$		$6 + 2 =$	
Week 2				
Week 3				
Week 4				
Week 5				
Week 6				
Week 7				
Week 8				
Week 9				
Week 10				
Week 11				
Week 12				
Week 13				
Week 14				

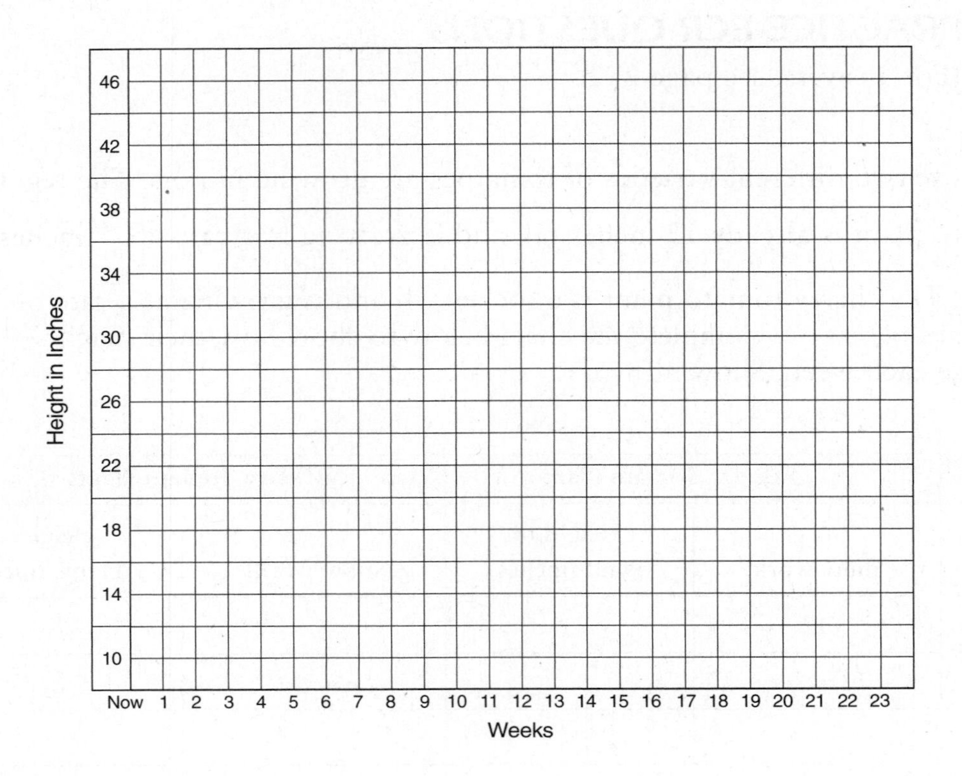

Use the coordinate grid above. Let the horizontal axis be the weeks; let the vertical axis be the height (inches).

- Create a double line graph. Label each line "regular tomatoes" or "cherry tomatoes." Graph each plant's height for at least the next 14 weeks.
- In how many weeks will both plants be the same height?
- What happens where the two lines intersect?
- What happens after this?

2. Ten cars were monitored driving down the hill in a 50-mile-per-hour zone. They were clocked at the following miles per hour:

44	58	50	62	54
48	55	60	52	48

a) Create a stem-and-leaf plot of this data.

b) Find the mean, median, mode, and range using your stem-and-leaf plot.

c) Use the same data and create a box-and-whisker plot and label the median.

d) Using your box-and-whisker plot, what is the value of the upper quartile?

e) What percent of cars were driving slower than 50 miles per hour?

f) What percent of cars were driving faster than 50 miles per hour?

3. Refer to the circle graph of "JB's Bakery: Annual Sales Breakdown" shown below.

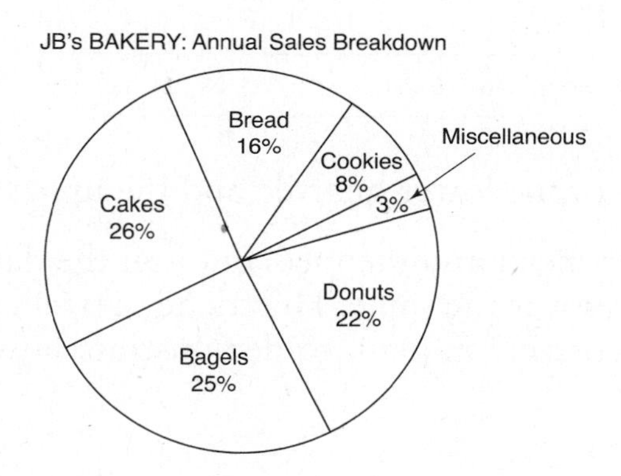

- If the bakery sold $23,769 in May 2009, how much money did it make from bread and bagels?
- After taking a survey of his customers, the owner expects to sell 50% more cookies than he sold during the year shown. If his annual total remains the same, how much money would he expect to make from cookies next year?

4. This is a fun "handshake" problem. Show your work and label your answers.

   - There are six students in a group, and each student shakes hands with each other student (only once). How many handshakes would there be altogether?
   - If three more students were added to the group, and each student shakes hands with each other student (only once), how many total handshakes would there be now?

5. - How do you find the probability that a card drawn at random from a full deck of 52 cards will be a 10?
   - If one card is chosen at random, how do you find the probability that you will choose a black King? What is that probability?
   - What is the probability of selecting a card that is a prime number? Show how you found your answer.

6. Use the box-and-whisker plot below to answer these questions.

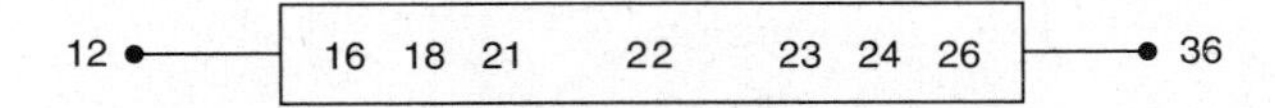

   - What is the value of the lower quartile and the upper quartile?
   - Jared says that the mean and the median of all the data shown in this box-and-whisker plot is the same. His friend, Terrell, says that is not possible. Who is correct? Explain or demonstrate how you arrived at your answer.

SOLUTIONS TO PRACTICE QUESTIONS

PROBABILITY (page 164)

1. A 0 There are no cards numbered 12 in a regular deck.
2. C 1 Any card you pick would be correct since all are even numbers.
3. C 1 Yes, there is a 100% chance that it will snow in Alaska each year.
4. A 0 There is no number less than 2; zero possibility of selecting that.

EXPERIMENTAL PROBABILITY (page 166)

1. B 30% $\frac{\text{3 samples}}{\text{10 total}} = \frac{3}{10} = 0.30 = 30\%$

2. D 50% $\frac{6}{12} = \frac{1}{2} = 0.50 = 50\%$

3. B 50% $\frac{4}{8} = \frac{1}{2} = 0.50 = 50\%$

4. B 14.29% $\frac{\text{1 Monday}}{\text{7 days in week}} = \frac{1}{7} = 0.1428571 \text{ or } 0.1429 = 14.29\%$

5. C in May $\frac{\text{1 May}}{\text{12 months in year}} = \frac{1}{12} = 0.083 \text{ or} \sim 8\%$

   $\text{Spring} = \frac{\text{1 spring}}{\text{4 seasons}} = \frac{1}{2} = 0.50 = 50\%$

   $\text{Wednesday} = \frac{\text{1 Wednesday}}{\text{7 days}} = \frac{1}{7} = 0.1428 = 14.28\%$

   $\text{Even numbers} = \frac{\text{1 even}}{\text{2 choices}} = \frac{1}{2} = 0.50 = 50\%$

6. A $\frac{5}{8} = 62.5\%$ $\frac{\text{NOT odd numbers}}{\text{All choices}} = \frac{5}{8} = 0.625 = 62.5\%$

PROBABILITY OF EVENTS (page 169)

1. B 0.6, 0.4 If 60% win, then 40% will lose; if 0.60, then 0.40.
2. B 55% Probability 1 run = 0.30, probability 2 runs = 0.25; since this is an "or" situation, you add the two probabilities: 0.30 + 0.25 = 0.55 or 55%

3a.	A	25%	Since they are 50% together, then Zak + Kyle = 50%. Since Zak and Kyle are equal in their abilities, Zak = 25%, Kyle = 25%.
b.	A	6–7%	Each has 0.25; since this is an "and" situation, you multiply the two (0.25)(0.25) = 0.0625 or 6.25%, which is between 6 and 7%.
4.	D	10%	This is an "and" situation, so you multiply: (0.50)(0.20) = 0.10 = 10%.
5a.	D	55%	In an "**or**" situation, you add the two probabilities: 0.05 + 0.50 = 0.55, which is 55%.
b.	A	28%	In an "**and**" situation, you multiply the two probabilities: (0.40)(0.70) = 0.28, which is 28%. (It is less likely that he will get into both classes.)

MEAN, MEDIAN, MODE, AND RANGE (page 172)

1a.	B	12	mean (Add all the numbers, then divide by 9.) $\frac{7+8+9+11+12+12+13+15+21}{9} = \frac{108}{9} = 12$
b.	B	12	median (Put the numbers in numerical order; this is the one in the middle.)
c.	A	12	mode (the one that appears most often); 12 appears two times.
2.	D	20	The range is 53 – 33 = 20.
3a.	D	$72.10	
b.	B	$60	
c.	A	$152.10	
4.	D		nothing would change
5.	B	37	100 – 63 = 37

THE FUNDAMENTAL COUNTING PRINCIPLE (page 174)

1.	D	48	bread, 3 choices; meat, 2 choices; cheese, 2 choices; spread, 4 choices $3 \times 2 \times 2 \times 4 = 48$
2.	D	2%	$\frac{\text{Favorable}}{\text{Total}} = \frac{1}{48} = 0.02 = 2\%$
3.	B	18	Large, 3 choices; medium, 3 choices; small, 2 choices; one slice, 1 choice $3 \times 3 \times 2 \times 1 = 18$
4.	C	45,697,600	$26 \times 26 \times 26 \times 10 \times 10 \times 26 = 45{,}697{,}600$ (26 letters each time A–Z; 10 numbers each time 0–9)

THE FUNDAMENTAL COUNTING PRINCIPLE AND TREE DIAGRAMS (page 176)

1. B 12 meat, 2; potatoes, 2; vegetables, 3; $2 \times 2 \times 3 = 12$ different variations

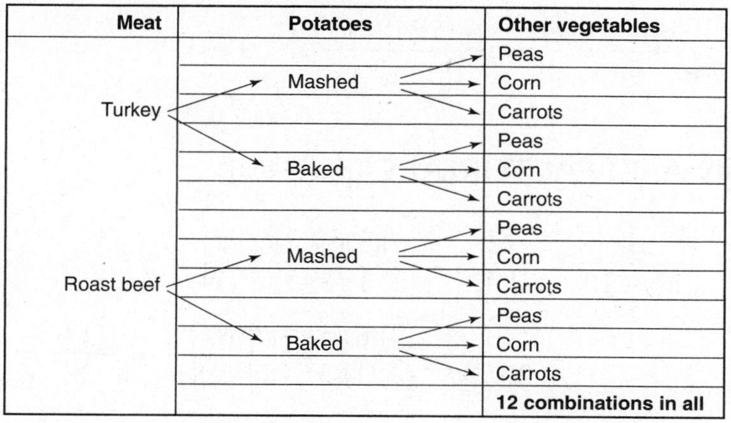

Meat	Potatoes	Other vegetables
Turkey	Mashed	Peas
		Corn
		Carrots
	Baked	Peas
		Corn
		Carrots
Roast beef	Mashed	Peas
		Corn
		Carrots
	Baked	Peas
		Corn
		Carrots
		12 combinations in all

2a.

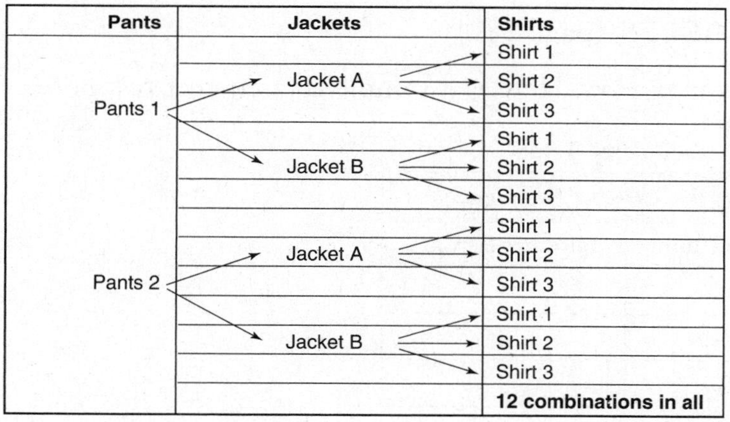

Pants	Jackets	Shirts
Pants 1	Jacket A	Shirt 1
		Shirt 2
		Shirt 3
	Jacket B	Shirt 1
		Shirt 2
		Shirt 3
Pants 2	Jacket A	Shirt 1
		Shirt 2
		Shirt 3
	Jacket B	Shirt 1
		Shirt 2
		Shirt 3
		12 combinations in all

b. B 12

3. C 25% $\frac{1}{4} = 25\%$

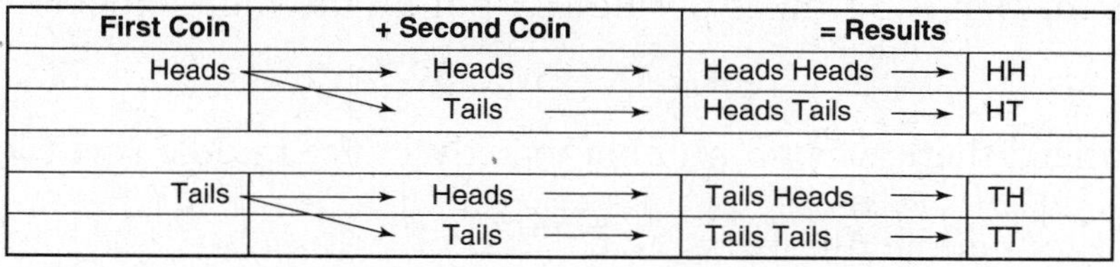

First Coin	+ Second Coin	= Results	
Heads	Heads	Heads Heads	HH
	Tails	Heads Tails	HT
Tails	Heads	Tails Heads	TH
	Tails	Tails Tails	TT

COMBINATIONS, PERMUTATIONS, AND FACTORIALS (page 179)

1. C 24 $\quad \dfrac{4!}{(4-4)!} = \dfrac{4!}{0!} = \dfrac{(4)(3)(2)(1)}{1} = 24$

2. D 720 $\quad \dfrac{6!}{(6-6)!} = \dfrac{6!}{0!} = \dfrac{(6)(5)(4)(3)(2)(1)}{1} = 720$

COMBINATIONS AND FACTORIALS (page 180)

1. B 15 $\quad \dfrac{6!}{2!(6-2)!} = \dfrac{6!}{2!(4!)} = \dfrac{(6)(5)\cancel{(4)}\cancel{(3)}\cancel{(2)}\cancel{(1)}}{(2)(1)\cancel{(4)}\cancel{(3)}\cancel{(2)}\cancel{(1)}} = \dfrac{(6)(5)}{2} = \dfrac{30}{2} = 15$

2. D 21 $\quad \dfrac{7!}{2!(7-2)!} = \dfrac{7!}{2!(5!)} = \dfrac{(7)(6)\cancel{(5)}\cancel{(4)}\cancel{(3)}\cancel{(2)}\cancel{(1)}}{(2)(1)\cancel{(5)}\cancel{(4)}\cancel{(3)}\cancel{(2)}\cancel{(1)}} = \dfrac{(7)(6)}{2} = \dfrac{42}{2} = 24$

FREQUENCY TABLES (page 182)

1. Answers may vary. Here is a sample of a correct response.

Frequency Table

Intervals	Frequency
Miles per hour	Number of Cars
40–44	1
45–49	2
50–54	3
55–59	2
60–64	2

2. The median is 53 miles per hour. I put all the miles clocked in numerical order:

44 48 50 52 | 54 55 58 60 62

I noticed there was no number exactly in the middle so I took the average of the two middle numbers: $\dfrac{52+54}{2} = 53$

The range is 18 miles. I subtracted the lowest mph from the highest: 62 – 44 = 18 miles.

STEM-AND-LEAF PLOTTING (page 183)

1. D 90
2. B 85
3. A 11

BAR GRAPHS (page 184)

1. B 20 $50 - 30 = 20$

HISTOGRAMS (page 185)

1a. 10 teachers
b. 45 teachers $15 + 30 = 45$
c. 80 $15 + 30 + 10 + 25 = 80$
d. 31.25% $\frac{35}{80} = 0.3125 = 32.25\%$ or $31\frac{1}{4}\%$
e. 100% all of the teachers are older than 22

BOX-AND-WHISKER PLOTS (page 187)

1. 30 is the least value; 61 is the greatest value.
2. 48 is the median (the number in the middle).
3. 38 is the first quartile; 30, 35, <u>38</u>, 42, 44
   55 is the third quartile; 52, 54, <u>55</u>, 58, 61
4. 31 is the range; $61 - 30 = 31$

SCATTER PLOTS (page 189)

1. B (–8, 2)
2. A Yes, the coordinate (5, 5) seems to be on the trend line.
3. C \$210 She saved \$230 but also withdrew \$20, which means she has $230 - 20 = \$210$.

4a. A positive
b. B yes ($\frac{1}{2}$ hr and 95%)
c. C ($\frac{1}{2}$, 95) $\frac{1}{2}$ hr studied and 95% on the test

5. B negative As the price of gasoline increases, the number of hours for Sunday drives decreases.

CIRCLE GRAPHS (page 192)

1. C 65% $35 + 30 = 65$
2. D 45 11% of 410 students = $(0.11)(410) = 45.1$
3. A $(410)(0.35) - (410)(0.07)$ = hamburger – chili
4. A chili, pizza, or hot dog
   chili + pizza + hot dog = 11% + 30% + 18% = 59%
5. A 19% Total number of students in all = 315; $\frac{60}{315} = 0.19$ or 19%
6. C 56% $90 + 85 = 175$ students had birthdays in April or May; $\frac{175}{315} = 0.56$ or 56%
7. A 0% No students surveyed had birthdays in the fall. The fall months are September, October, and November.

VENN DIAGRAMS (page 194)

1a. C $180 + 45$ (180 took both + 45 who took only the PSAT)

b. B $\frac{180}{255}$ = about 71%

c. B $\frac{255}{396}$ = about 64%

2a. A 90

b. C 570 Add all the students: $50 + 210 + 220 + 90 = 570$

c. A $\frac{50+210+220}{570} = \frac{\text{Football Game only + Both + Dance only}}{\text{All students interviewed}}$

NETWORKS (page 197)

1. D from 1 to 2 to 12 to 6 (notice that the path from 12 to 6 is the hypotenuse of a triangle. The hypotenuse is shorter than going to 8 and then to 7 (the legs of that triangle).
2. B

SCR NONCALCULATOR QUESTIONS (page 198)

1. There is no change. All remain the same. 75, 80, **80**, 80, 85.
   Mean = 80, mode = 80, and median = 80

2. There is a positive correlation. As they use more KWH, their cost increases.

3. 9 combinations (3 flavors) × (3 kinds of sprinkles) = 3 × 3 = 9

4. $\frac{2}{52}$ or $\frac{1}{26}$ There is a red 5 of diamonds and a red 5 of hearts.

   To find probability you put $\frac{\text{What you are looking for}}{\text{Over the total number}} = \frac{2}{52}$

5. 120 students 60% of 100 students = 60 students; therefore,
   60% of 200 students = 120 students

6. 2 lawns $4 + 2 + 1 + 1 + 2 = 10$ $\frac{10}{5} = 2$ is the mean (average)

7. $150 First, arrange the numbers in order. Then select the middle number.

   100 125 150 **150** 150 200 225

8. $\frac{25}{100} = \frac{1}{4}$ $\frac{\text{Area small shaded square}}{\text{Area large square}} = \frac{(5)(5)}{(10)(10)} = \frac{25}{100} = \frac{1}{4}$

9. $\frac{1}{2}$ $\frac{\text{Favorable outcomes}}{\text{Total possible outcomes}} = \frac{1(\text{head})}{2(\text{heads or tails})} = \frac{1}{2}$

10. 110 students Wednesday + Friday = 30 + 80 = 110

ECR QUESTIONS (page 201)

1.

	Regular Tomato Plants Show work	Regular, Height of Plants (inches)	Cherry Tomato Plants Show work	Cherry, Height of Plants (inches)
Now	12	12	6	6
Week 1	12 + 1.5 =	13.5	6 + 2 =	8
Week 2	13.5 + 1.5 =	15	8 + 2 =	10
Week 3	15 + 1.5 =	16.5	10 + 2 =	12
Week 4	16.5 + 1.5 =	18	12 + 2 =	14
Week 5	18 + 1.5 =	19.5	14 + 2 =	16
Week 6	19.5 + 1.5 =	21	16 + 2 =	18
Week 7	21 + 1.5 =	22.5	18 + 2 =	20
Week 8	22.5 + 1.5 =	24	20 + 2 =	22
Week 9	24 + 1.5 =	25.5	22 + 2 =	24
Week 10	25.5 + 1.5 =	27	24 + 2 =	26
Week 11	27 + 1.5 =	28.5	26 + 2 =	28
Week 12	**28.5 + 1.5 =**	**30**	**28 + 2 =**	**30**
Week 13	30 + 1.5 =	31.5	30 + 2 =	32
Week 14	31.5 + 1.5 =	33	32 + 2 =	34

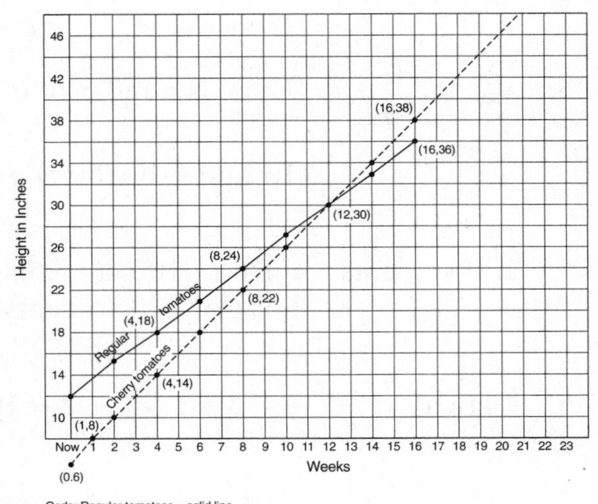

- At week #12 both plants will be 30 inches tall.
- The two plants are the same height where the lines intersect.
- The cherry tomatoes begin to grow taller than the regular tomatoes. At week #13 the cherry tomatoes are 32 inches tall and the regular tomatoes are only 31.5 inches tall.

2.

a. See stem-and-leaf plot below.

6	0 2
5	0 2 4 5 8
4	4 8 8
Stems	Leaves

b. Mean = 53.1 miles per hour
Median = 53
Mode = 48 miles per hour
Range = 18 miles per hour [difference between the fastest and slowest car (62 – 44)]

c. See box-and-whisker plot below:

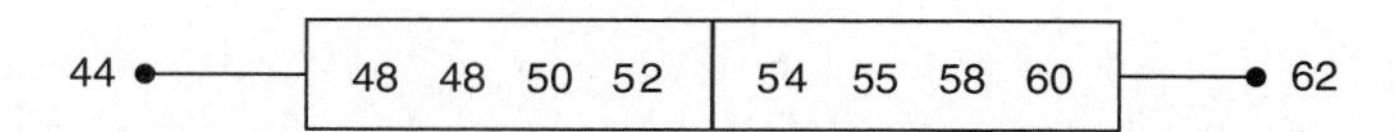

53 is the median. The median is the number in the middle, or in this case, it is the average of 52 and 54, which is 53.

d. 58.0 is the value of the upper quartile. Here you can see that 58 is the middle number of the data in the upper quartile (including the highest number):

54 55 58 60 62

e. 30% were driving slower than 50 miles per hour: $\frac{3}{10} = 0.30 = 30\%$.

f. 60% were driving faster than 50 miles per hour: $\frac{6}{10} = 0.60 = 60\%$.

3.

- 41% of his sales came from bread and bagels.

  (% bread) + (% bagels) = 16% + 25% = 41%

- He could expect to get $2,852.28 next year on cookie sales.

  This year, cookies are 8% of the total: (.08)(23,769), which equals $1901.52. Next year, cookies would be 8% + 4% (50% more), or 12% of the total. (0.12)($23,769) = $2,852.28

4.

- With six students there would be a total of 15 handshakes.

  #1 – shakes hands with 2, 3, 4, 5, and 6
  #2 – shakes hands with 3, 4, 5, and 6
  #3 – shakes hands with 4, 5, and 6
  #4 – shakes hands with 5 and 6
  #5 – shakes hands with 6
  #6 – does not shake any more hands

A shortcut way would be to think: 5 + 4 + 3 + 2 + 1 = 15 handshakes.

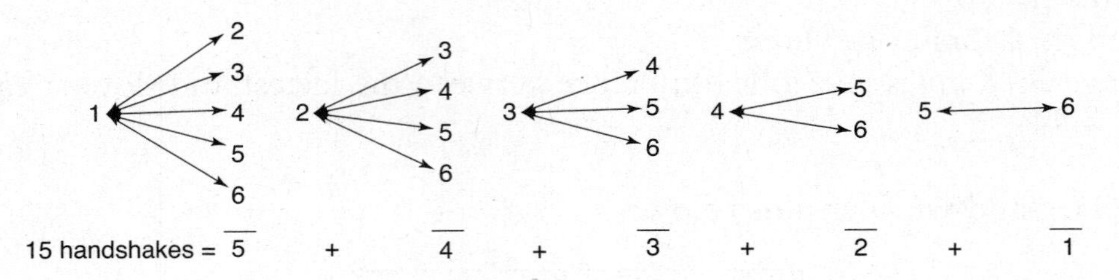

- When three more students are added, you now have a total of nine students. There would be a total of 36 handshakes: 8 + 7 + 6 + 5 + 4 + 3 + 2 + 1 = 36 handshakes.

5. - $\frac{1}{13}$ There are four 10s in a deck of 52 cards. The probability would be

   $\frac{4}{52} = \frac{2}{26} = \frac{1}{13}$.

   - $\frac{1}{26}$ There are only two black Kings: the King of spades and the King of clubs;

   so, the probability of selecting a black King is $\frac{2}{52} = \frac{1}{26}$.

   - $\frac{4}{13}$ The following are the numbered cards that are prime numbers: 2, 3, 5, 7.
   There are four of each one, so there are (4)(4) = 16 cards that are prime numbers. The probability of selecting 16 cards out of the total of 52 is
   $\frac{16}{52} = \frac{4}{13}$.

6. - Lower quartile is 17.0 $\frac{16+18}{2} = \frac{34}{2} = 17$

   - Upper quartile is 25.0 $\frac{24+26}{2} = \frac{50}{2} = 25$

   - Jared is correct. The mean of all the numbers is $\frac{198}{9} = 22$.
   The median, the number in the middle, is also 22.

Name: ______________________ Date: ______________

Cluster III Test

35 minutes
(Use the *NJ ASK 8 Mathematics Reference Sheet* on page 265.)

SHORT CONSTRUCTED RESPONSE AND MULTIPLE-CHOICE QUESTIONS

DIRECTIONS FOR QUESTIONS 1 THROUGH 12: Each of the questions or incomplete statements below is followed by four suggested answers. Select the one that is the best in each case, and fill in the corresponding lettered circle. Be sure the circle is filled in completely so you cannot see the letter. Unless you are told to do so in the question, do NOT include sales tax in your answer to questions involving purchases.

1. After 500 spins of the spinner, the following information was recorded. Estimate the probability of the spinner landing on blue.

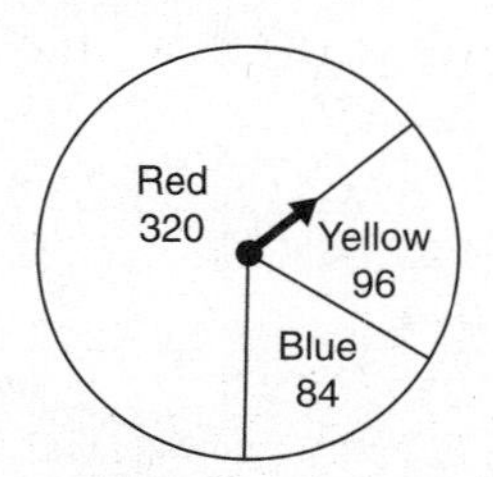

Outcome	Red	Yellow	Blue
Spins	320	96	84

A. approximately 84%
B. approximately 17%
C. approximately 8.4%
D. approximately 1.7%

2. What is the probability of landing on a perfect square number using a six-sided die with the following numbers: 4, 5, 6, 7, 8, 9 ?

A. $\frac{1}{6}$ or approximately 17%

B. $\frac{2}{6}$ or approximately 33%

C. $\frac{3}{6}$ or approximately 50%

D. $\frac{4}{6}$ or approximately 67%

Ⓐ Ⓑ Ⓒ Ⓓ

---

3. If two coins are tossed, what is the probability of getting exactly two heads $P(2H)$?

Answer: __________

---

4. What are the number of possible outcomes of choosing a dinner from 4 main courses (meat, chicken, fish, or pasta), 3 salads, and 2 beverages?

A. 4 + 3 + 2 + 1
B. (4)(5)
C. 4!
D. 25

GO ON TO THE NEXT PAGE

5. A family homemade ice cream shop offers some 8 different items that can be blended into your ice cream. They offer: chocolate chips, walnuts, wet-nuts, M & Ms, marshmallows, raisins, cherries, and strawberries. How many different *pairs* of items can be blended into your ice cream?

Answer: ________

6. What is the *median* of the following data:

Stem	Leaves			
14	2	6	9	
13	0	0	2	
12	5	5	7	9
11	2	4	6	
10	0	3	6	

Answer: ________

7. Use the box-and-whisker plot below. What is the value of the *upper quartile*?

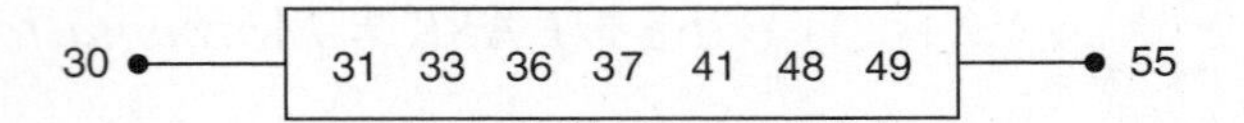

A. 48.25
B. 48.5
C. 48.0
D. 49.0

8. In a scatter plot, a trend line is also called

A. a parallel line
B. a diagonal line
C. an average line
D. the line-of-best-fit

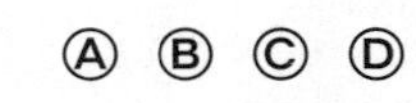

9. Use the *histogram* below. Approximately how many recorded accidents occurred during the past year between 4:00 P.M. and 4:00 A.M. according the histogram data?

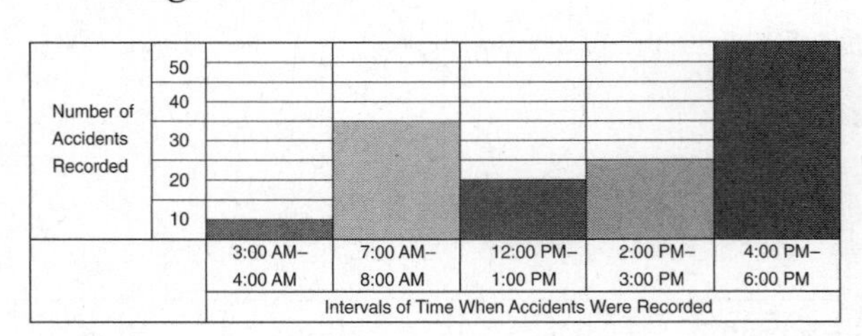

A. 60
B. 55
C. 50
D. 45

GO ON TO THE NEXT PAGE

10. Find the approximate mean average of the following data set:

42.6 41.3 35.8 23.1
21.5 20.4 15.7

A. 27.2
B. 58.3
C. 28.62
D. 29.15

Ⓐ Ⓑ Ⓒ Ⓓ

11. The local girl's softball team scored the following runs in their first 5 games: 2, 1, 4, 2, 3. If the score of the next game is 3 runs, what will change in this team's average?

A. Only the mean will change.
B. Only the median will change.
C. Only the mode will change.
D. All of them—the mean, median, and mode—will change.

Ⓐ Ⓑ Ⓒ Ⓓ

12. What is the probability you will select a red heart from a regular deck of cards?

A. $\frac{13}{52}$ or 25%

B. $\frac{4}{20}$ or 20%

C. $\frac{4}{52}$ or about 7.6%

D. $\frac{4}{50}$ or 8%

Ⓐ Ⓑ Ⓒ Ⓓ

GO ON TO THE NEXT PAGE ➡

EXTENDED CONSTRUCTED RESPONSE QUESTIONS

DIRECTIONS FOR QUESTIONS 13 AND 14: Respond fully to the ECR questions that follow. Show your work and clearly explain your answer. You will be graded on the correctness of your method as well as the accuracy of your answer.

13. The following are NJ ASK 8 scores of 9 students randomly selected from the 8th grade. Use the stem-and-leaf graph below to answer the following questions. Be sure to answer each bulleted question.

Stem	Leaves
18	2 5
19	0 5
20	6
22	0
25	0 0
26	8

- What is the range of scores?
- What is the median score?
- Draw and label a box-and-whisker plot using the information above.
- Use the box-and-whisker-plot to answer the following two questions: What is the lower quartile? What is the upper quartile?

14. Use the circle graph below to answer the following questions:

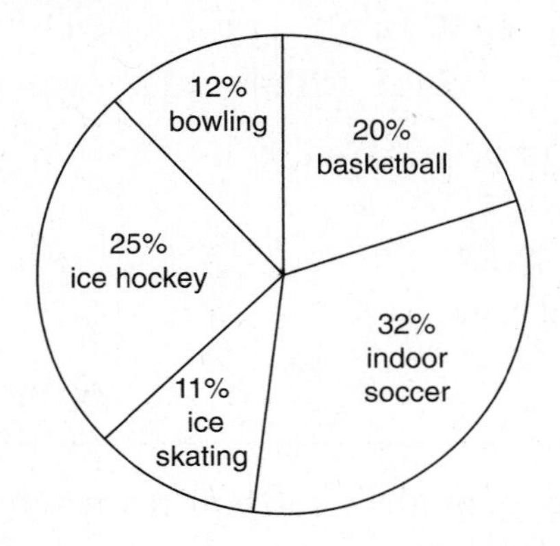

- Of the 520 students surveyed, how many said their favorite winter sport was ice hockey?
- How many students chose bowling or basketball as their favorite winter sport?
- How many more students selected indoor soccer than ice skating?
- One of the students who collected the survey data said that all was correct except the percentage of students selecting ice skating. He said that should have been 20%. Explain why this could not be possible if all the other data is true.

SOLUTIONS TO CLUSTER III TEST

SHORT CONSTRUCTED RESPONSE AND MULTIPLE-CHOICE QUESTIONS

1. B	2. B	3. 25%	4. C	5. 28	6. 126
7. B	8. D	9. B	10. C	11. D	12. A

EXTENDED CONSTRUCTED RESPONSE QUESTIONS

QUESTION 13

■ 86 The range is the difference between the highest and lowest.

$$268 - 182 = 86$$

■ 206 The median score is the score in the middle when all are arranged from lowest to highest.

■ Box-and-whisker plot

182	185	190	195	**206**	220	250	250	**268**
lowest				median				greatest

■ Lower quartile (the median of the lower numbers)

182 185 190 195

$$\underbrace{185 \quad 190}$$

$$\frac{185+190}{2}$$

$$\frac{375}{2}$$

187.5

is the lower quartile value (the number in the middle)

■ Upper quartile (the median of the upper numbers)

220 250 250 268

$$\underbrace{250 \quad 250}$$

the middle number here would be 250

250

is the upper quartile value (the number in the middle)

QUESTION 14

Using the circle graph (Total students surveyed = 520; 20% basketball, 12% bowling, 32% indoor soccer, 25% ice hockey, 11% ice skating)

- 130 students selected ice hockey   (0.25)(520) = 130
- 166 students selected basketball or bowling

  Basketball: (0.20)(520) = 104 students

  Bowling: (0.12)(520) = 62.4 students

  Basketball + Bowling = 104 + 62.4 or approximately 166 students

- Approximately 109 more students selected indoor soccer over ice skating.

  Indoor soccer: (0.32)(520) = 166.4 students

  Ice skating (0.11)(520) = 57.2 students

  166 − 57 = 109 students

- If the percent for ice skating was increased, the total percent would be greater than 100%, and that is not possible if all the other information is correct.

  20% + 12% + 32% + 25% + 11% = 100% now

Chapter 4

Cluster IV: Patterns, Functions, and Algebra

WHAT DO ASK 8 PATTERNS, FUNCTIONS, AND ALGEBRA QUESTIONS LOOK LIKE?

MULTIPLE-CHOICE QUESTIONS (MC)

Example 1: Patterns
What is the next number in the sequence?

4   −8   16   −32   ____

A. −49
B. 46
C. 64
D. −64

Example 1: Strategies and Solutions
Look at the sequence; notice if the "numbers" are getting larger or smaller. If they are getting larger, you probably need to add or multiply to get from one number to the next. If the sign changes every other number it is probably because each number is multiplied by a negative number. Remember, when multiplying

$$(-)(+) = - \quad \text{and} \quad (+)(-) = -$$

$$(-)(-) = + \quad \text{and} \quad (+)(+) = +$$

$$(-32)(-2) = 64$$

Correct choice is C.

Example 2:
Is this an *arithmetic* or *geometric* sequence?

6   2   −2   −6   −10   −14

A. arithmetic
B. geometric
C. both
D. neither

Example 2: Strategies and Solutions
Here you see the numbers begin as positive and then all become negative. By trial and error, you see that the pattern is to add a −4 to each number to get the next number. When you *add* to continue a sequence this is called an *arithmetic* sequence.

Correct choice is A.

SHORT CONSTRUCTED RESPONSE QUESTION (SCR)

<table>
<tr>
<td>

Example 3:
(No calculator permitted.)
What is the slope of the line represented by the following equation?

$$4y = -2x + 16$$

Answer: __________

</td>
<td>

Example 3:
Explanation: First, you need to divide all terms by 4 so it is in the correct form to easily see the slope.

$$4y = -2x + 16$$

$$\frac{4y}{4} = \frac{-2x}{4} + \frac{16}{4} \text{ or } y = -\frac{1}{2}x + 4$$

Therefore, the slope is $\frac{-1}{2}$

Answer: $\frac{-1}{2}$

</td>
</tr>
<tr>
<td>

Example 4:
(No calculator permitted.)
What value of x will make the following equation a true statement?

$$3x + 4(x - 6) = 4$$

Answer: __________

</td>
<td>

Example 4:
Explanation: $3x + 4(x - 6) = 4$

$$3x + 4x - 24 = 4$$
$$7x - 24 = 4$$
$$7x = 28$$
$$x = 4$$

Answer: $x = 4$

</td>
</tr>
</table>

Remember, no partial credit is given for short constructed response answers and no calculators are permitted.

EXTENDED CONSTRUCTED RESPONSE QUESTION (ECR)

Example 5: Functions and Algebra

- Describe the two lines drawn on the graph below; describe two ways they are the same and two ways they are different. Be specific.
- Explain how you know that the equation $y = \frac{1}{2}x$ is the equation that matches line A.

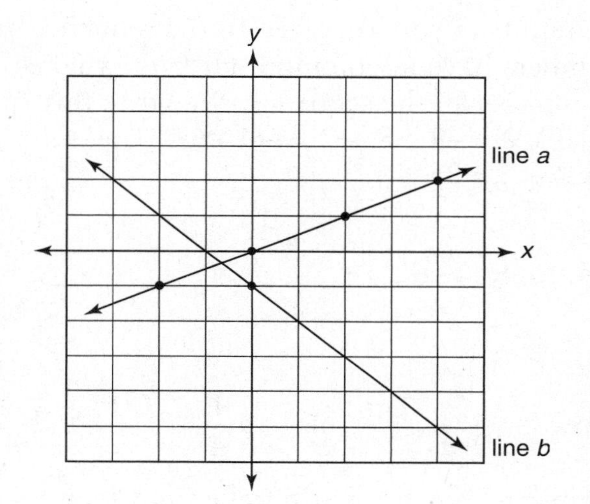

Example 5: Strategies and Solutions

Underline or circle important information in the question. Make a table to help you organize data. Remember, almost every open-ended NJ ASK 8 question can be organized with a chart or table. Show your work even if you use a calculator.

Similarities:

- Both are straight lines.
- Both continue forever in two directions. (They are lines, not line segments.)
- Both are slanted (not vertical or horizontal).
- Both are drawn on a coordinate grid. (They are oblique lines.)
- Both have points on them that you can describe with coordinates (x, y).

Differences:

- Line a has a positive slope while line b has a negative slope.
- Going from left to right, line a slants upward, but line b slants downward.
- Line a will never be in quadrant II; Line b passes through quadrant II.
- Line b will never be in quadrant I; Line a passes through quadrant I.
- Line a crosses the origin [passes through (0, 0)]; line b does not.

Equation $y = \frac{1}{2}x$ matches line a because

- It has a slope of $\frac{1}{2}$ (see diagram for my sketch) and that matches the coefficient of x.
- The point (0, 0) makes the equation true and it is a point on the line.
- The point (2, 1) ($x = 2$, $y = 1$) makes the equation true, and I can see from the graph that it is a point on the line.

PATTERNS

In this section, you will learn how to recognize, describe, extend, and create patterns. You will work with whole numbers (i.e., 1, 2, 3, 4, 5), rational numbers (i.e., $\frac{4}{5}$, $\frac{2}{5}$, 0.2, 0.10), and positive and negative integers (i.e., –5, –2, 1, 4, 7).

The following are examples of patterns based on *arithmetic sequences*. In an arithmetic sequence, there is some rule based on *adding* that creates the pattern. Sometimes you add a positive number; other times you add a negative number.

Examples

A. What are the next three numbers in this sequence?

–10, –8, –6, –4, ___, ___, ___.

Look for a pattern to find out how the numbers you are given in the sequence are found. Look for something that the numbers have in common. Here the rule is: add +2 to each number to get the next number in the sequence. We write this rule as $x + 2$. (We see this is true because –10 + 2 = –8; –8 + 2 = –6; – 6 + 2 = –4; therefore, – 4 + 2 = –2; –2 + 2 = 0, and 0 + 2 = 2).

Answer: –2, 0, 2

B. Continue the pattern: 19, 14, 9, 4, –1, – 6 ___, ___, ___, ___.

Here the rule is to add –5.

Answer: –11, –16, –21, –26

C. What is the 9th number in this sequence? $\frac{1}{2}$ $\frac{3}{4}$ 1 $1\frac{1}{4}$ $1\frac{1}{2}$ $1\frac{3}{4}$

Here the rule is to add $+\frac{1}{4}$ to each number to get the next number and then to simplify fractions when possible. ($1\frac{3}{4} + \frac{1}{4} = 1\frac{4}{4}$, which = 2; $2 + \frac{1}{4} = 2\frac{1}{4}$, $2\frac{1}{4} + \frac{1}{4} = 2\frac{2}{4}$ or $2\frac{1}{2}$)

Answer: The ninth term would be $2\frac{1}{2}$.

Another type of sequence is called a *geometric sequence*. In a geometric sequence, *multiplication* is used to generate the next number(s).

D. What is the rule to continue this pattern?

100 80 64 51.2 10.24 _____ _____

A. multiply by a whole number
B. multiply by a fraction or decimal number
C. subtract a decimal number
D. add a negative number

Choice B is the correct answer.

Now, use the rule you just discovered and determine the 7th number in the sequence.

The specific rule is to multiply by 0.80.

(100)(0.80) = 80, (80)(0.80) = 64, ..., (10.24)(0.80) = **8.192**, (8.192)(0.80) = **6.5536**

Answer: 6.5536 is the 7th number in the sequence.

E. Fill in the missing numbers in this *geometric* sequence:

$$\frac{1}{2} \quad \frac{3}{2} \quad \frac{9}{2} \quad \frac{27}{2} \quad \frac{?}{?} \quad \frac{?}{?} \quad \frac{729}{2}$$

The rule here is to *multiply* the numerator by 3. The two missing numbers are $\frac{81}{2}$ and $\frac{243}{2}$.

F. At other times you may find patterns that you recognize by just looking and noticing a repetition of a particular sequence of numbers. Let your calculator help here.

Use the following pattern to find the *units* digit of 3^{10}.

$3^1 = 3$ $\quad$ $3^2 = 9$ $\quad$ $3^3 = 27$ $\quad$ $3^4 = 81$ $\quad$ $3^5 = 243$

A. 3 $\quad$ B. 9 $\quad$ C. 7 $\quad$ D. 1

The correct answer is B.

Once you notice that the sequence in the units is 3, 9, 7, 1, you will notice that this repeats so that the tenth number in the sequence will have a 9 in the units place.

3^6	has a 9 in the units place	279
3^7	has a 7 in the units place	2,187
3^8	has a 1 in the units place	6,561
3^9	has a 3 in the units place	19,683
3^{10}	has a 9 in the units place	59,049
3^{11}	will have a 7 in the units place	
3^{12}	will have a 1 in the units place	
3^{13}	will have a 3 in the units place; and so on.	

What number will be in the units digit in the 20th place?

There will be a 1 as the units digit in the 20th place.

G. What are the missing numbers in the sequence?

2 $\quad$ 8 $\quad$ 18 $\quad$ 32 $\quad$ 50 $\quad$ 72 $\quad$ ___ $\quad$ ___ $\quad$ 162

Look for a pattern to find out how the numbers you are given in the sequence are found. Look for something that the numbers have in common. Here you can see that the numbers seem to be doubles of perfect square numbers in order (1, 4, 9, 16, 25, 36, etc.)

You see	2		8		18		32		50		72		?		?
You think!	2	$2 \times 4 =$	8	$2 \times 9 =$	18	$2 \times 16 =$	32	$2 \times 25 =$	50	$2 \times 36 =$	72	$2 \times ? =$		$2 \times ? =$	

You realize the next numbers are doubles of 49 and 64. You compute; the next numbers are: 98 and 128. [(2) (49) = 98 and (2) (64) = 128]

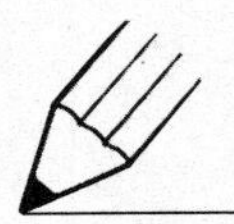

PRACTICE: Patterns and Numbers

(For answers, see page 249.)

1. Which value is the missing number in the sequence?

61 57 53 ___ 45 41

A. 52 B. 51 C. 49 D. 47

2. Use the pattern to find the units digit of 4^{12}.

$4^1 = 4$ $4^2 = 16$ $4^3 = 64$ $4^4 = 256$ $4^5 = 1{,}024$

A. 2 B. 4 C. 6 D. 18

3. If $x + 3$ is the rule, what are the next three numbers?

7, 10, 13, ___, ___, ___

4. If $a - 2$ is the rule, what are the missing numbers?

100, 98, 96, ___, 92 , ___, 88

5. What is the next number in the sequence?

3, 15, 75, 375, ___

6. Continue the sequence; fill in the next three numbers:

1,000, 500, 250, ___, ___

7. What is the rule? 3, −6, 12, −24

8. What is the rule? 25, −12.5, 6.25

9. What is the rule? 2, −8, 32, −128

10. $3x - 2$ 10, 28, ___, ___, ___, ___

11. x^2 2, 4, 16, ___, ___

12. $x^2 + 1$ 2, 5, 26, ___

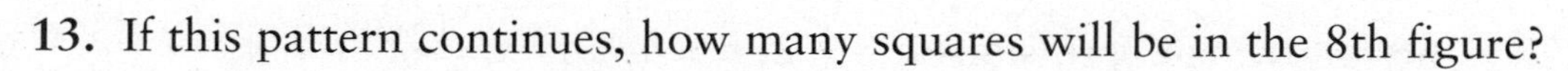

13. If this pattern continues, how many squares will be in the 8th figure?

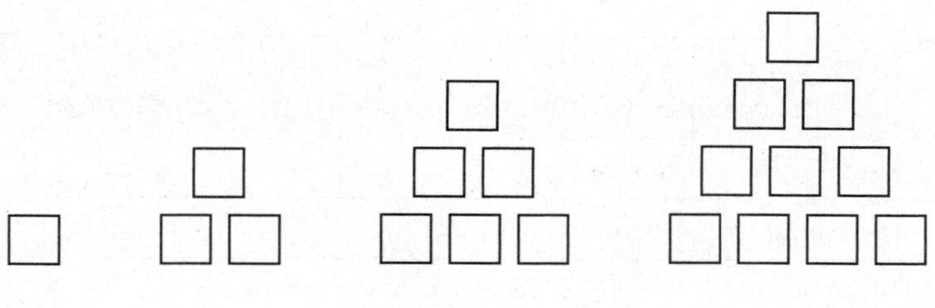

A. 36 B. 38 C. 45 D. 48

14. If this pattern continues, how many triangles will be in the 6th figure?

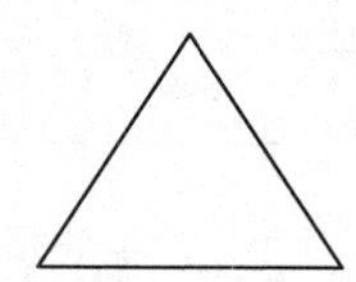

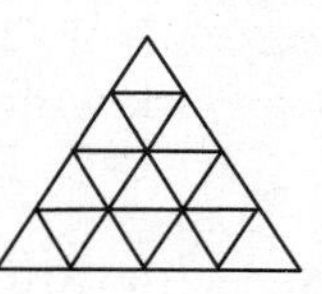

A. 128 B. 256 C. 532 D. 1,024

MONOMIALS, TERMS, AND EXPRESSIONS

In algebra you work with *numbers* and *variables* and different combinations of them. Here are some terms to remember and recognize:

- **Monomial:** A number, variable, or product of numbers, variables, or numbers and variables.

  A number: 5 -4.5 $\frac{1}{2}$ -0.008

  A variable: a b x m

  Different ways to write the *product* of monomials, variables, or numbers and variables:

  $(5)(9)$ $(\frac{1}{2})(-4)(5)$ $5(n)$ $-16x$ $20ab$ xyz $25abx^2$

- **Expressions:** Combinations of monomials (numbers and variables)

$3x + 2.5$ $16 - 19a$ $24a + (3b - \frac{3}{4})$ $-12 + \frac{1}{2}x - 6ab$ $xy + 2a - 3(4b)^2$

SIMPLIFY AND EVALUATE EXPRESSIONS

Evaluate each of the following expressions if $a = 2$, $b = 4$, and $x = -2$.

The Expression	Work Shown (replace variables with numbers and do the computation)	The Evaluation
$3x + 2.5$	$3(-2) + 2.5 = -6 + 2.5 =$	-3.5
$16 - 19a$	$16 - 19(2) = 16 - 38 =$	-22
$24a + 3b - 4$	$24(2) + 3(4) - 4 = 48 + 12 - 4 = 60 - 4 =$	56
$-12 + \frac{1}{2}b(8)$	$-12 + \frac{1}{2}(4)(8) = -12 + 2(8) = -12 + 16 =$	4
$3(x - 4)^2 + b$	$3(-2 - 4)^2 + 4 = 3(-6)^2 + 4 = 3(36) + 4 = 108 + 4$	112
	Remember: PEMDAS (work inside Parentheses first, then work with Exponents.)	

Combine like terms. Sometimes you do not know the value of a variable. Often you are asked to just simplify an expression by combining like terms:

- 8 apples + 15 cherries + \$3 + 4 cherries – \$2;
  simplified: 8 apples, 19 cherries, and \$1
- $2a + 15b - 6 + 3a + 4b - 2$; put like terms next to each other:
  $\underline{2a + 3a} + \underline{15b + 4b} - \underline{6 - 2}$ and then combine like terms: $5a + 19b - 8$ (answer)
- $10a - 2b + 15a + 6b - 4$; put like terms next to each other:
  $\underline{10a + 15a} - \underline{2b + 6b} - \underline{4}$ and then combine like terms: $25a + 4b - 4$ (answer)

Use the distributive property:

- $2(4x + 5)$ $8x + 10$ This is the same as $2(4x) + 2(5) = 8x + 10$ (answer)
- $-6(3x - 3)$ $-18x + 18$ This is the same as $-6(3x) - 6(-3) = -18x + 18$ (answer)

EQUATIONS

After learning how to combine like terms to simplify expressions, you now can learn how to solve basic linear equations.

SOLVING LINEAR EQUATIONS WITH ONE VARIABLE ON ONE SIDE

Remember, do the same thing to both sides of the equation to keep the equation balanced. Study the following examples and remember the correct order of operations (PEMDAS) and the rules about working with positive and negative integers.

Solving One-Step Equations

Study each chart. The original equations are in bold, with steps and explanations below.

$\mathbf{10 = x - 2}$ $+2 \quad +2$ Undo the subtraction; add 2 to both sides. $12 = x$	$\mathbf{b + 5 = 8}$ $-5 \quad -5$ Undo the addition; subtract 5 from both sides. $b = 3$	$\mathbf{3x = 15}$ $\frac{3x}{3} = \frac{15}{3}$ Undo the multiplication; divide both sides by 3. $x = 5$	$\mathbf{w + 12 = -10}$ $-12 \quad -12$ Undo the addition; subtract 12 from both sides. $w = -22$
$\mathbf{-15 = 4 + w}$ $-4 \quad -4$ Undo the addition; subtract 4 from both sides. $-19 = w$	$\mathbf{\frac{1}{4}m = 5}$ $\left(\frac{4}{1}\right)\frac{1}{4}m = 5\left(\frac{4}{1}\right)$ Multiply both sides by the reciprocal of $\frac{1}{4}$; use $\frac{4}{1}$. $m = \frac{20}{1}$ $m = 20$	$\mathbf{-5a = 15}$ $\frac{-5a}{-5} \quad \frac{15}{-5}$ Undo the multiplication; divide both sides by -5. $a = -3$	$\mathbf{10 = x + 2.5}$ $-2.5 \quad -2.5$ Undo the addition; subtract 2.5 from both sides. $7.5 = x$

Solving Two-Step Equations

Study each chart. The original equations are in bold, with steps and explanations below. The equations have been solved working one step at a time.

$\mathbf{3x + 6 = 36}$ $-6 \quad -6$ Subtract 6 from both sides. $3x = 30$ Divide both sides by 3 $x = 10$	$\mathbf{100 - 2x = 60}$ $-100 \quad -100$ Subtract 100 from both sides. $-2x = -40$ Divide both sides by -2 $x = 20$	$\mathbf{\frac{1}{3}x - 8 = -4}$ $+8 \quad +8$ Add 8 to both sides. $\frac{1}{3}x = 4$ Multiply both sides by 3 $x = 12$	$\mathbf{16 = 11 + 2x}$ $-11 \quad -11$ Subtract 11 from both sides. $5 = 2x$ Divide both sides by 2 $\frac{5}{2} = x$

Solving Linear Equations with Variable on Both Sides

$\mathbf{5x = -75 + 2x}$ $-2x \quad -2x$ $3x = \quad -75$ $x = \quad -25$	$\mathbf{16 + 4w = -48 + 3w}$ $-3w \quad -3w$ $16 + w = -48$ $-16 \quad -16$ $w = -64$	$\mathbf{2x + 8 = 6x + 80}$ $-2x \quad -2x$ $8 = 4x + 80$ $-80 \quad -80$ $-72 = 4x$ $-18 = x$	$\mathbf{12 + 2(x - 2) = 3x}$ $12 + 2x - 4 = 3x$ $8 + 2x = 3x$ $-2x \quad -2x$ $8 = x$

PRACTICE: Solving Equations

(For answers, see page 249.)

1. $3x = -21$

A. $x = 7$
B. $x = -7$
C. $x = 18$
D. $x = -18$

2. $12 = 2a - 4$

A. $a = 16$
B. $a = -8$
C. $a = 8$
D. $a = -16$

3. $2b - 9 = 5b$

A. $b = -3$
B. $b = 3$
C. $b = -6$
D. $b = -\frac{9}{7}$

4. $3(4 + x) = 24$

A. $x = -12$
B. $x = 12$
C. $x = 9$
D. $x = 4$

5. $3w - (5)^2 = 20$

A. $w = -15$
B. $w = 15$
C. $w = 10$
D. $w = -10$

6. $3x + 4 = 12x - 8$

A. $x = \frac{3}{2}$
B. $x = \frac{4}{3}$
C. $x = \frac{3}{4}$
D. $x = \frac{-3}{2}$

7. $2(a - 3) = 3(2a + 10)$

A. $a = -1$
B. $a = -4$
C. $-9 = a$
D. $-8 = x$

8. $\frac{1}{4}z = 20$

A. $z = 5$
B. $z = 16$
C. $z = 80$
D. $z = -16$

9. $x(5 - 3)^2 = -64$

A. $x = 4$
B. $x = -4$
C. $x = 16$
D. $x = -16$

10. $16 + \frac{1}{2}y = -24$

A. $y = -80$
B. $y = 80$
C. $y = 20$
D. $y = -20$

FUNCTIONS AND RELATIONS

In this section, we will review graphing simple functions (seen as various lines) and discuss their general behavior (how they slant, where they are placed on the grid, and how they relate to other lines); we'll recognize if they are parallel, perpendicular, or intersecting lines.

- In Chapter 1, you reviewed graphing *integers* on a number line and sometimes connected them to see a range of points that created a *line segment.*

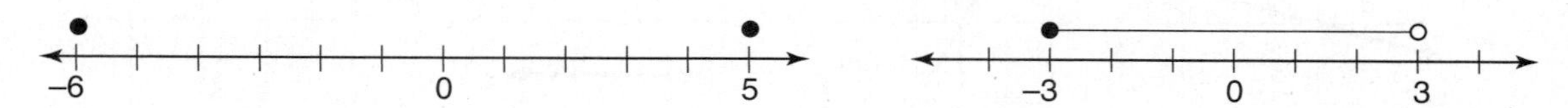

- In Chapter 2, you extended this knowledge and graphed *points* on a coordinate plane and sometimes connected the points to make a *line segment* and eventually, a *geometric shape.*

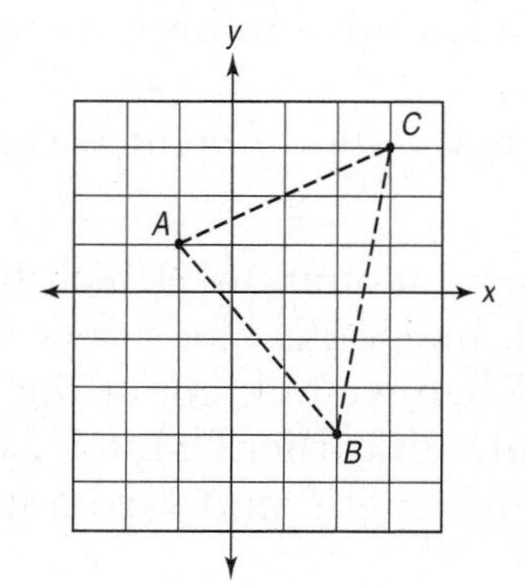

- In this section, you will review how to determine which x and y values match a particular *equation* of a line and what that means. You also will learn how to match equations to lines on the coordinate plane. How does the line described by $y = x$ compare to the line $y = -x$? How does the line $y = 4x$ compare to the line $y = \frac{1}{2}x$?

LINES, SLOPES, AND LINEAR EQUATIONS

There are four basic areas you will review in this section:

- Lines with positive and negative slopes
- Intersecting, parallel, and perpendicular lines
- The slope–intercept form of a linear equation ($y = mx + b$)
- Finding the slope and y-intercept of a line if you are given
  - the slope-intercept form of the equation; for example, $y = \frac{-2}{3}x + b$
  - two points on the line; for example, (2, 3) and (4, –8)

SLOPE

First we'll just LOOK at the lines on the different coordinate grids below. Think of how they are alike and how they are different.

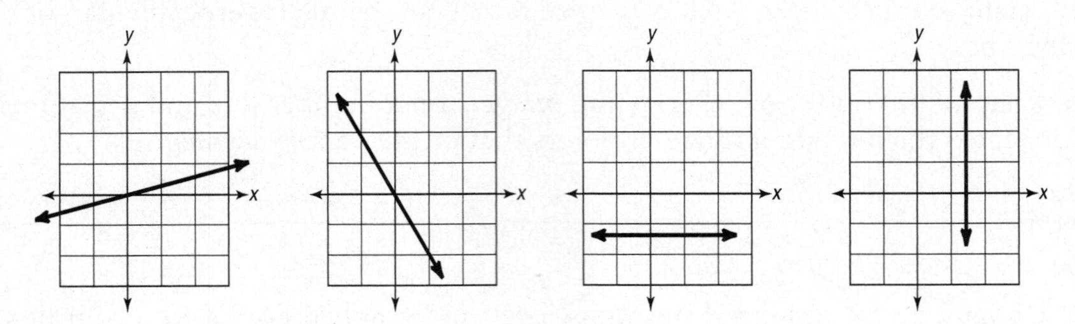

- How are they alike? They are all straight lines, two of them cross the origin (0, 0), and all lines seem to continue in both directions. (Notice the arrows at both ends.)
- How are they different? They *slant* different ways; two of them do not cross the origin (0, 0).

Now, think of each line as a hill or a mountain. Which line looks like a small hill that would be easy to walk up? Which line looks like a very steep hill, one that would be difficult to ride up with a bike? When we talk about the slant or steepness of lines, we describe this as the *slope* of the line. In earlier chapters, you measured a specific line segment, used a ruler to measure its length, and used a number line to describe its location.

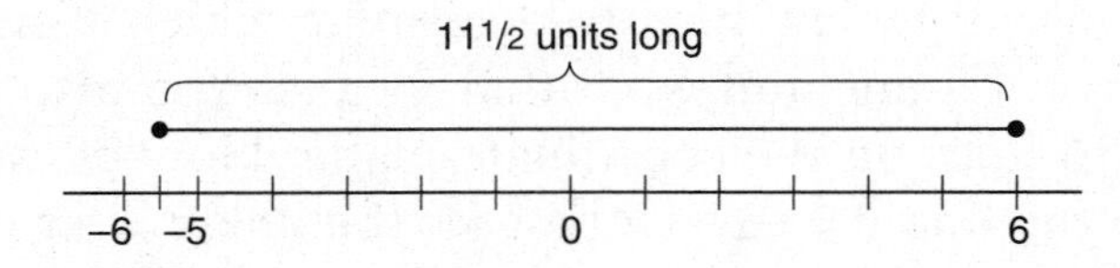

But when a line is on a coordinate plane it is not always a horizontal line and not usually a segment. We must know its slope to describe it accurately. We use a grid to help us be exact.

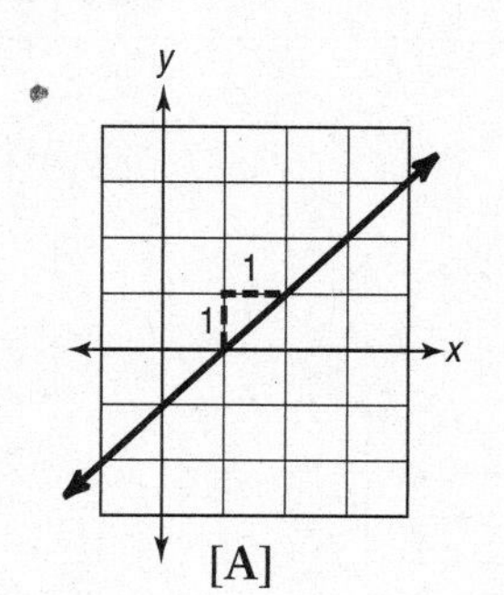

[A]

Slope here is 1/1 = 1.

The equation of this line is $y = 1x$ or $y = x$.

This is like a small hill.

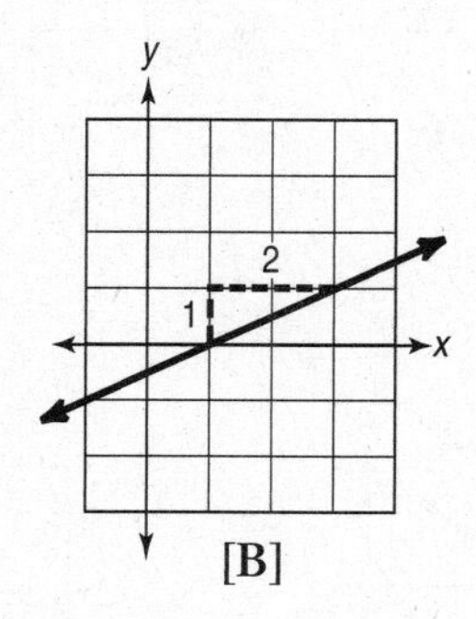

[B]

Slope here is 1/2.

The equation of this line is $y = \frac{1}{2}x$.

This is a much smaller hill.

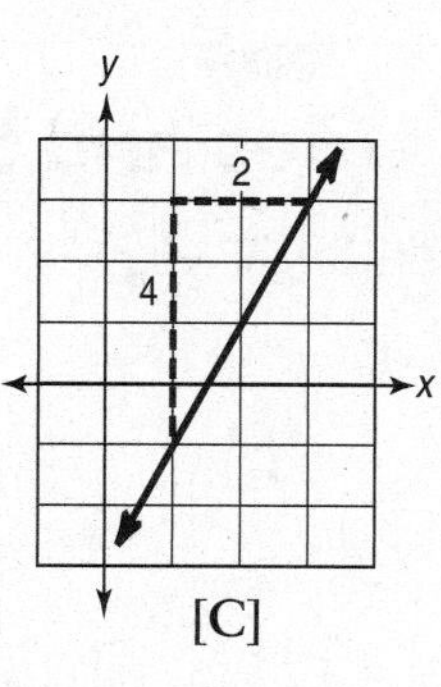

[C]

Here, the slope is 4/2, which reduces to 2/1.

The equation of this line is $y = 2x$.

This is a steeper hill.

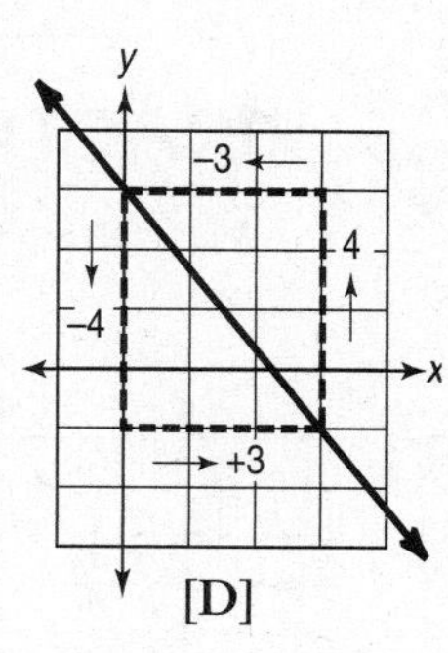

[D]

What is the slope of this line?

Slope = $\frac{-4}{3}$ or slope = $\frac{4}{-3}$.

The equation of this line is $y = -\frac{4}{3}x$.

To determine the slope of any line, begin at *any* point on the line and move up or down (count the number of boxes you moved (a y movement); then move right or left (an x movement) until you reach the line (count the number of boxes you moved left or right). The slope is written as the fraction:

$$\text{Slope} = \frac{\text{The } y \text{ movement}}{\text{The } x \text{ movement}}$$

$$\text{Remember } \frac{+}{+} = + \qquad \frac{2}{5} = \frac{2}{5}$$

$$\text{Remember } \frac{-}{-} = + \qquad \frac{-2}{-5} = \frac{2}{5}$$

- For A, B, and C above, our movements were all "positive" movements (up and right) or (down and left). The slopes of these lines were all positive.
- For D above, we see a line that slants in a different direction. From a point on this line we moved down (a negative y movement), and then right (a positive x movement). The slope of this line is $\frac{-}{+}$ or − (negative).

Here are two facts to remember:

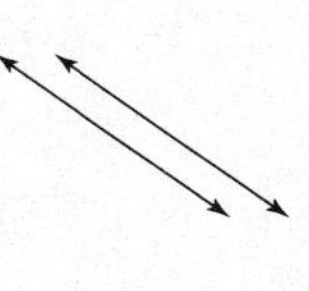

- If two lines have the same slope, they are parallel lines.

$$y = \frac{3}{4}x + 12 \text{ is parallel to the line } y = \frac{3}{4}x - 8$$

- If two lines have slopes that are the *negative reciprocals* of each other, they are perpendicular lines (lines that meet at right angles).

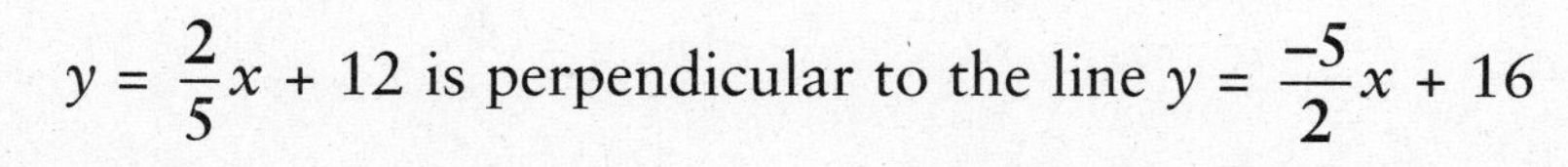

$$y = \frac{2}{5}x + 12 \text{ is perpendicular to the line } y = \frac{-5}{2}x + 16$$

PRACTICE: Lines and Slope

(For answers, see page 249.)

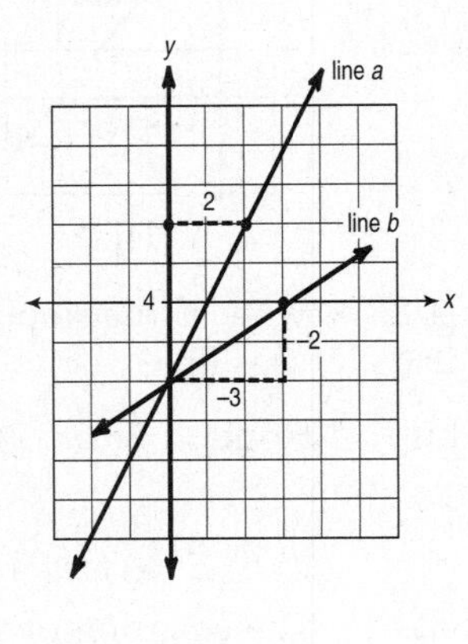

A

B

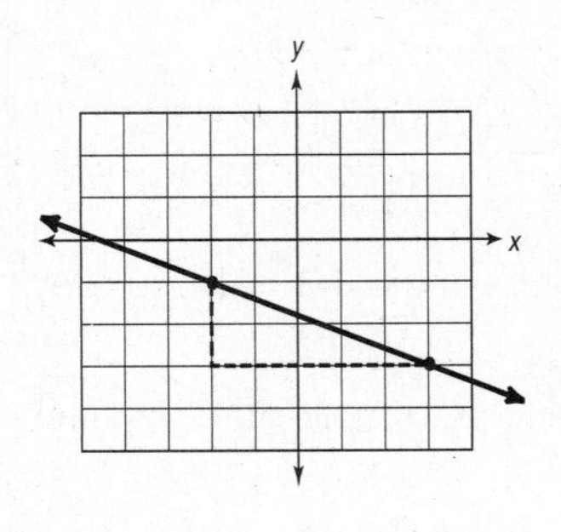

C

1. Look at Figure A above. What is the slope of the line labeled "line *a*"?

   A. $\frac{1}{3}$ B. $\frac{3}{1}$ C. $\frac{2}{4}$ or $\frac{1}{2}$ D. $\frac{4}{2}$ or 2

2. Look at Figure A above. What is the slope of the line labeled "line *b*"?

   A. $\frac{2}{3}$ B. $\frac{3}{2}$ C. $\frac{2}{1}$ D. $\frac{3}{1}$

3. Look at Figure B above. What do you notice about the two lines (*c* and *d*)? They seem to be

   A. parallel B. perpendicular
   C. intersecting D. the same line

4. Look at Figure B above. What is the slope of line *c*? What is the slope of line *d*?

   A. slope of line $c = \frac{-3}{2}$; slope of line $d = \frac{-3}{2}$

   B. slope of line $c = \frac{-2}{3}$; slope of line $d = \frac{-2}{3}$

   C. slope of line $c = \frac{2}{3}$; slope of line $d = \frac{2}{3}$

   D. slope of line $c = \frac{3}{2}$; slope of line $d = \frac{3}{2}$

5. From what you have observed complete this statement. If the slope of one line *equals* the slope of another line then

   A. the two lines are intersecting lines
   B. the two lines are perpendicular lines
   C. the two lines are parallel lines
   D. the two lines always have very steep slopes

6. Look at Figure C on the previous page. What is the slope of the line here?

7. In the following two equations the coefficient of x is the slope of each line.

   Line m: $y = \frac{-3}{2}x + 12$ Line n: $y = \frac{2}{3}x + 12$

   Just by looking at the slope of these two equations, what can you tell about their lines?

   A. They are parallel.
   B. Line n has a negative slope and line m has a positive slope.
   C. They are perpendicular.
   D. They have the same slope.

8. Which of the lines below has a negative slope?

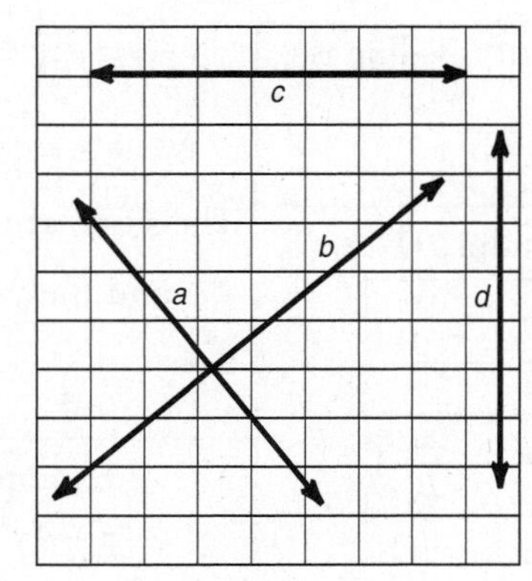

There are four basic ways to get information about a line so that you can determine its slope. The table below outlines these four ways. Study each example and then do the practice examples on the next pages.

FINDING THE SLOPE OF A LINE

Remember: It is always best to leave the slope as a fraction: $\frac{3}{2}$ not $1\frac{1}{2}$.

Graph

Example A

Start at any point on the line; count up (y) and then count across (x) until you reach the line again.

$\frac{y}{x}$ = Slope

In this figure the slope is $\frac{1}{3}$.

Example B

Start at any point on the line; count up or down (y) and then count across, left or right, (x) until you reach the line again.

$\frac{y}{x}$ = Slope

In this figure the slope is $= -\frac{1}{2}$.

Table

Example C

This is a list of different coordinate points (x, y)

x	y
0	2
1	3
2	4
3	5
4	6

Look at the change in the y values.

$2 + 1 = 3$
$3 + 1 = 4$
$4 + 1 = 5$
$5 + 1 = 6$

The slope of this line is $+1$ or $\frac{1}{1}$.

Example D

x	y
0	0
1	−2
2	−4
3	−6
4	−8

Look at the change in the y values.

$0 + -2 = -2$
$-2 + -2 = -4$
$-4 + -2 = -6$
$-6 + -2 = -8$

The slope of this line is -2 or $-\frac{2}{1}$.

Equation in slope–intercept form*

Example E

$y = 4x + 6$

The slope of this line is 4 or $\frac{4}{1}$, and the y-intercept (where the line crosses the y-axis) is 6.

Example F

$4y = -2x + 16$

(Here we need to divide by 4 to get the equation in the correct form.)

$y = -\frac{2}{4}x + \frac{16}{4}$

$y = -\frac{1}{2}x + 4$

The slope of this line is $-\frac{1}{2}$ and the y-intercept is 4.

Example G

$y + 4x = 12$

(Here we need to add $-4x$ to both sides.)

$\quad -4x \quad -4x$

$y = -4x + 12$

The slope of this line is -4 or $-\frac{4}{1}$.

Two Points

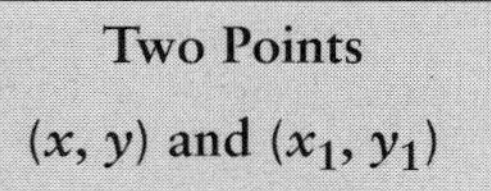

(x, y) and (x_1, y_1)

Example H

Use the formula

$\frac{y_1 - y}{x_1 - x}$ = Slope

(1, 0) and (5, 3)
(x, y) (x_1, y_1)

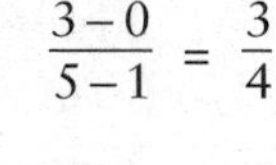

$\frac{3-0}{5-1} = \frac{3}{4}$

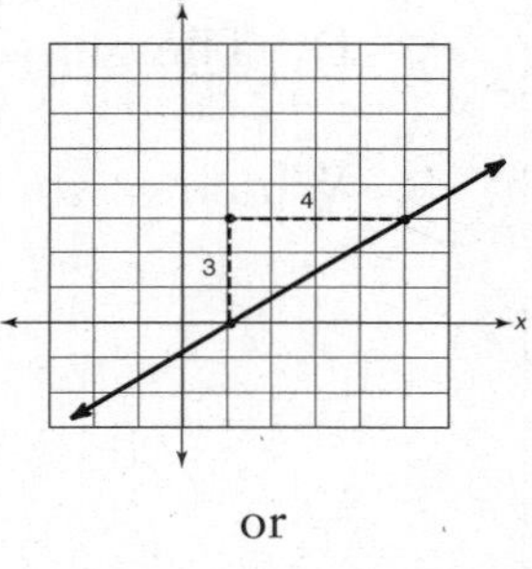

or

(5)(3) and (1)(0)
(x, y) (x_1, y_1)

$\frac{0-3}{1-5} = \frac{-3}{-4} = \frac{3}{4}$

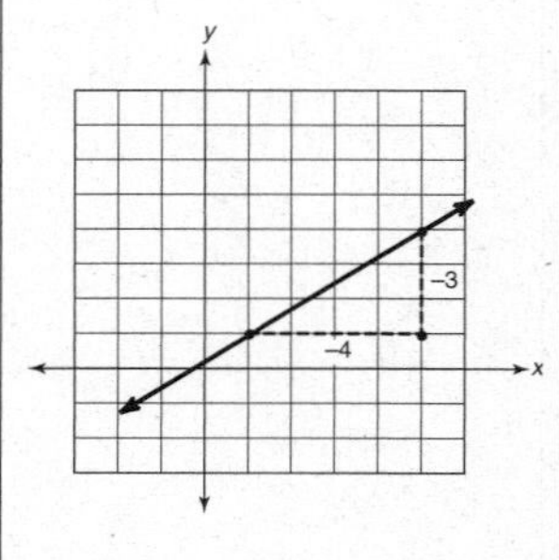

Remember: The slope–intercept form of an equation is $y = mx + b$. The m is the slope of the line, and the b is the point where the line intersects the y-axis.

PRACTICE: Slope

(For answers, see page 250.)

1. Look at the graph and determine the slope of the line.

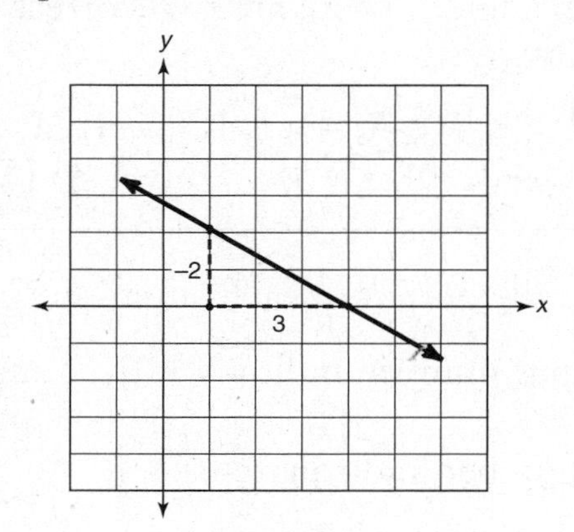

A. $\frac{2}{3}$ B. $\frac{-2}{3}$

C. $\frac{3}{2}$ D. $\frac{-3}{2}$

2. Look at the graph and determine the slope of the line.

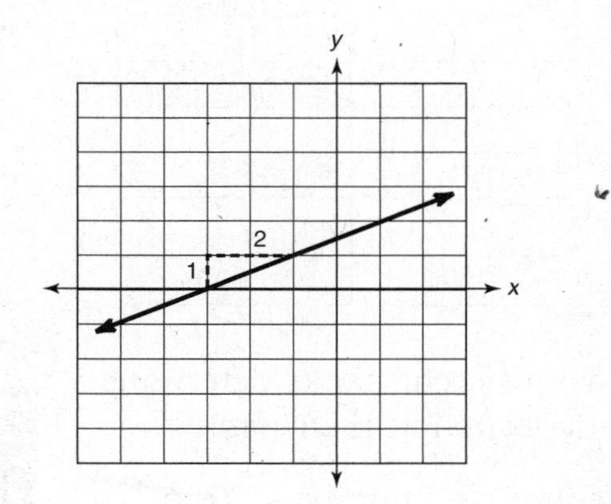

A. $\frac{1}{2}$ B. $\frac{2}{1}$

C. $\frac{-1}{2}$ D. $\frac{-2}{1}$

3. Look at the following linear equation. It already is in the required form. What is the slope of the line represented by this equation?

$$y = \frac{2}{3}x + 5$$

A. 2 B. 3

C. $\frac{2}{3}$ D. 5

4. Look at the following linear equation. It is not yet in the correct form. Put it in the correct form; then determine the slope of the line represented by this equation?

$$2y = 6x + 10$$

A. 6 B. 3
C. 10 D. 5

5. Here is another linear equation that is not yet in correct form. Put it in correct form; then determine the slope of the line it represents.

$$y - 6 = 4x$$

A. −6 B. 6

C. $\frac{-2}{3}$ D. 4

6. This linear equation is not yet in correct form. What is the slope of this line?

$$y - 2x = 14$$

A. −2 B. 2
C. 14 D. 7

7. The following two points are on the same line: (3, 0) and (2, 4). What is the slope of this line?

A. $\frac{4}{-1}$ or -4 B. $\frac{-1}{4}$

C. $\frac{1}{2}$ D. $\frac{4}{1}$ or 4

8. You are told that the slope of a line is $\frac{-1}{3}$. Which one of the following pairs of coordinate points are on this line? (There are two correct answers.)

A. (2, 4) (–5, –3) B. (2, 4) (5, 3)
C. (–2, –4) (5, 3) D. (–4, 4) (5, 1)

9. In both tables there is a rule that determines the output in each situation.

- In Table A, the rule is $2x + 1$ (take the input number, multiply it by 2 and add 1).
- In Table B, the rule is x^2 (take the input number and square it).

Table A Shows Linear Growth	
Input	*Output*
x	$2x + 1$
0	1
1	3
2	5
3	7
4	?
5	11
6	13
7	?
↓	↓
100	201

Table B Shows Exponential Growth	
Input	*Output*
x	x^2
0	0
1	2
2	4
3	?
4	16
5	?
6	36
7	?
↓	↓
100	10,000

Let's say these tables represent the way two different banks determine the interest you would receive on money deposited at their bank.

- Use the rule for each table and fill in the missing numbers (?) in the shaded cells.
- According to Table A, how much money would you receive if you deposited $50?
- According to Table B, how much money would you receive if you deposited $50?
- If you deposit $50 in each bank explain why you receive so much more interest in one bank than in the other?

COMBINING ALGEBRA AND GEOMETRY

Examples

A. The perimeter of the rectangle drawn is 42 centimeters.

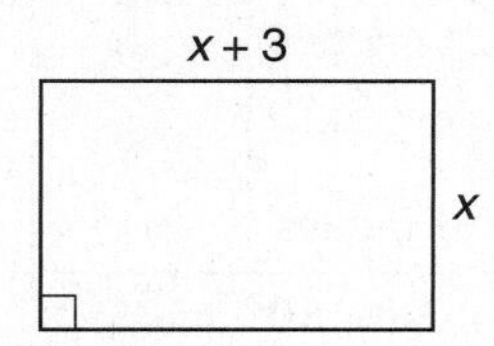

- Write an equation to represent this perimeter.
- Find the value of x.
- Find the length and width of the rectangle.
- Find the area of the rectangle.

	Show All Work	Reminders
Equation	Perimeter $= x + 3 + x + 3 + x + x = 42$ or Perimeter $= 2(x + 3) + 2(x) = 42$	For perimeter, add all sides.
Solve for x	$4x + 6 = 42$ $-6 \quad -6$ $4x = 36$ $x = 9$	Combine like terms. Undo addition. Undo multiplication.
Length of each side	Side $a = x + 3 = 9 + 3 = 12$ cm Side $b = x = 9$ cm	Substitute the value for x and solve each equation.
Area of rectangle	Area rectangle $= (12)(9) = 108$ sq. cm	Area = (Side)(Side)

B. The perimeter of the triangle drawn is 80 inches.

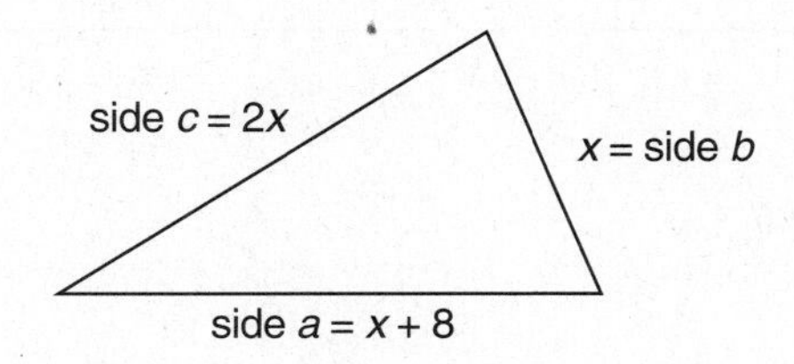

- Write an equation to represent this perimeter.
- Find the value of x.
- Find the value of each side.
- Check your work.

Equation	$2x + x + 8 + x = 80$ inches	
Solve for x	$4x + 8 = 80$ $4x = 72$ $x = 18$	
Length of each side	Side $a = 26$ inches Side $b = 18$ inches Side $c = 36$ inches	side $a = x + 8 = 18 + 8 = 26$ inches side $b = x = 18$ inches side $c = 2x = 2(18) = 36$ inches
Check by adding sides	$26 + 18 + 36 = 80$	Correct, the perimeter is 80!

C. The area of the rectangle drawn is 36 sq. units

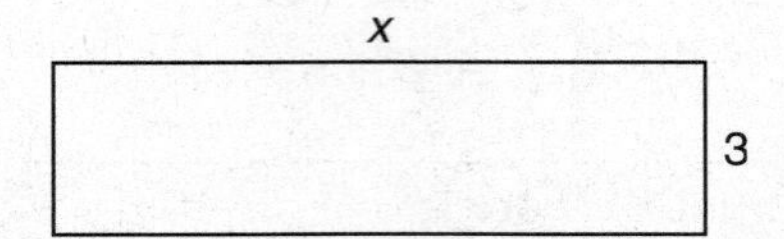

- Write an equation to represent the perimeter of this rectangle.
- Find the value of x.
- Find the value of each side.
- Check your work.

Equation	$3x = 36$	Area = Length × Width; $(3)(x)$ = Area
Solve for x	$x = 12$	$3x = 36$ (undo multiplication); $x = 12$
Length of each side	Length = x = 12 units Width = 3 units	Top = 12 units long, Bottom = 12 units long Each side = 3 units long
Perimeter of rectangle	30 units	Side + Side + Side + Side = 6 + 6 + 3 + 3

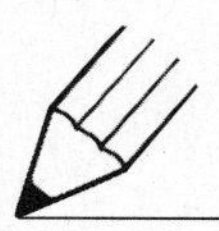

PRACTICE: Combining Algebra and Geometry

(For answers, see page 250.)

The following are multi-step problems. Show all steps and label and circle each of your answers. Each bullet requires an answer. Work on a separate sheet of paper.

1. The perimeter of the rectangle drawn is 22 inches.

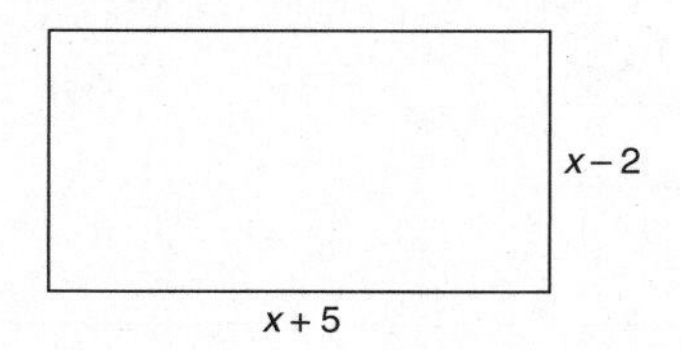

- Write an equation to represent this perimeter.
- Find the value of x.
- Find the value of each side.
- Check your work.

2. The perimeter of the isosceles triangle drawn is 100 feet.

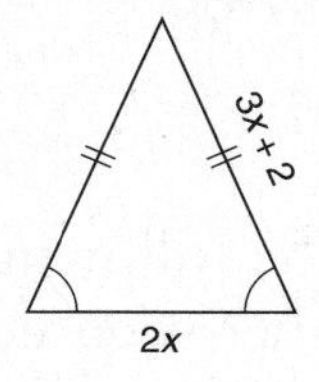

- Write an equation to represent this perimeter.
- Find the value of x.
- Find the value of each side.
- Check your work.

3. The area of the rectangle drawn is 18 square yards.

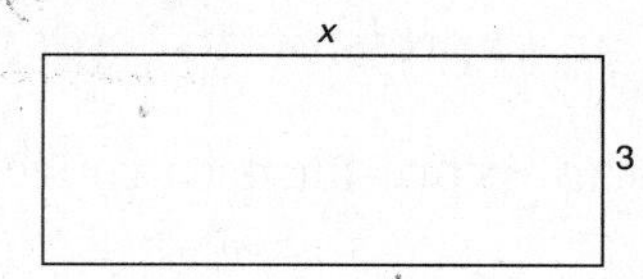

- Find the value of x.
- Write an equation to represent the perimeter of this rectangle.
- Find the perimeter of this rectangle.
- Check your work.

4. The perimeter of this equilateral triangle is 108 meters.

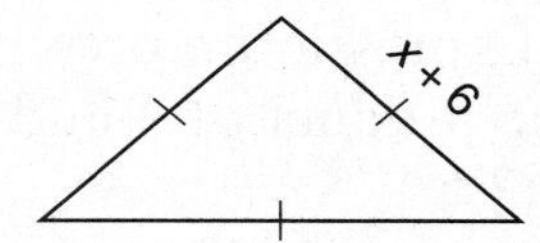

- Write an equation to represent the perimeter of this equilateral triangle.
- Find the value of x.
- Find the length of each side.
- Check your work.

5. The triangle below shows that $\angle A = x°$, $\angle B = 2x°$, $\angle C = 3x°$.

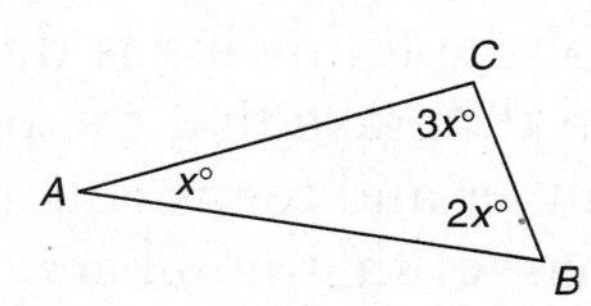

- Write an equation to represent the total of these three angles.
- Find the value of x.
- Find the value of $\angle B$.
- Find the value of $\angle C$.

WRITING EXPRESSIONS AND EQUATIONS

Examples

A. Write an expression to represent the product of a number and 6. $(6)(n)$ or $6n$

B. Write an expression to represent the absolute value of a number. $|n|$

C. Write an expression to represent twice a number divided by five. $\frac{2n}{5}$

D. The sum of 16 and a number is five less than twice that number. $16 + n = 2n - 5$

E. The product of 4 and some number is less than 25. $4n < 25$

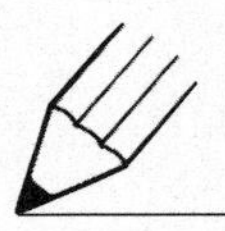

PRACTICE: Writing Expressions and Equations

(For answers, see page 251.)

1. Which expression represents six less than the product of 4 and some number?

A. $6 - 4 + n$ B. $4n - 6$
C. $6 - 4n$ D. $4 - 6n$

2. Which of the following says that the square root of a number is greater than 4 less than the number?

A. $\sqrt{n} > n - 4$ B. $\sqrt{n} < n - 4$
C. $4 - n \geq \sqrt{n}$ D. $\sqrt{4 - n} \leq \sqrt{n}$

3. Which of the following is the equation that says that the product of a number and ten is equal to 20 percent of that number?

A. $10n = 2.0n$
B. $10 + n = 0.20n$
C. $n + 0.20n = 10$
D. $10n = 0.20n$

4. Which of the following expressions is most likely *not* the perimeter of a rectangle?

A. $2(x + 3) + 2(x + 2)$
B. $x + 3 + x + 3$
C. $2(x + 3)$
D. $9x$

5. Which of the following represents that a number squared is equal to twenty more than nine times that number.

A. $x^2 = 9x + 20$
B. $x^2 = (20)(x) + 9$
C. $x^2 > 9 + 20x$
D. $x^2 = 9 + x + 20$

REAL-LIFE APPLICATIONS

Very often it is easiest to solve some basic everyday situations using algebraic equations and special formulas.

Example

If Christine put \$2,000 into a savings account that received 3% interest *compounded annually,* how much money would she have in that account after 5 years?

The formula to determine compound interest is

$$A = P(1 + r)^t$$

If you do not remember this formula, use the *NJ ASK 8 Mathematics Reference Sheet* on page 265. A is the amount she would have after five years, P is the principal amount in the account now, r is the rate (or the percent) of interest she would receive, and t is the time (in this case, $t = 5$ years).

$$A = 2{,}000\ (1 + 0.03)^5 = 2{,}000\ (1.03)^5$$

$$A = 2{,}000\ (1.159274) = 2318.548 \text{ or } \$2{,}318.55$$

At the end of five years Christine would have \$2,318.55 in her savings account.

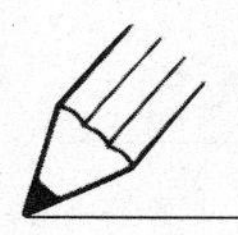

PRACTICE: Real-Life Applications

(For answers, see page 251.)

1. The freshman class just collected money from their major fund-raiser for the year. They have \$5,300 and plan to put this money into a savings account for 3 years; the bank gives 2.9% interest annually.

How much money can they expect to have at the end of the four years? (Use the formula $A = P\ (1 + r)^t$.)

A. \$5,329
B. \$5,453.70
C. \$5,774.60
D. \$5,940.07

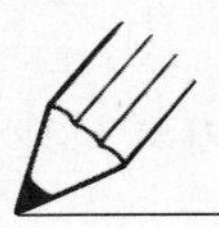

PRACTICE SCR NONCALCULATOR QUESTIONS

Each question is worth 1 point. No partial credit is given.
(For answers, see page 252.)

1. Simplify the expression: $-8 - 5$ Answer: ____________

2. Simplify the expression: $12x + 6x - 4x$ Answer: ____________

3. Combine like terms: $10x + 6 - 3x + 5$ Answer: ____________

4. Distribute the 2: $2(x - 10)$ Answer: ____________

5. Multiply: $(-3)(-2)(2)$ Answer: ____________

6. Divide and simplify: $\frac{-6}{-12}$ Answer: ____________

7. Reduce to lowest terms: $-\frac{10}{5}$ Answer: ____________

8. Distribute the -3: $-3(3 - 2x)$ Answer: ____________

9. Evaluate the expression if $x = 2$.

   $3x + 4 + 2x - 1$ Answer: ____________

10. Evaluate the expression when $x = -3$

    $4x - 6\ (x + 2)$ Answer: ____________

11. If $x + 2$ is the rule, what are the next three numbers in the sequence?

    11 ___ ___ ___ Answer: ____________

12. What is the slope of the line represented by the equation below?

    $y = -\frac{1}{2}x + 4$ Answer: ____________

13. What is the slope of the line represented by this equation?

    $y - 3x = 16$ Answer: ____________

14. What is the slope of a line that is *parallel* to the line represented by this equation?

$$5x = y + 4$$

Answer: ____________

15. What is the slope of a line that is *perpendicular* to the line represented by this equation?

$$y = \frac{2}{3}x - 10$$

Answer: ____________

16. What value of x will make this equation true?

$$2x - 4 = 10$$

Answer: ____________

17. What value of y will make the following equation true?

$$-3(x - 2) = 18$$

Answer: ____________

18. The perimeter of the rectangle drawn below is 28 feet. Solve for x.

2x

x + 2

Answer: ____________

19. If a line goes through the points (1, 3) and (4, 8), does it have a positive or a negative slope?

Answer: ____________

20. Write an equation that matches the input–output information given in the table below?

Answer: ____________

x	y
2	5
1	3
0	1
–1	–1
–2	–3

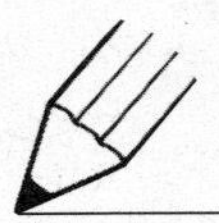

PRACTICE ECR QUESTIONS

(For answers, see page 253.)

1. Use the information provided on the table below.

   a) Make a double line graph on the grid provided. One line should represent the electric bills; the other should show the gas bills. Label each axis.

   b) What month were the electric and gas bills approximately the same?

   c) What is the difference between the month with the highest total bills and the month with the lowest total bills? Show and label your work.

   d) Mr. and Mrs. B. use gas for heating their home and electric for air conditioning. Use this information to explain the major changes in their gas and electric bills throughout the year shown.

Average Monthly Energy Bills of Mr. and Mrs. B's house in central New Jersey		
Month	**Electric**	**Gas**
December	180.00	415.00
January	140.00	330.00
February	150.00	400.00
March	130.00	270.00
April	140.00	110.00
May	330.00	75.00
June	450.00	30.00
July	300.00	20.00
August	300.00	20.00
September	180.00	30.00
October	130.00	80.00
November	110.00	280.00

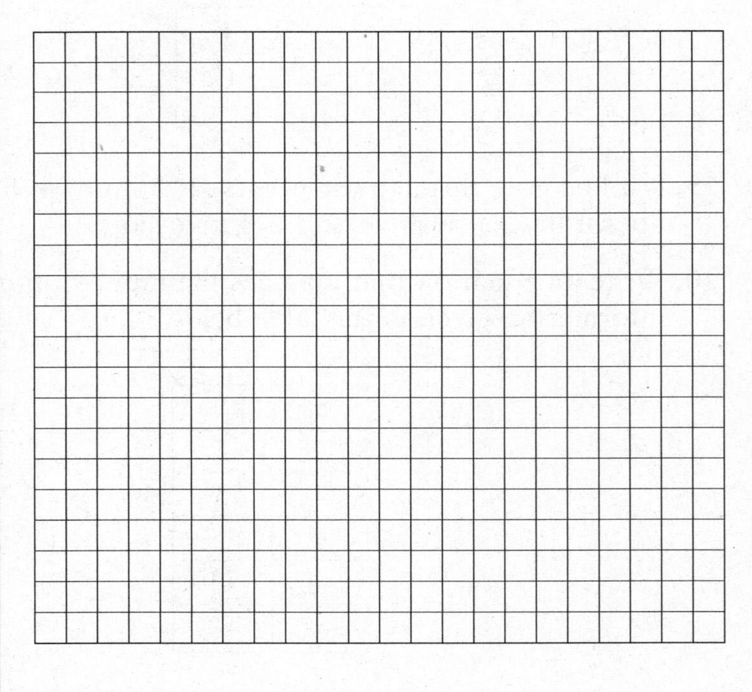

2. Nicholas and Miguel solved the same equation. One is correct, and one is incorrect.

Nicholas	Miguel
$2x - 3(4 + 5x) = 2x + 18$	$2x - 3(4 + 5x) = 2x + 18$
$2x - 12 - 15x = 2x + 18$	$2x - 12 + 15x = 2x + 18$
$-12 - 13x = 2x + 18$	$+12 \qquad + 12$
$+ 13x \qquad +13x$	$2x + 15x = 2x + 30$
$-12 = 15x + 18$	$17x = 2x + 30$
$-18 \qquad -18$	$-2x \qquad -2x$
$-30 = 15x$	$15x = 30$
$-2 = x$	$x = 2$

a) Which student has the correct answer?

b) Replace this correct value for x in the following equation to show that it is the correct solution: $2x - 3(4 + 5x) = 2x + 18$

c) How would you explain to the student who was incorrect what he did wrong?

3. If x = positive, and y = negative will the sum of x and y be positive, negative or zero? Explain your answer.

4. We use the equation $A = P(1 + r)^t$ when calculating compound interest.

A represents the accumulated value of the investment.
P represents the principal (the amount originally invested).
R represents the rate.
T represents the time (in years).

If Tim and his wife MaryJo saved \$8,000 at 4% compounded annually, what would be the value of their savings after one year? After 5 years? Set up the equations and show your work.

5. The sequence given is 3, 7, 15, 31, and it continues indefinitely. Study the sequence and look for a pattern.

    - What is the 8th term in the sequence?
    - Describe the pattern suggested by the sequence.

6.

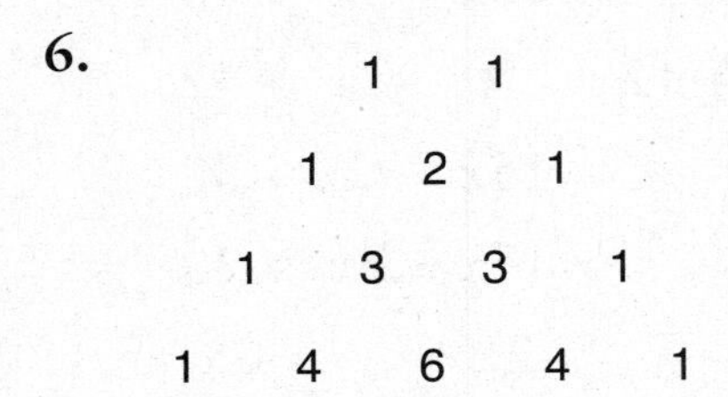

The triangular array of numbers above is called Pascal's triangle. Each number is found by adding the pair of numbers to its left and right above. For example:

$2 = 1 + 1 \qquad 3 = 1 + 2 \qquad 4 = 1 + 3 \qquad 5 = 1 + 4$

- Find the sum of the numbers in each row shown in the diagram above.
- Write the numbers in the next row (the 6th row) and explain how you selected them.
- Predict the sum of the numbers in the 7th row. Explain or show how you arrived at your answer.

7. - Write an expression to represent three consecutive integers if x is the first integer.
    - If the sum of the three consecutive integers is 1,356, set up an equation and find the value of each integer.

SOLUTIONS TO PRACTICE QUESTIONS

PATTERNS AND NUMBERS (page 226)

1. C 49 $x - 4$ is the rule $61 - 4 = 57$; $57 - 4 = 53$; $53 - 4 = 49$
2. C 6
3. 16, 19, 22
4. 94 and 90
5. 1,875 Multiply by 5. (The next number would be 9,375.)
6. 125, 62.50, 31.25
7. multiply by –2
8. multiply by $-\frac{1}{2}$
9. multiply by –4
10. 82, 244, 730, 2,188 $(28)(3) = 84 - 2 = 82$; $(82)(3) = 246 - 2 = 244$; $(244)(3) = 732 - 1 = 730$; $(730)(3) = 2{,}190 - 2 = 2{,}188$
11. 256, 65,536 $(16)^2 = 256$; $(256)^2 = 65{,}536$
12. 677 $(26)^2 = 676$, $676 + 1 = 677$
13. A 36 squares is the 8th figure; 45 squares is the 9th figure
14. D 1,024 triangles

SOLVING EQUATIONS (page 230)

1. B $x = -7$
2. C $a = 8$
3. A $b = -3$
4. D $x = 4$
5. B $w = 15$
6. B $x = \frac{4}{3}$
7. C $-9 = a$
8. C $z = 80$
9. D $x = -16$
10. A $y = -80$

LINES AND SLOPE (page 234)

1. D line with slope of $\frac{4}{2}$ or 2
2. A line with slope of $\frac{2}{3}$
3. A parallel line with slope of $\frac{2}{3}$
4. C slope of line c is $\frac{2}{3}$; slope of line d is $\frac{2}{3}$
5. C The two lines are parallel; they have the same slope.
6. $\frac{-2}{5}$

7. C Perpendicular (the slope of one is the negative reciprocal of the other) $\frac{-3}{2}$ and $\frac{2}{3}$
8. Line *a*

SLOPE (page 237)

1. B slope = $\frac{-2}{3}$
2. A slope = $\frac{1}{2}$
3. C $\frac{2}{3}$
4. B 3 $2y = 6x + 10$ becomes $y = 3x + 10$; slope is 3
5. D 4 $y - 6 = 4x$ becomes $y = 4x + 6$; slope is 4
6. B 2 $y - 2x = 14$ becomes $y = 2x + 14$; slope is 2
7. A $\frac{4}{-1}$ or -4 $\frac{4-0}{2-3} = \frac{4}{-1}$
8. B Choice B shows that $\frac{(3-4)}{(5-2)} = \frac{-1}{3}$.
9.

Table A	
x	$2x + 1$
4	$2(4) + 1 = 9$
7	$2(7) + 1 = 15$

Table B	
x	x^2
3	$(3)(3) = 9$
5	$(5)(5) = 25$
7	$(7)(7) = 49$

- Table A: \$50 would give you (2)(50) + 1 = \$101.00
- Table B: \$50 would give you (50)(50) = \$2,500.00
- Answers will vary: In the Table B bank the money amount is squared, which increases your money much more quickly than if you just doubled it and added \$1.00 (as in Table A). In Table A, the money is just multiplied by the number 2, but in Table B the number is multiplied by a higher and higher number each time.

COMBINING ALGEBRA AND GEOMETRY (page 241)

1.
- Equation: $P = 2(x + 5) + 2(x - 2) = 22$ or $P = x + 5 + x + 5 + x - 2 + x - 2 = 22$
- $x = 4$

$2x + 10 + 2x - 4 = 22$
$4x + 6 = 22$
$4x = 16$
$x = 4$

or

$4x + 10 - 4 = 22$
$4x + 6 = 22$ (Combine like terms.)
$4x = 16$ (Undo multiplication.)
$x = 4$

- One side is 9 units long. $x + 5 = 4 + 5 = 9$
  The other side is 2 units long. $x - 2 = 4 - 2 = 2$ units long
- Check: Perimeter $= 2(x + 5) + 2(x - 2)$
  $= 2(4 + 5) + 2(4 - 2)$
  $= 2(9) + 2(2)$
  $= 18 + 4 = 22$, which is correct!

2. *Remember:* In an isosceles triangle, two sides are equal in length.
- Equation: $2(3x + 2) + 2x = 100$ or $3x + 2 + 3x + 2 + 2x = 100$
- $x = 12$
  $6x + 4 + 2x = 100$
  $8x + 4 = 100$
  $8x = 96$
  $x = 12$

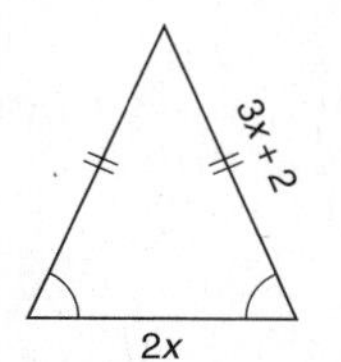

- Side: $x + 2 = 3(12) + 2 = 36 + 2 = 38$ units long
- Base: $2x = 2(12) = 24$ units long
- Check: Side + Side + Base should = 100; 38 + 38 + 24 = 100, correct!

3.
- $x = 6$ yards Area = (Side)(Side) $18 = (3)(x)$; $6 = x$
- $P = 3 + 3 + 6 + 6$ or $P = 2(3) + 2(6)$
- Perimeter = 18 yards

4.
- $P = 3(x - 6)$ or $P = x - 6 + x - 6 + x - 6$ (*Remember:* In an equilateral triangle all sides are the same length.)
- $x = 30$ $3(x + 6) = 108$; $3x + 18 = 108$; $3x = 90$; $x = 30$
- Each side: $x + 6 = 30 + 6 = 36$
- Check: Perimeter = Side + Side + Side = 36 + 36 + 36 = 108, correct!

5. *Remember*: The sum of the three angles in any triangle will be 180°.
- Equation: $x° + 2x° + 3x° = 180°$
- $6x = 180°$, $x = 30°$
- $\angle B = 2x = (2)(30°) = 60°$
- $C = 3x = (3)(30°) = 90°$

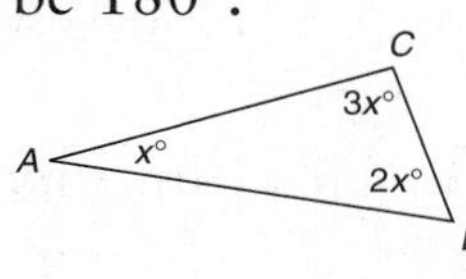

WRITING EXPRESSIONS AND EQUATIONS (page 242)

1. B $4n - 6$
2. A $\sqrt{n} > n - 4$
3. D $10n = 0.20n$
4. D $9x$
5. A $x^2 = 9x + 20$

REAL-LIFE APPLICATIONS (page 243)

1. C $A = 5{,}300\ (1 + .029)^3 = 5{,}300\ (1.029)\ (1.029)\ (1.029)$
   $= 5{,}300\ (1.0895473) = \$5{,}774.60$

SCR NONCALCULATOR QUESTIONS (page 244)

1. -13
2. $14x$
3. $7x + 11$ — Group the like terms: $10x + 6 - 3x + 5$ becomes $10x - 3x + 6 + 5$
4. $2x - 20$
5. 12 — $(-3)(-2)(2) = (6)(2) = 12$
6. $\frac{1}{2}$ — Remember: a negative divided by a negative is a positive.
7. -2 — Remember: a negative divided by by a positive is a negative.
8. $-9 + 6x$
9. 13

$$\begin{aligned} 3x + 4 + 2x - 1 &= \\ 3(2) + 4 + 2(2) - 1 &= \\ 6 + 4 + 4 - 1 &= \\ 14 - 1 &= 13 \end{aligned}$$

10. -6

$$\begin{aligned} 4x - 6(x + 2) &= \\ 4x - 6x - 12 &= \\ -2x - 12 &= \\ (-2)(-3) - 12 &= \\ 6 - 12 &= -6 \end{aligned}$$

11. 13, 15, 17 — $11 + 2 =$ **13**; $13 + 2 =$ **15**; $15 + 2 =$ **17**
12. $-\frac{1}{2}$
13. 3 or $\frac{3}{1}$

$$\begin{aligned} y - 3x &= 16 \\ +3x &\quad +3x \\ \hline y &= 3x + 16 \end{aligned}$$

First put the equation in *slope-intercept* form.

The slope is 3 or $\frac{3}{1}$

14. 5 or = $\frac{5}{1}$ — In slope-intercept form, $5x = y + 4$ would be $5x - 4 = y$ or $y = 5x - 4$.
15. $-\frac{3}{2}$ — The slope if the line $y = \frac{2}{3}x - 10$ is $\frac{2}{3}$.

A line perpendicular to this would have a slope that is the opposite reciprocal of the slope of that line.

16. 7

$$\begin{aligned} 2x - 4 &= 10 \\ +4 &\quad +4 \\ \hline \frac{2x}{2} &= \frac{14}{2} \\ x &= 7 \end{aligned}$$

17. −4

$$
\begin{aligned}
-3(x-2) &= 18 \\
-3x + 6 &= 18 \\
-6 \quad & \quad -6 \\
\hline
\frac{-3x}{-3} &= \frac{12}{-3} \\
x &= -4
\end{aligned}
$$

18. $x = 4$ $\quad 2(x+2) + 2(2x) = 2x + 4 + 4x = 6x + 4 = 28;\ 6x = 24;\ x = 4$

19. It has a positive slope. You can tell just by plotting the line and seeing how it increases from left to right. You can also use the slope formula.

$$\frac{y_1 - y_2}{x_1 - x_2} = \frac{8-3}{4-1} = \frac{5}{3}$$

20. $y = 2x + 1$

ECR QUESTIONS (page 246)

1. a)

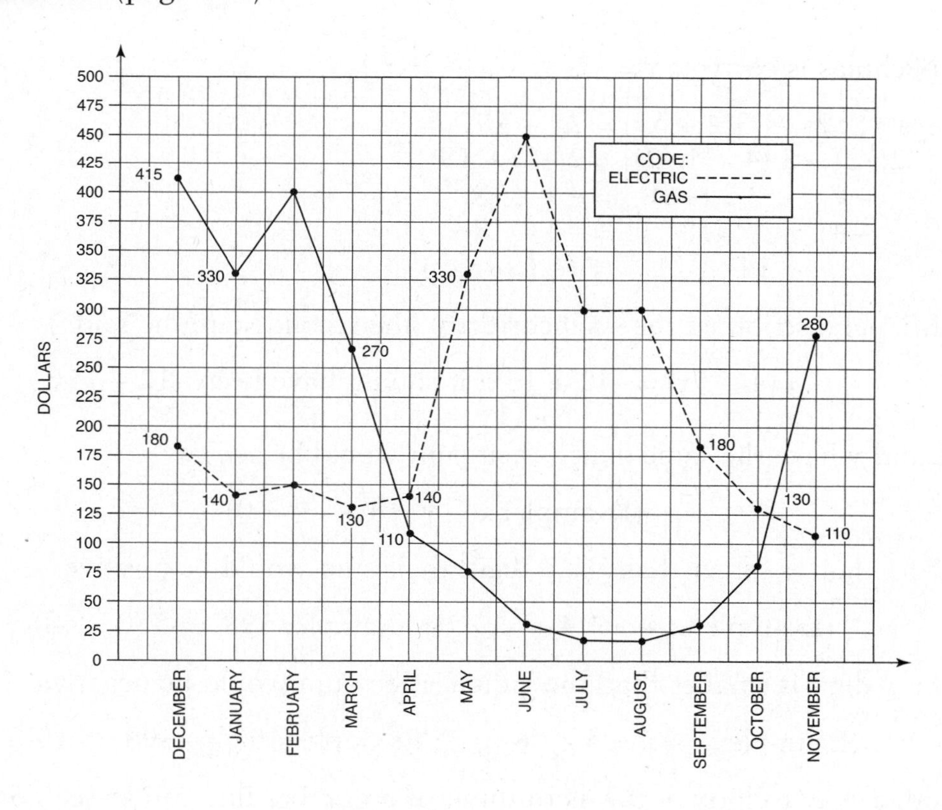

b) **April** If you look at the numbers on the chart, you see that in April the electric bill was \$140 and the gas bill was \$110; there was only a \$30 difference (140 – 110 = 30).

However, if you look at the double line graph above you notice that the two lines intersect twice; the gas and electric usage between March and April at one point were each about \$140, and the usage between October and November were each about \$125.

c) **July** In July the electric bill was \$450 and the gas bill was only \$30; this was a \$420 difference (450 – 30 = 420).

d) The electric bills were high (over \$300) in May, June, July, and August; these were the hotter months when Mr. and Mrs. B. had their air conditioning working more often. The colder months, when Mr. and Mrs. B. used their heating system often, seems to be from November to March, when the gas bills were at least \$130 per month.

2. a) Nicholas is correct; $x = -2$

b)
$$\begin{aligned} 2x - 3(4 + 5x) &= 2x + 18 \\ 2(-2) - 3\,(4 + 5(-2)) &= 2(-2) + 18 \\ -4 \quad -3\,(-6) &= -4 + 18 \\ -4 \quad +18 &= -4 + 18 \\ 14 &= 14 \text{ (correct)} \end{aligned}$$

c) Miguel distributed the –3 incorrectly (the second step); he said:

$-3\,(4 + 5x) = -12 + 15x$; it should have been $-12 - 15x$

3. If x and y have the same digits, their sum would be zero.

(Example: $x + y = 4 + -4 = 0$)

If the x digit is larger than the y digit, their sum would be positive.

(Example: $x + y = 4 + -1 = 3$; Example: **598** + –550 = 48)

If the x digit is smaller than the y digit, their sum would be negative.

(Example: $x + y = 4 + \mathbf{-6} = -2$; Example: 598 + **–698** = **–100**)

Another way to look at this is to think of a number line and to see how far each number is from zero. If x and y are the same distance from zero, their sum would be 0. If x is further away from zero than y, their sum would have the sign of x (positive). If y is further away from zero than x, their sum would have the sign of y (negative).

4. After one year their savings would equal $8,320.

$$A = P\,(1 + r)^t$$
$$A = \$8{,}000(1 + 0.04\,)^1 = 8{,}320$$

After five years their savings would equal $9,733.22

$$A = \$8{,}000(1 + 0.04)^5 = 9{,}733.22$$

5. ▪ 511 is the 8th term.

(**3**)(2) + 1 = 7   (**7**)(2) + 1 = 15   (**15**)(2) + 1 = 31   (**31**)(2) + 1 = 63
(**63**)(2) + 1 = 127   (**127**)(2) + 1 = 255   (**255**)(2) + 1 = 511

▪ The pattern suggested is that each term is multiplied by two and then 1 is added.

6. ▪ 1 6 15 20 15 6 1

▪ The sum of the numbers in the 5th row would be **32**.

1 + 1 = **2**
1 + 2 + 1 = **4**
1 + 3 + 3 + 1 = **8**
1 + 4 + 6 + 4 + 1 = **16**
1 + 5 + 10 + 10 + 5 + 1 = **32**

▪ The sum of the numbers in the 7th row would be **128**.

1 1
1 2 1
1 3 3 1
1 4 6 4 1
1 5 10 10 5 1

The sum of each row is double the row above: 2 + 4 + 8 + 16 + 32 + 64 = 128.

7. ▪ $x \quad + \quad x + 1 \quad + \quad x + 2$

▪ $x \quad + \quad x + 1 \quad + \quad x + 2 = 1{,}356$

$$3x + 3 = 1{,}356$$
$$3x = 1{,}353$$

$x =$ **451**, $x + 1 =$ **452**, and $x + 2 = 453$

Check: 451 + 452 + 453 = 1,356 (correct).

Name: ______________________________ Date: ______________

Cluster IV Test

35 minutes
(Use the *NJ ASK 8 Mathematics Reference Sheet* on page 265.)

MULTIPLE-CHOICE QUESTIONS

DIRECTIONS FOR QUESTIONS 1 THROUGH 12: Each of the questions or incomplete statements below is followed by four suggested answers. Select the one that is the best in each case, and fill in the corresponding lettered circle. Be sure the circle is filled in completely so you cannot see the letter. Unless you are told to do so in the question, do NOT include sales tax in your answer to questions involving purchases.

1. Which of the following is a geometric series?

A. $\frac{1}{2}$ $\frac{1}{4}$ $\frac{1}{8}$ $\frac{1}{16}$

B. 12 10 8 6

C. 0.5 1.0 1.5 2.0

D. $\frac{-1}{2}$ $\frac{-5}{8}$ $\frac{-3}{4}$ $\frac{-7}{8}$

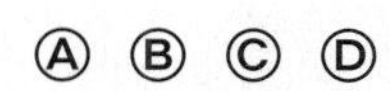

2. In the following equation what is the first step in isolating the variable?

$$-8x - 34 = 14$$

A. Subtract 34 from both sides
B. Divide both sides by –8
C. Add 34 to both sides
D. Multiply both sides by 8

3. Which equation does *not* have a solution of 12?

A. $4a + 3 = 51$
B. $14 - 2b = -10$
C. $3(x - 5) = 21$
D. $3(6 + y) = 30$

Ⓐ Ⓑ Ⓒ Ⓓ

4. Jacob's mom is paid twice her usual hourly wage for each hour she works over 40 hours a week. This week she worked 50 hours and earned $873.00. What is her hourly wage?

A. $9.96 **B.** $14.55
C. $ 17.46 **D.** $ 21.83

Ⓐ Ⓑ Ⓒ Ⓓ

5. Judy and Janet took their 3 young children to the movies in nearby Franklin Township. An adult ticket costs 3 times more than a child's ticket. If they paid a total of $27.00, what is the cost of each adult ticket?

A. $3.00 **B.** $4.00
C. $6.00 **D.** $9.00

Ⓐ Ⓑ Ⓒ Ⓓ

GO ON TO THE NEXT PAGE ➡

6. Which equation matches the following: Six less than some number is five squared.

   A. $5^2 = n - 6$
   B. $n + 6 = 5^2$
   C. $6 - n = 5^2$
   D. $5^2 - 6 = n$

Ⓐ Ⓑ Ⓒ Ⓓ

7. Continue the pattern. How many □ will be in the 8th figure?

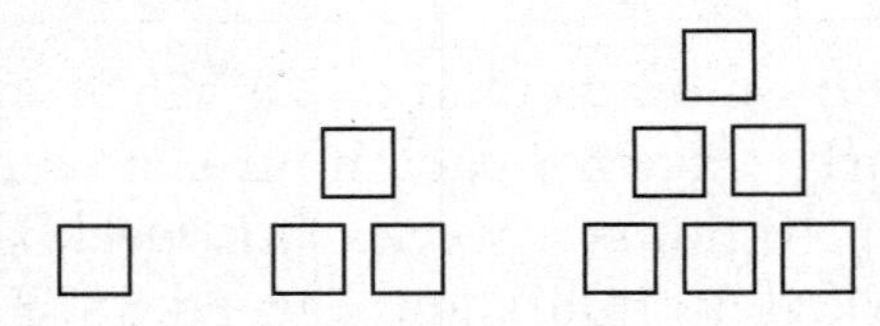

   A. 18
   B. 21
   C. 32
   D. 36

8. Refer to this equation:

$$y = \frac{-2}{5}x + 10$$

   A. When you graph this line, it will have a very steep slope (like a tall mountain).
   B. When graphed, this line will slant to the right.
   C. When graphed, this line will slant to the left.
   D. When graphed, this line will be parallel to the line $y = \frac{5}{2}x + 10$

9. Write the following equation in slope–intercept form:

$$3x - 9y = 12$$

   A. $3x = 9y + 12$
   B. $y = \frac{1}{3}x - \frac{4}{3}$
   C. $x = 9y + 12$
   D. $x - 3y = 4$

GO ON TO THE NEXT PAGE

10. Solve for c in the following equation:

$$ac + 4 = b$$

A. $c = \dfrac{b+4}{a}$

B. $c = \dfrac{b}{a} + 4$

C. $c = \dfrac{b-4}{a}$

D. $c = 4b - a$

Ⓐ Ⓑ Ⓒ Ⓓ

11. Solve the following equation:

$$2x + 3(x - 2) = 24$$

A. $x = 6$
B. $x = -6$
C. $x = 25$
D. $x = \dfrac{4}{5}$

Ⓐ Ⓑ Ⓒ Ⓓ

12. How many of the following lines have a slope of 4?

$$y = 4x + 3$$
$$y = 4x$$
$$4y = x$$
$$16x - 4y = 24$$
$$y + 4x = 6$$

A. 1
B. 2
C. 3
D. 4

GO ON TO THE NEXT PAGE

EXTENDED CONSTRUCTED RESPONSE QUESTIONS

DIRECTIONS FOR QUESTIONS 13 AND 14: Respond fully to the ECR questions that follow. Show your work and clearly explain your answer. You will be graded on the correctness of your method as well as the accuracy of your answer.

13. Refer to the diagram of a triangle below.

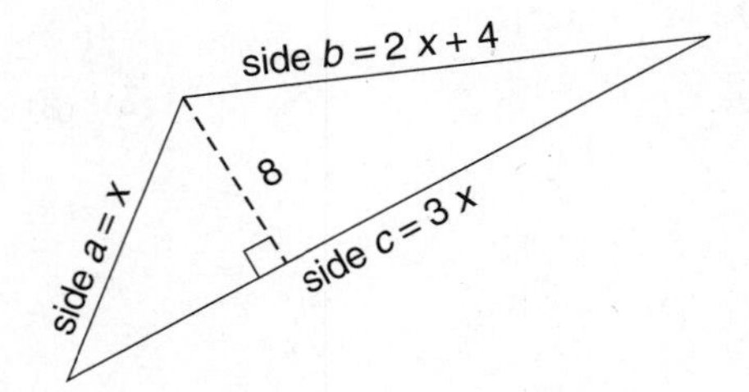

The perimeter of the triangle shown is 100 cm.

- Write an equation to represent the perimeter.
- Solve for x.
- Find the length of side a, side b, *and* side c.
- If the height of this triangle is 8 cm long, what is the area of the large triangle?

14. The following equation was given to three different students to solve:

$$2(x - 4) = 16$$

Each student made at least one mistake.

- Describe each student's mistake.
- Then solve the equation properly.

Student A	Student B	Student C
$2(x - 4) = 16$	$2(x - 4) = 16$	$2(x - 4) =16$
$2x + 8 = 16$	$2x - 4 = 16$	$2x - 8 = 16$
$-8 \quad -8$	$+4 \quad +4$	$+8 \quad +8$
$2x = 8$	$2x = 22$	$2x = 24$
$\frac{2x}{2} = \frac{8}{2}$	$\frac{2x}{2} = \frac{22}{2}$	$\frac{2x}{-2} = \frac{24}{-2}$
$x = 4$	$x = 11$	$x = -12$

SOLUTIONS TO CLUSTER IV TEST

MULTIPLE-CHOICE QUESTIONS

1. A	2. C	3. D	4. B	5. D	6. A
7. D	8. C	9. B	10. C	11. A	12. C

EXTENDED CONSTRUCTED RESPONSE QUESTIONS

QUESTION 13

- Equation of Perimeter: $x + 2x + 4 + 3x = 100$ cm or $6x + 4 = 100$ cm
- $x = 16$ — $6x + 4 = 100$, $6x = 96$, $x = 16$
- *Side a* = 16 cm — Side $a = x$, $x = 16$ cm
  Side b = 36 cm — Side $b = 2x + 4 = 2(16) + 4 = 36$
  Side c = 48 cm — Side $c = 3x = 3(16) = 48$
- Area = 192 sq. cm — Area of triangle = $\frac{1}{2}$(Base)(Height)

  Area = $\frac{1}{2}(8)(48) = 192$ sq. cm

QUESTION 14

- Student A multiplied $2(-4)$ incorrectly. $2(-4) = -8$ not $+8$; $(+)(-)$ = a negative.
- Student B did not distribute the 2 correctly. He only multiplied $2(x)$; he did not multiply $2(-4)$; $2(x - 4) = 2(x) + 2(-4)$ or $2x - 8$.
- Student C did not solve for x correctly in the last step. She should have divided both sides by 2 not by -2.
- Here is the equation solved correctly.

$2(x - 4) = 16$

$2x - 8 = 16$ — distribute the 2 to both the x and the -4

$+8 \quad +8$ — add 8 to both sides (to undo the -8) to get the $2x$ by itself

$2x = 24$

$\frac{2x}{2} = \frac{24}{2}$ — divide both sides by 2 (to undo the multiplication)

a positive divided by a positive is a positive

$x = 12$

Chapter 5

Practice Tests

A SCAVENGER HUNT TO DO BEFORE TAKING THE PRACTICE TESTS

Use the *NJ ASK 8 Mathematics Reference Sheet* on page 265 to help you answer these questions. This scavenger hunt will help you remember what information will be available to you when you take the official ASK 8 in the spring. All the formulas you need are on the *ASK 8 Mathematics Reference Sheet.*

1. What is the area and circumference of a circle with radius 4 ft.? _____ and ______

2. If one side of a rectangle measures 3 feet, and another side measures 4.2 feet, what is the perimeter of the rectangle? _______

3. If a rectangular swimming pool is 25 feet long, 12 feet wide, and 6 feet deep, how many cubic feet of water would fill the pool to the top? (*Hint*: Find the volume of the pool; its shape is a rectangular prism.) _______

4. What is the total area of the shape on the right? (*Hint*: Find the area of each shape and add them together.) _______

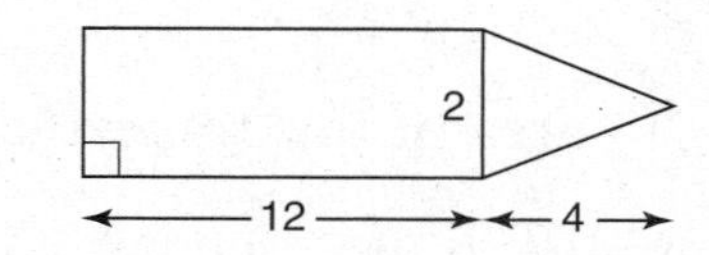

5. What is the area of a trapezoid that is 3 inches high, with one base 9 inches long, the other base 5 inches long? _______

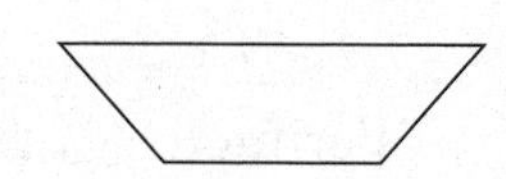

6. Judy worked one month in her grandfather's store for 3 hours a day (including weekends). How much money would she make at $7 per hour? _______

7. What is the perimeter of a right triangle with an 8 cm base, and a 10 cm hypotenuse? (*Hint*: First use the Pythagorean formula to find the height; then find the perimeter.) _______

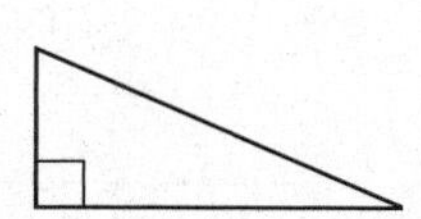

8. The recipe for a small cheese cake requires 8 ounces (oz) of cream cheese. How many **pounds** are needed to make 15 cakes for the PTA bake sale? ______

9. If a room measures 12 feet by 18 feet, how many square yards of carpeting will be needed to cover the entire floor? ______

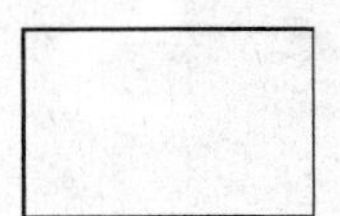

10. A New York City office building has a very large elevator that can hold a maximum of 1.5 tons. What is the maximum number of people who should be on that elevator if they each weigh 150 pounds (lb)? ______

SOLUTIONS TO SCAVENGER HUNT

1. Area of circle = πr^2; area of this circle = $\pi(4)(4) = 3.14(16) \sim 50.24$ or ~50 sq. feet circumference of circle = $2\pi r$; circumference of this circle = $2(3.14)(4) = 25.12$ or ~25 feet
2. Perimeter of rectangle = $2(l + w)$; perimeter of this rectangle = $2(3 + 4.2) = 2(7.2) = 14.4$ feet
3. Volume of rectangle = lwd; volume of this rectangle = $(25)(12)(6) = 1{,}800$ cubic feet
4. Area of rectangle = lw; area of this rectangle = $(12)(2)$ = 24 sq. units

   Area triangle = $\frac{1}{2}bh$; area of this triangle = $\frac{1}{2}(2)(4)$ = <u>4 sq. units</u>

   Total area of this shape = 28 sq. units
5. Area of the trapezoid = $\frac{1}{2}h(b_1+b_2)$;

   area of this trapezoid = $\frac{1}{2}(3)(5 + 9) = (3)(7) = 21$ sq. feet
6. We'll use 30 days = 1 month; (3 hr)($7) (30 days) = (3)(7)(30) = $630. If you used 31 days = 1 month; (3 hr)($7)(31 days) = (3)(7)(31) = $651
7. $a^2 + b^2 = c^2$; $a^2 + 8^2 = 10^2$; $a^2 + 64 = 100$, $a^2 = 36$, a = 6
   Perimeter of triangle = Side + Side + Side; in this case, it is 8 + 10 + 6 = 24 cm
8. 16 oz = 1 pound; 15×8 oz = 120 oz needed; $\frac{120 \text{ oz}}{16 \text{ oz}}$ = 7.5 pounds needed
9. 3 feet = 1 yard; a room 12 ft by 18 ft is $\frac{12}{3} \times \frac{18}{3}$ = (4 yd)(6 yd) = 24 sq. yd
10. 1 ton = 2,000 pounds (or lb); 1.5 tons = 3,000 lb, $3000 \div 150 = 20$ people

Remember: Use the *ASK 8 Mathematics Reference Sheet* each time you practice or take an ASK 8 test!

New Jersey Assessment of Skills and Knowledge
Grade 8
MATHEMATICS REFERENCE SHEET

Use the information below to answer questions on the
Mathematics section of the Grade Eight Assessment of Skills and Knowledge (NJ ASK 8)

The sum of the measures of the interior angles of any n-sided polygon = $(n - 2)180°$
Distance = rate × time
Simple Interest Formula $A = P + Prt$ **Compound Interest Formula** $A = P(1 + r)^t$
A = amount after t years; P = principal; r = annual interest rate; t = number of years

The number of combinations of r objects you can choose from a set of n objects is $\frac{n!}{(n-r)!r!}$

The number of permutations of size r taken from n objects is $\frac{n!}{(n-r)!}$

Pythagorean Formula $c^2 = a^2 + b^2$	Use the following equivalents for your calculations: $\pi \approx$ **3.14 or** $\frac{22}{7}$	**Rectangle** Area = lw Perimeter = $2l + 2w$
Triangle Area = $\frac{1}{2}bh$	**Parallelogram** Area = bh	**Trapezoid** Area = $\frac{1}{2}h(b_1 + b_2)$
Circle Area = πr^2 Circumference = $2\pi r$ = πd	**Sphere** Volume = $\frac{4}{3}\pi r^3$ Surface Area = $4\pi r^2$	**Cylinder** Volume = $\pi r^2 h$ Surface Area = $2\pi rh + 2\pi r^2$
Cone Volume = $\frac{1}{3}\pi r^2 h$	**Rectangular Prism** Volume= lwh Surface Area= $2lw + 2wh + 2lh$	**Pyramid** Volume = $\frac{1}{3}lwh$

USE THE FOLLOWING EQUIVALENTS FOR YOUR CALCULATIONS

60 seconds = 1 minute	12 inches = 1 foot	10 millimeters = 1 centimeter
60 minutes = 1 hour	3 feet = 1 yard	100 centimeters = 1 meter
24 hours = 1 day	36 inches = 1 yard	10 decimeters = 1 meter
7 days = 1 week	5,280 feet = 1 mile	1000 meters = 1 kilometer
12 months = 1 year	1,760 yards = 1 mile	
365 days = 1 year		

8 fluid ounces = 1 cup	16 ounces = 1 pound
2 cups = 1 pint	2,000 pounds = 1 ton
2 pints = 1 quart	
4 quarts = 1 gallon	1000 milligrams = 1 gram
	100 centigrams = 1 gram
1000 milliliters (mL) = 1 liter (L)	10 grams = 1 dekagram
	1000 grams = 1 kilogram

centimeters 1 2 3 4 5 6 7 8 9 10 11 12 13 14 15
inches 1 2 3 4 5

DIRECTIONS FOR THE NJ ASK 8 MATHEMATICS TEST

The following has been taken directly from the directions provided by the NJ Department of Education. When you take the official NJ ASK 8 test, these directions will be read to you by the proctor before you begin.

The Mathematics section of the ASK 8 Assessment is made up of three parts consisting of short constructed response questions, multiple-choice questions, and extended constructed response questions.

Work as rapidly as you can without sacrificing accuracy. Do not spend too much time puzzling over a question that seems too difficult for you. Answer the easier questions first; then return to the harder ones. Try to answer every question, even if you have to guess.

YOU MUST RECORD ALL OF YOUR ANSWERS IN THE SEPARATE ANSWER FOLDER. No credit will be given for anything written in your test booklet. Your responses must be in English in order to be scored.

For multiple-choice questions, mark only one answer for each question by filling in the corresponding circle on the answer folder. MAKE SURE THAT EACH MARK IS HEAVY AND DARK AND COMPLETELY FILLS THE CIRCLE. If you change an answer, be sure to erase your first choice completely. Incomplete erasures may be read as intended answers.

Respond FULLY to the extended constructed response questions in the area provided in the answer folder. Specific directions with each question will refer you to the page in your answer folder where your response is to be written. For each of these questions, provide enough explanation so that the scorer can understand your solution. You will be graded on the correctness of your methods as well as the accuracy of your answer.

In addition to a ruler and geometric shapes, the Mathematics Reference Sheet provides formulas and other information you may find useful. You may use the information on the reference sheet and a calculator to help you solve problems on the test.

You will have 2 hours and 8 minutes to complete the five parts of the Mathematics test.

Part 1—10 SCR items, 20 minutes
Part 2—8 MC and 1 ECR item, 22 minutes
Part 3—8 MC and 1 ECR item, 22 minutes
Part 4—10 MC and 1 ECR item, 25 minutes
Part 5—10 MC and 1 ECR item, 25 minutes
Part 6—6 MC and 1 ECR item, 19 minutes

Name: ______________________ Date: ______________

Practice Test 1
Part I (20 minutes)

Use the *NJ ASK 8 Mathematics Reference Sheet* on page 265.
NO CALCULATOR PERMITTED

DIRECTIONS FOR QUESTIONS 1 THROUGH 10: Write your answer neatly on the line provided. No partial credit is given for these questions.

1. Which line segment is longer, $\overline{AB}$ or $\overline{BC}$?

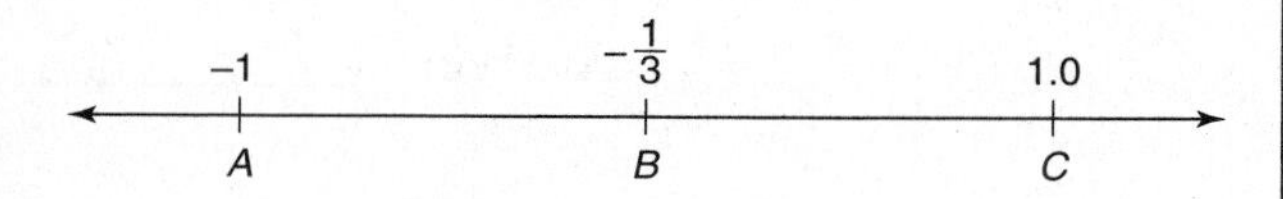

(Figure not drawn to scale.)

Answer: ________

2. What is the sum of all the prime factors of 24?

Answer: ________

3. If the perimeter of the rectangle below is 50 cm, how long is side *a*?

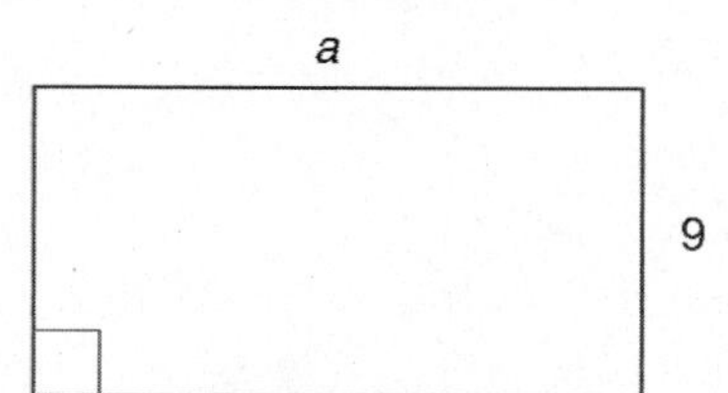

Answer: ________

4. If one side of a rectangle is 14 inches long, and the perimeter of that rectangle is 36 inches, what is the area of the rectangle?

Answer: ________

5. Below is a chart showing the points our town basketball team scored.

72	39	43
29	72	50
53	59	67
69	42	63

If one game is randomly chosen, what is the probability that the team scored fewer than 50 points? (Write your answer as a fraction.)

Answer: ________

6. What type of figure is polygon *ABCD* if its vertices are as listed below?

A(1, 1) *B*(2, 5) *C*(6, 5) *C*(8, 1)

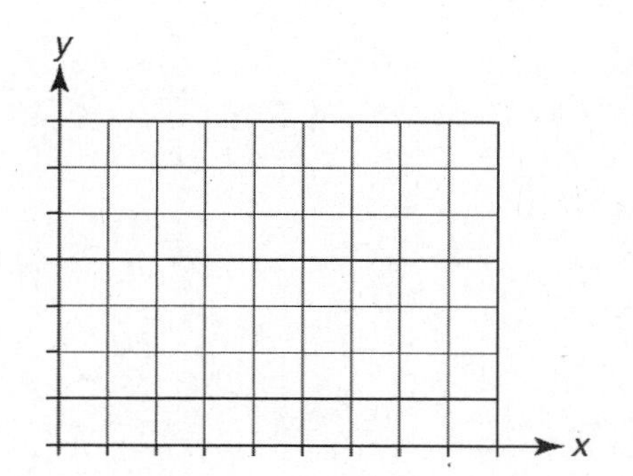

Answer: ________

GO ON TO THE NEXT PAGE 

7. Using the correct order of operations, what is the first step you should do when evaluating this expression?

$$3 \times 4 + (4 - 3)^2 - 6$$

Answer: __________

8. In a rectangular array, 36 seats can be arranged in six rows. Give two other rectangular arrays that are possible for 36 seats.

Answer: __________

9. If this pattern continues, what letter will be in the 100th place?

P I Z Z A P I E P I Z Z A P I E

Answer: __________

10. Solve for x in the given equation.

$$52 = 2x + 4$$

Answer: __________

Part II (22 minutes)

Use the *NJ ASK 8 Mathematics Reference Sheet* on page 265.
NO CALCULATOR PERMITTED

DIRECTIONS FOR QUESTIONS 11 THROUGH 18: Each of the questions or incomplete statements below is followed by four suggested answers. Select the one that is the best in each case, and fill in the corresponding lettered circle. Be sure the circle is filled in completely so you cannot see the letter. Unless you are told to do so in the question, do NOT include sales tax in your answer to questions involving purchases.

11. Evaluate the following expression.

$$3 + 4 \times 2(6 - 2)^2$$

A. 24
B. 131
C. 67
D. 147

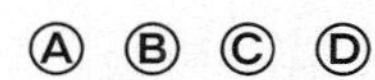

12. What diagram below illustrates an *acute* angle?

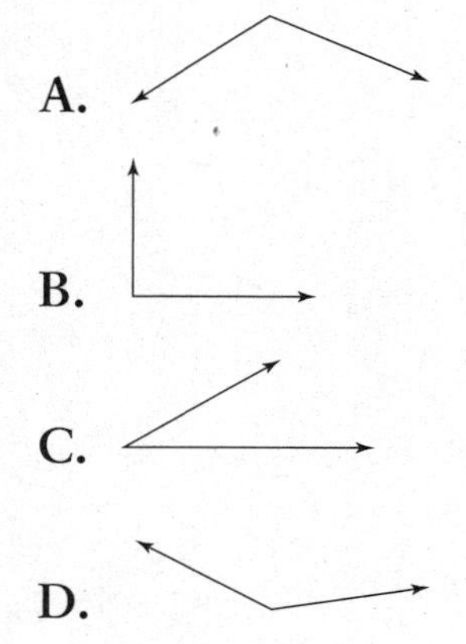

Ⓐ Ⓑ Ⓒ Ⓓ

13. What number comes next in the pattern

5 13 29 61 125 ?

A. 253
B. 250
C. 225
D. 186

Ⓐ Ⓑ Ⓒ Ⓓ

14. What value of n makes the equation true?

$$69 + n = 87$$

A. 17
B. 156
C. −18
D. 18

Ⓐ Ⓑ Ⓒ Ⓓ

15. Evaluate the following algebraic express if $a = -2$, $b = 6$, and $c = 4$:

$$2a + b^2 - 2\,(c + 14)$$

A. 0
B. −4
C. −35
D. 52

Ⓐ Ⓑ Ⓒ Ⓓ

16. Evaluate the following expression:

$$|-4| + |9|$$

A. 5
B. −5
C. 12
D. 13

Ⓐ Ⓑ Ⓒ Ⓓ

GO ON TO THE NEXT PAGE

17. Look at the network below. It describes Luis's bike ride from his house (1), to school (2), to the library (3), to the store (4), and then back home (5). What is this type of network called?

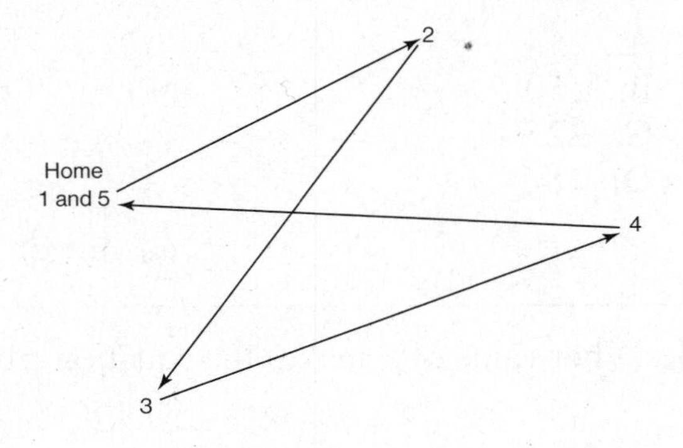

Answer: ________

18. Find the value of x in the following equation.

$$2x - 4(x - 3) = 10$$

A. $x = -4$
B. $x = 1$
C. $x = 4$
D. $x = -6$

DIRECTIONS FOR QUESTION 19: Respond fully to the extended constructed response question that follows. Show your work and clearly explain your answer. You will be graded on the correctness of your method as well as the accuracy of your answer.

19. There are two swimming pools that are being filled. They each need to be filled to the 4-foot mark (48 inches). Mrs. Brown has been filling Pool B. It already has been filled up to the 20-inch mark. With Mrs. Brown's hose, she can add 2 inches per hour. Mrs. Carol has been filling Pool C. It already has been filled up to the 10-inch mark. With Mrs. Carol's new hose she can add 4 inches per hour. Show all work on chart below.

	Mrs. Brown's Pool B Show work	Pool B, Height of Water (inches)	Mrs. Carol's Pool C Show work	Pool C, Height of Water (inches)
Now				
Hour 1				
Hour 2				
Hour 3				
Hour 4				
Hour 5				
Hour 6				
Hour 7				
Hour 8				
Hour 9				
Hour 10				
Hour 11				
Hour 12				
Hour 13				
Hour 14				

- How many hours will it take for the water in both pools to be the same height?
- How many hours will it take for the water in each pool to reach the 4-foot mark?

GO ON TO THE NEXT PAGE

Create a double-line graph. Let the horizontal axis be the hours; let the vertical axis be the height (inches). Label each line. Graph each pool's height for at least the next 14 hours. Connect the points; use a solid line for Mrs. Brown's data and a dotted line for Mrs. Carol's data.

- Complete the graph below and label each line (Mrs. Brown or Mrs. Carol).
- What happens where the two lines intersect?

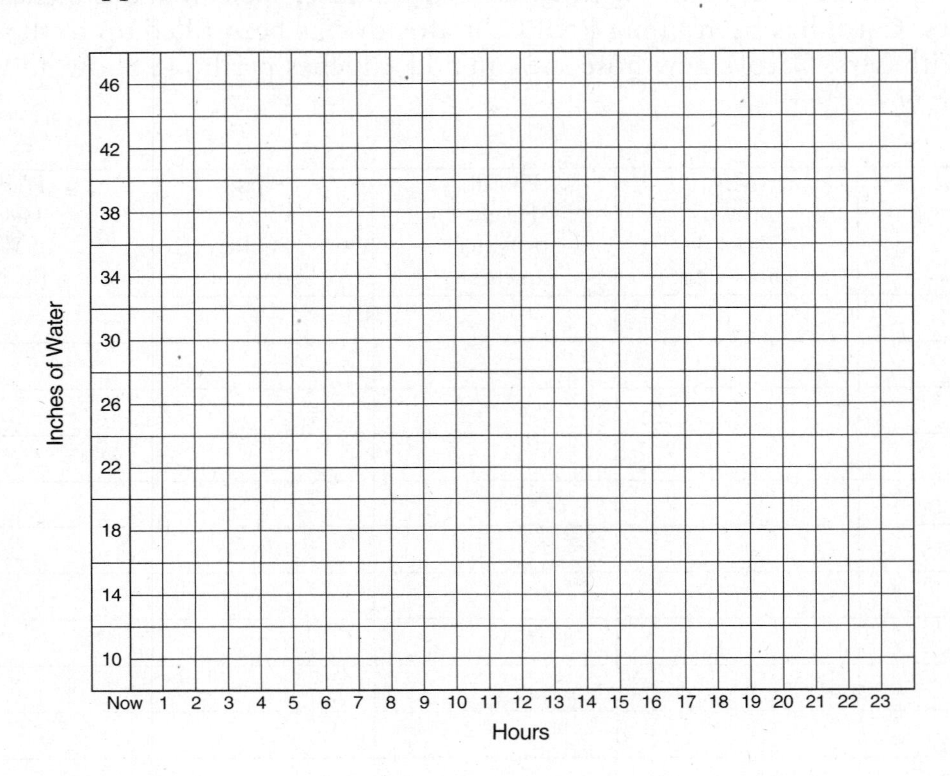

Part III (22 minutes)

Use the *NJ ASK 8 Mathematics Reference Sheet* on page 265.
NO CALCULATOR PERMITTED

DIRECTIONS FOR QUESTIONS 20 THROUGH 27: Each of the questions or incomplete statements below is followed by four suggested answers. Select the one that is the best in each case, and fill in the corresponding lettered circle. Be sure the circle is filled in completely so you cannot see the letter. Unless you are told to do so in the question, do NOT include sales tax in your answer to questions involving purchases.

20. How many prime factors does the number 64 have?

A. 0
B. 1
C. 6
D. 8

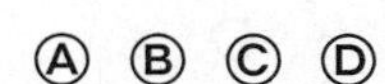

21. Each pair of students received a stack of cards numbered 1 through 12. What is the probability of selecting a prime number from this stack of cards?

A. $\frac{1}{3}$

B. $\frac{1}{4}$

C. $\frac{1}{2}$

D. $\frac{5}{12}$

Ⓐ Ⓑ Ⓒ Ⓓ

22. What is the probability that the spinner will land on a 2?

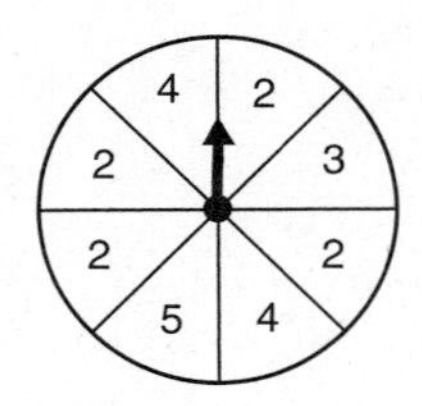

A. $\frac{1}{3}$ **B.** $\frac{1}{2}$

C. $\frac{3}{5}$ **D.** $\frac{3}{4}$

Ⓐ Ⓑ Ⓒ Ⓓ

23. If the right triangle shown below is reflected over the x-axis, which graph shows its new location?

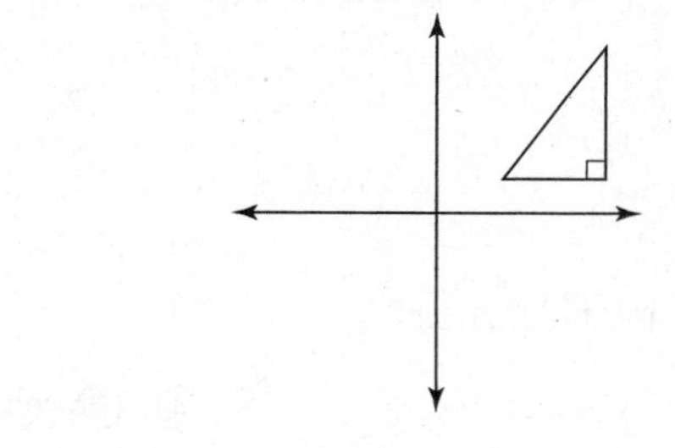

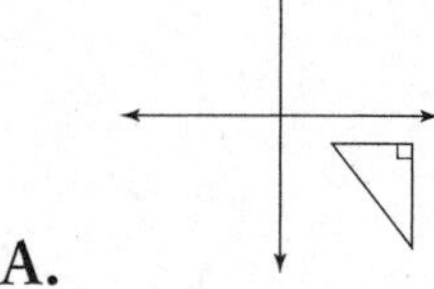

A.

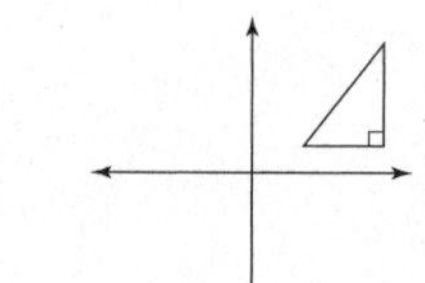

B.

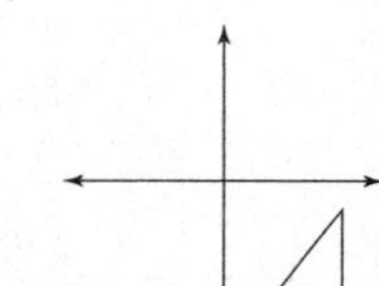

C.

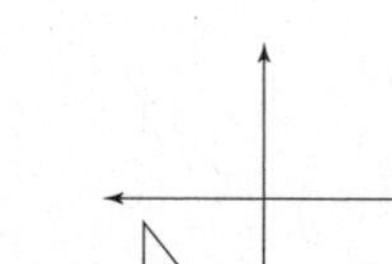

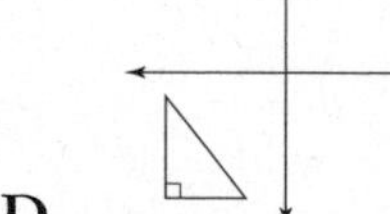

D.

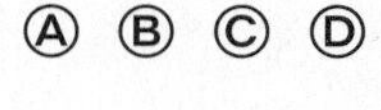

GO ON TO THE NEXT PAGE ➡

24. Which fraction is less than $\frac{1}{2}$?

A. $\frac{7}{10}$

B. $\frac{4}{5}$

C. $\frac{2}{3}$

D. $\frac{1}{3}$

25. If this is a rectangular box, which face is parallel to the front face *ABCD*?

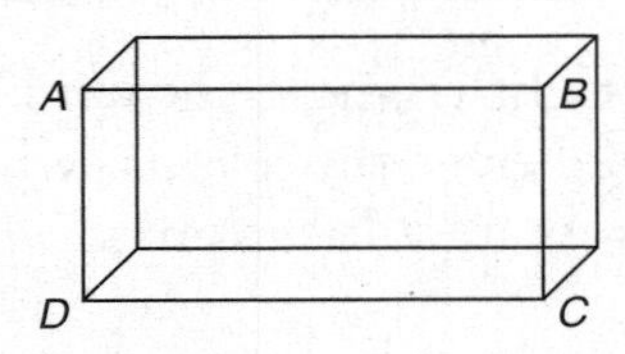

A. top
B. bottom
C. back
D. one of the sides

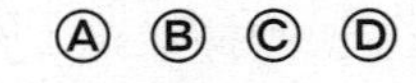

26. The distance from Earth to a nearby planet is 679,000,000 miles away. Use scientific notation to write the distance an astronaut would travel on a roundtrip from Earth to this planet.

A. 1.358×10^9
B. 13.58×10^8
C. 135.8×10^7
D. $1{,}358 \times 10^6$

27. Which of the following is an *outlier* in this data set?

51 35 48 34 65 54
46 38 42 50 52

A. 31
B. 34
C. 44
D. 65

GO ON TO THE NEXT PAGE

DIRECTIONS FOR QUESTION 28: Respond fully to the extended constructed response question that follows. Show your work and clearly explain your answer. You will be graded on the correctness of your method as well as the accuracy of your answer.

CALCULATOR ACTIVE

28. In November a computer costs $1,000. Each month it is reduced by 8%.

- If Allison has $725 saved, in what month will she have enough money to buy that computer?
- Show your work and explain how you arrived at your answer.

Part IV (25 minutes)

Use the *NJ ASK 8 Mathematics Reference Sheet* on page 265.

CALCULATOR ACTIVE

DIRECTIONS FOR QUESTIONS 29 THROUGH 38: Each of the questions or incomplete statements below is followed by four suggested answers. Select the one that is the best in each case, and fill in the corresponding lettered circle. Be sure the circle is filled in completely so you cannot see the letter. Unless you are told to do so in the question, do NOT include sales tax in your answer to questions involving purchases.

29. Cindy bought a coat that was on sale for $249.60. This was the sale price after a 20% discount. What was the original price?

A. $254.59 **B.** $312.00
C. $299.52 **D.** $199.68

Ⓐ Ⓑ Ⓒ Ⓓ

30. What is the student-to-teacher ratio in grade 7 at this New Jersey middle school?

Grade 7 Homeroom Teachers	Boys	Girls	Total Number of Students per Class
Mr. Perkins	10	14	24
Ms. Crowley	12	10	22
Mrs. Williams	14	14	28
Mr. Kakar	10	11	21
Mr. Martin	15	11	26
Mrs. Vargas	10	13	23

A. 1 : 24 **B.** 6 : 144
C. 24 : 16 **D.** 24 : 1

Ⓐ Ⓑ Ⓒ Ⓓ

31. Which list correctly shows the fractions in order from least to greatest?

A. $\frac{1}{4}$ $\frac{1}{2}$ $\frac{2}{3}$ $\frac{5}{6}$

B. $\frac{1}{50}$ $\frac{3}{10}$ $\frac{6}{10}$ $\frac{2}{5}$

C. $\frac{1}{50}$ $\frac{6}{10}$ $\frac{2}{5}$ $\frac{4}{5}$

D. $\frac{2}{3}$ $\frac{5}{6}$ $\frac{5}{9}$ $\frac{7}{8}$

Ⓐ Ⓑ Ⓒ Ⓓ

32. Judy and Janet took their 4 young children to the movies in nearby Washington Township. An adult ticket costs 3 times a child's ticket. If they paid a total of $30.00, what is the cost of each child's ticket?

A. $3.00
B. $4.00
C. $6.00
D. $9.00

Ⓐ Ⓑ Ⓒ Ⓓ

GO ON TO THE NEXT PAGE

33. Miguel exercises 45 minutes a day, 3 days a week. How many hours does Miguel exercise in 6 weeks?

A. 13.5
B. 14
C. 135
D. 810

Ⓐ Ⓑ Ⓒ Ⓓ

---

34. On six tests of 100 points each, Tim has a 90 average. What is the lowest score he could have made on any one of these tests?

A. 40
B. 60
C. 80
D. 90

Ⓐ Ⓑ Ⓒ Ⓓ

---

35. DiDi and John each drive 60 mph. DiDi drives for 4 hours and John drives for 2 hours. How many more miles does DiDi drive than John?

A. 120 miles
B. 130 miles
C. 140 miles
D. 150 miles

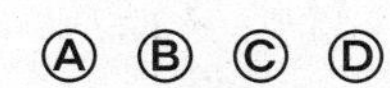

36. In the drawing shown, there is a large triangle and a small triangle. These two triangles are *similar.* What proportion would you use to find the value of x (the base of the small triangle)?

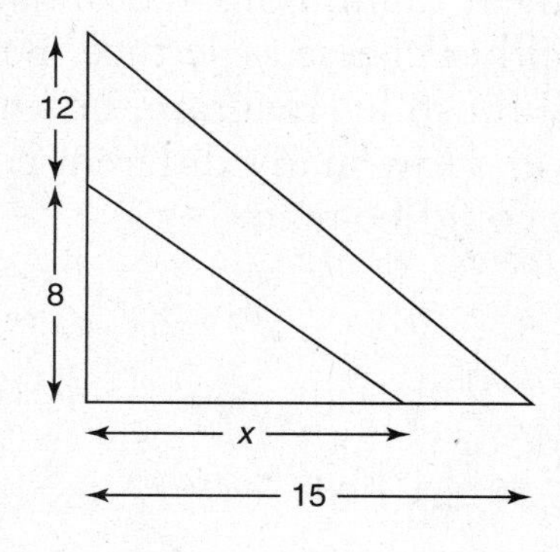

A. $\frac{8}{12} = \frac{x}{15}$

B. $\frac{-8}{x} = \frac{20}{15}$

C. $\frac{x}{8} = \frac{3}{4}$

D. $\frac{8}{20} = \frac{x}{15}$

Ⓐ Ⓑ Ⓒ Ⓓ

37. The owner of Marc's Deli asked David to make a tree diagram to see how many different combinations of sandwiches could be made from the following choices: one bread (roll or rye), one meat (ham, bologna, or salami), and either cheese or lettuce. Before he could finish his diagram, he knew his answer. How many different combinations could be made?

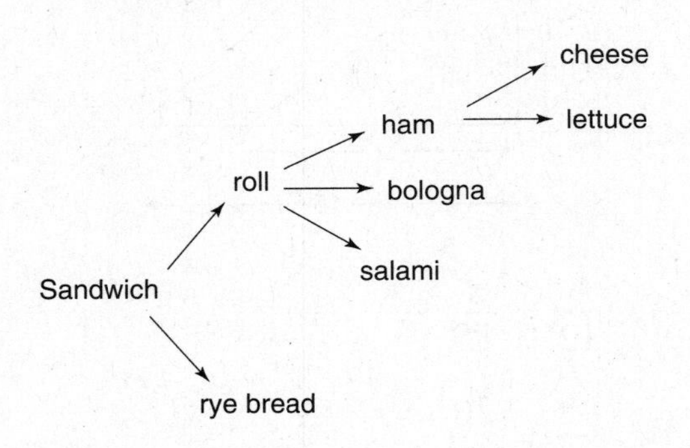

A. 18
B. 12
C. 9
D. 6

Ⓐ Ⓑ Ⓒ Ⓓ

38. On a map, $\frac{1}{4}$ inch represents 16 feet.

If a driveway is 40 feet long, what is its length, in inches, on the map?

A. $\frac{3}{8}$ inches

B. $\frac{5}{8}$ inches

C. $\frac{3}{4}$ inches

D. $2\frac{1}{2}$ inches

GO ON TO THE NEXT PAGE

DIRECTIONS FOR QUESTION 39: Respond fully to the extended constructed response question that follows. Show your work and clearly explain your answer. You will be graded on the correctness of your method as well as the accuracy of your answer.

39. Geometry: Use the diagram of a boxed present below to answer the following questions.

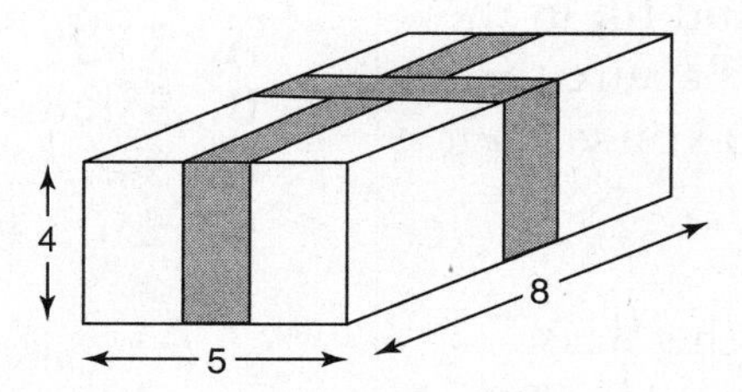

- What is the volume of this box?
- What is the surface area of this box?
- How long a piece of ribbon is needed to go around the box as shown? Assume there is no overlapping.

Part V (25 minutes)

Use the *NJ ASK 8 Mathematics Reference Sheet* on page 265.

CALCULATOR ACTIVE

DIRECTIONS FOR QUESTIONS 40 THROUGH 49: Each of the questions or incomplete statements below is followed by four suggested answers. Select the one that is the best in each case, and fill in the corresponding lettered circle. Be sure the circle is filled in completely so you cannot see the letter.

40. There are 28 students in the class. Their grades on the last test showed that five students scored 90, fifteen students scored 85, six students scored 78, and two students scored 70. Approximately what percent of students scored 85 on the test?

A. 18%
B. 15%
C. 54%
D. 71%

Ⓐ Ⓑ Ⓒ Ⓓ

41. What is the probability that it will snow on Sunday and Tuesday, according to the following January weather forecast chart?

Day of Week	Sunday	Monday	Tuesday	Wednesday
Probability of snow	50%	20%	10%	5%

A. 5%
B. 10%
C. 40%
D. 60%

Ⓐ Ⓑ Ⓒ Ⓓ

42. A new clothing store sells four pairs of socks for $16.88. If you bought 11 pairs of socks, how much change would you get back from $50.00?

A. $3.58
B. $3.68
C. $46.42
D. $50.64

Ⓐ Ⓑ Ⓒ Ⓓ

43. At a local middle school, 131 of the 7th graders were on the high honor roll in June last year. If there were a total of 616 seventh graders at the school, approximately what percentage of 7th grade students were **not** on the high honor roll?

A. 20%
B. 21%
C. 79%
D. 89%

Ⓐ Ⓑ Ⓒ Ⓓ

44. John needs to find the approximate square root of 60. What information will help him make a good estimate?

A. If he knows the value of 6^2 and 7^2
B. If he knows the value of 7^2 and 8^2
C. If he knows that 36 and 81 are perfect square numbers
D. If he knows that half of 60 = 30

Ⓐ Ⓑ Ⓒ Ⓓ

GO ON TO THE NEXT PAGE

45. In the triangle below, which is the smallest angle?

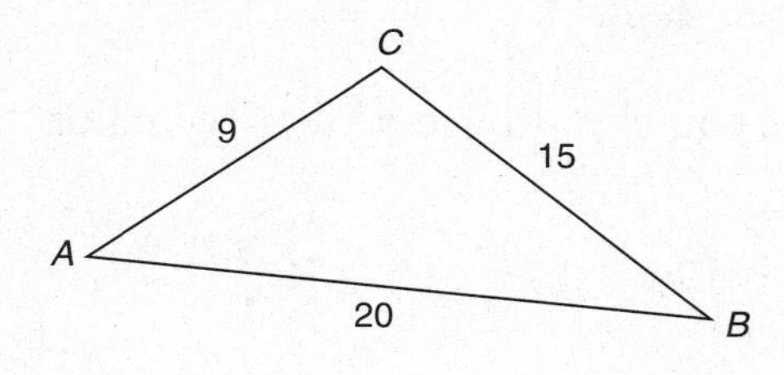

(Figure not drawn to scale.)

A. Angle *A*
B. Angle *B*
C. Angle *C*
D. Not enough information given

46. Michelle works 42 hours per week and earns $15.45 per hour. Which is the best estimate of her gross pay per year?

A. $45,420
B. $16,871
C. $33,743
D. $32,445

47. If you were to measure the length of your foot, which of the following would be the most appropriate to use?

A. meters
B. centimeters
C. grams
D. millimeters

48. This year's 9th grade class has 220 students. Each student was able to select an elective if they were not assigned a remedial course. Using the chart below, approximately what percentage of students selected art or cooking?

Number of Students	Elective Course
75	Art
40	Cooking
80	Music
20	Remedial Studies

A. 7%
B. 16%
C. 53.5%
D. 57.5%

Ⓐ Ⓑ Ⓒ Ⓓ

49. The table below shows a relationship between x and y. Which value for y completes the table correctly?

x	−1	0	1	2	3	4
y	−5	−3	−1	1	3	?

A. 4
B. 5
C. 6
D. 0

GO ON TO THE NEXT PAGE ➡

DIRECTIONS FOR QUESTION 50: Respond fully to the extended constructed response question that follows. Show your work and clearly explain your answer. You will be graded on the correctness of your method as well as the accuracy of your answer.

50. For this extended constructed response question, you should use a ruler to make straight lines.

- Make a stem-and-leaf plot for the data listed below.

Ages of Middle School Principals in One New Jersey County								
52	55	38	42	52	50	36	38	45
48	42	59	56	41	60	40	42	50
45	58	40	51	59	37	35	46	49

- Now, make a second stem-and-leaf plot with the numbers in numerical order.
- What is the median age of middle school principals in this New Jersey County?
- What is the mean age of these principals?
- What is the mode of their ages?

Part VI (19 minutes)

Use the *NJ ASK 8 Mathematics Reference Sheet* on page 265.
CALCULATOR ACTIVE

DIRECTIONS FOR QUESTIONS 51 THROUGH 56: Each of the questions or incomplete statements below is followed by four suggested answers. Select the one that is the best in each case, and fill in the corresponding lettered circle. Be sure the circle is filled in completely so you cannot see the letter. Unless you are told to do so in the question, do NOT include sales tax in your answer to questions involving purchases.

51. If isosceles triangle $A \cong$ isosceles triangle B, what is the area of triangle B?

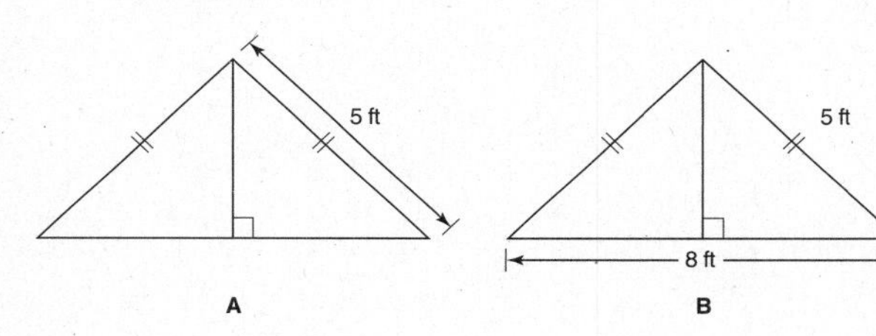

A. 15 square feet
B. 6 square feet
C. 24 square feet
D. 12 square feet

Ⓐ Ⓑ Ⓒ Ⓓ

52. A group of students were hiking for an hour, they stopped to rest for about $\frac{1}{2}$ hour, and then continued hiking for another hour. Which graph best describes this?

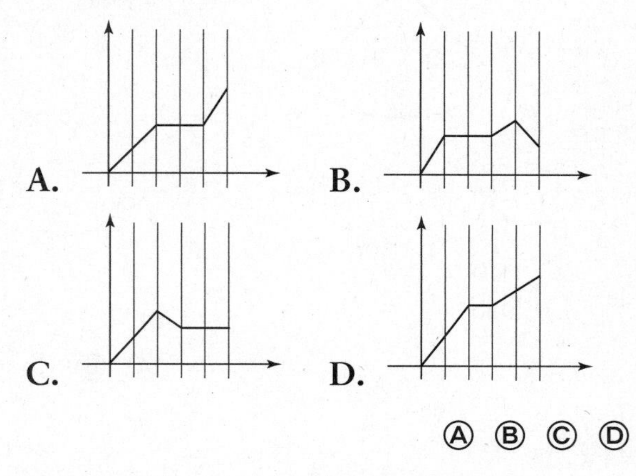

Ⓐ Ⓑ Ⓒ Ⓓ

53. These are Joey's grades for this marking period: 84, 92, 90, 82, 87. What would he have to get on his next test to have a mean average of 88?

A. 90
B. 92
C. 93
D. 95

Ⓐ Ⓑ Ⓒ Ⓓ

54. Estimate to find the approximate value of $\sqrt{79}$.

A. about 7.5
B. almost 8
C. greater than 9
D. between 8 and 9

Ⓐ Ⓑ Ⓒ Ⓓ

GO ON TO THE NEXT PAGE 

55. If I throw a dart at the dartboard shown, what is the probability that the dart will land in a red or yellow space?

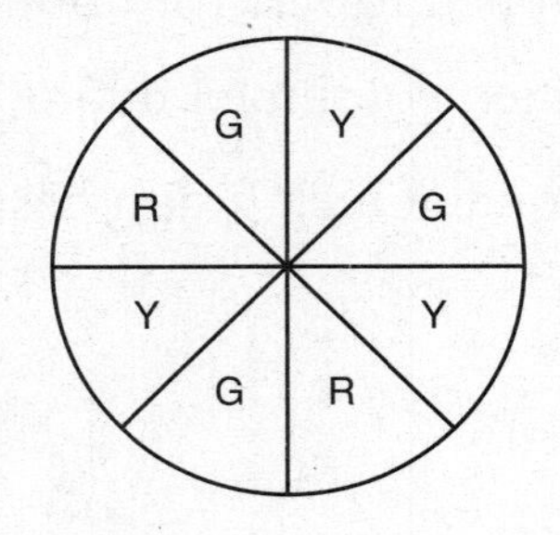

R = red
G = green
Y = yellow

A. 2:3
B. 3:5
C. 5:6
D. 5:8

Ⓐ Ⓑ Ⓒ Ⓓ

56. A computer randomly generates a 5-character serial number that must contain two letters followed by three single-digit numbers. How many serial numbers can be generated? (Letters and numbers can be repeated.)

A. 78
B. 82
C. 468,000
D. 676,000

GO ON TO THE NEXT PAGE

DIRECTIONS FOR QUESTION 57: Respond fully to the extended constructed response question that follows. Show your work and clearly explain your answer. You will be graded on the correctness of your method as well as the accuracy of your answer.

57. In the figure below, four congruent circles fit side by side in a rectangle.

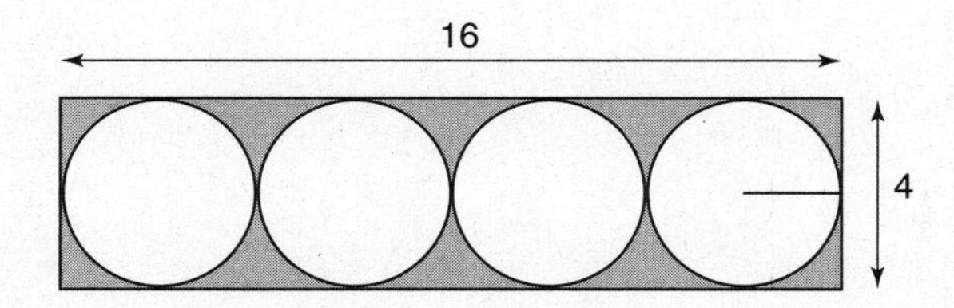

- How long is the radius of each circle?
- What is the area of each circle? (Use 3.14 for π.)
- What is the area of the rectangle?
- Describe how you would find the area of the shaded region. What is its area?

SOLUTIONS TO PRACTICE TEST 1

PART I

Short Constructed Response Questions

1. *BC* is longer   *AB* goes from –1 to $\frac{-1}{3}$ and is only $\frac{2}{3}$ units long

   BC goes from 1 to $\frac{-1}{3}$; (from 1 to 0 is 1 unit long and from 0 to $\frac{-1}{3}$ is $\frac{1}{3}$ units long), so *BC* is 1 + $\frac{1}{3}$ units long, or $1\frac{1}{3}$ units long.

2. 9   The prime factors of 24 are 2, 2, 2, 3; (2)(2)(2)(3) = 24; the sum is 2 + 2 + 2 + 3 = 9.

3. 16   Total perimeter = sum of all 4 sides. See diagram below.

   $50 - 2(9) = 50 - 18 = 32$

   $\frac{32}{2} = 16 =$ length side *a*

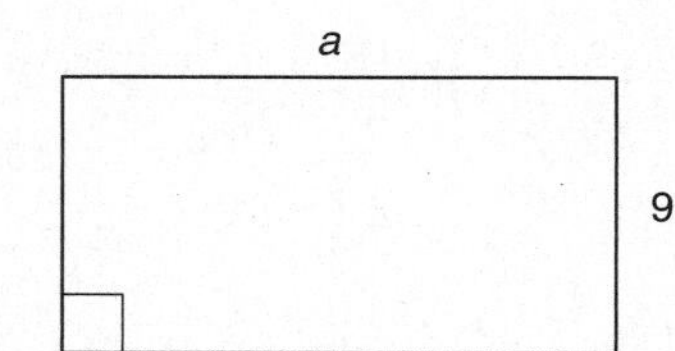

4. 56 sq. inches is the area of the rectangle. The sides of the rectangle are 14, 14, 4, and 4 since you know the perimeter is 36 and one side is 14. Therefore, the area is (14)(4), or 56 sq. inches.

5. $\frac{1}{3}$ 4 scores were less than 50; $\frac{\text{scores less than 50}}{\text{total number of scores}} = \frac{4}{12} = \frac{1}{3}$

72	39	43
29	72	50
53	59	67
69	42	63

6. a trapezoid

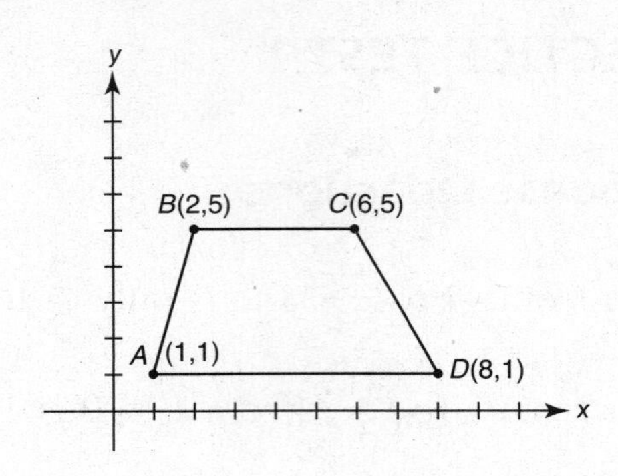

7. 4 – 3 Work inside parentheses first. Remember the algebraic order of operations.
8. There are four other possible arrays: 4×9, 3×12, 2×18, or 1×36.
9. "Z" would be in the 100th place. PIZZAPIE contains 8 letters; 100/8 = 12. (8)(12) = 96, with four letters remaining, "P," "I," "Z," and the 100th letter, which would be "Z."
10. $x = 24$

$$52 = 2x + 4$$
$$-4 \qquad -4$$
$$\frac{48}{2} = \frac{2x}{2}$$
$$24 = x$$

PART II

Multiple-Choice Questions

11. B 131 $3 + 4 \times 2(4)^2 = 3 + 4 \times 2(16) = 3 + 8(16) = 3 + 128 = 131$
Use the correct order of operations.
12. C An acute angle is less than 90°.
13. A 253 $5 \times 2 + 3 = 13$; $13 \times 2 + 3 = 29$; $29 \times 2 + 3 = 61$; $125 \times 2 + 3 = 253$
14. D 18 Subtract 69 from both sides and you get $n = 18$.
15. B −4 $2a + b^2 - 2(c + 14) = 2(-2) + (6)^2 - 2(4 + 14) = -4 + 36 - 2(18) = 32 - 36 = -4$
16. D 13 The absolute value of a number is always positive. $|-4| + |9| = 4 + 9 = 13$
17. B A circuit begins and ends at the same point.
18. B $x = 1$

$$2x - 4(x - 3) = 10$$
$$2x - 4x + 12 = 10; \qquad -2x + 12 = 10$$
$$-12 \quad -12$$
$$\frac{-2x}{-2} = \frac{-2}{-2}$$
$$x = 1$$

Extended Constructed Response Question

19. (See the completed chart that follows.)

- 5 hours — The water in both pools will be 30 inches high after 5 hours.
- 9.5 hours — The water in Mrs. Carol's pool will reach 48 inches or 4 feet in 9.5 hours.
- 13 hours — The water in Mrs. Brown's pool will reach 48 inches in 13 hours.
- The point where the two lines intersect is when the water in both pools is the same height.

	Mrs. Brown's Pool B Show work	Pool B, Height of Water (inches)	Mrs. Carol's Pool C Show work	Pool C, Height of Water (inches)
Now		20		10
Hour 1	20 + 2 =	22	10 + 4 =	14
Hour 2	22 + 2 =	24	14 + 4 =	18
Hour 3	24 + 2 =	26	18 + 4 =	22
Hour 4	26 + 2 =	28	22 + 4 =	26
Hour 5	**28 + 2 =**	**30**	**26 + 4 =**	**30**
Hour 6	30 + 2 =	32	30 + 4 =	34
Hour 7	32 + 2 =	36	34 + 4 =	38
Hour 8	36 + 2 =	38	38 + 4 =	42
Hour 9	38 + 2 =	40	42 + 4 =	46*
Hour 10	40 + 2 =	42	46 + 4 =	50*
Hour 11	42 + 2 =	44		
Hour 12	44 + 2 =	46		
Hour 13	**46 + 2 =**	**48**		
Hour 14				

*Note 9.5 hr = 48 in. or 4 ft

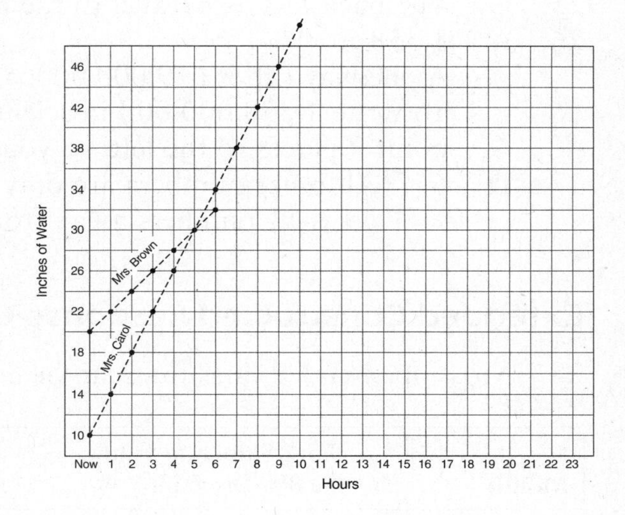

PART III

Multiple-Choice Questions

20. C 6 prime factors $2 \times 2 \times 2 \times 2 \times 2 \times 2 = 64$

21. D $\frac{5}{12}$ $\frac{\text{5 numbers are prime}}{\text{12 numbers in all}}$ 1 (2) (3) 4 (5) 6 (7) 8 9 10 (11) 12

22. B $\frac{1}{2}$ $\frac{\text{The number of 2s}}{\text{The total number of choices}} = \frac{4}{8} = \frac{1}{2}$

23. A A reflection is flipped. The new image looks like a mirror-image of the original shape. In this case it is flipped over the x-axis into the fourth quadrant.

24. D $\frac{1}{3} = 0.\overline{333}$ and $\frac{1}{2} = 0.50$

25. C The back face is parallel to the front face. They never touch.

26. A 1.358×10^9
A one-way trip is 679,000,000; a round trip is $679{,}000{,}000 \times 2 = 1{,}358{,}000{,}000$. To write 1,358,000,000 in scientific notation form, you move the decimal point 9 places to the left, so you have 1.358×10^9.

27. D 65 All other numbers are only 1 to 4 numbers away from any other number; 65 is 11 numbers away from its closest number, 54.

Extended Constructed Response Question

28. Most open-ended questions can be answered with a table. Study this one carefully.

Month	Work Shown to Find the 8% Discount	Last Month's Price Less the 8% Discount	New Price
November			$1,000.00
December	$1000 × 0.08 = $ 80.00	$1,000 – 80 = $920.00	$920.00
January	$ 920 × 0.08 = $ 73.60	$920 – 73.60 = $846.40	$846.40
February	$ 846.40 × 0.08 = $ 67.71	$846.40 – 67.71 = $778.69	$778.69
March	$ 778.69 × 0.08 = $ 62.30	$778.69 – 62.30 = $716.39	$716.39

From the chart, you can see that Allison can buy the computer in March. In March it would cost only $716.39, and she has $725.

PART IV

Multiple-Choice Questions

29. **B** $312.00 The sales price is 80% of the regular price. Let x = regular price, so 80% of x = $249.60, or $(.80)(x) = 249.60$; then $\frac{(.80)(x)}{.80} = \frac{249.60}{.80}$ and $x = 312$

30. **D** $\frac{24}{1}$ or 24 : 1 $\frac{\text{total students}}{\text{total teachers}} = \frac{144}{6} = \frac{24}{1}$

31. **A** $\frac{1}{4}\ \frac{1}{2}\ \frac{2}{3}\ \frac{5}{6}$ written as decimal numbers is 0.25 0.5 0.66 0.83

32. **A** $3.00 Let x represent a child's ticket. Then write an equation with the information you have.
$4x$ = (4 children's tickets) + $2(3)x$ = (2 adult tickets) = $30.00
$4x + 2(3)x = \$30.00$, $4x + 6x = \$30.00$, $10x = \$30$, $x = \$3.00$

33. **A** 13.5 hours (0.75)(3) = 2.25 hours/week; (2.25)(6) = 13.5 hours in 6 weeks.

34. **A** 40 points $90 \times 6 = 540$ points = total points he received on all six of his tests. The maximum he could get on five tests would be 5(100 points) = 500 points. Therefore, the lowest he could have scored on the 6th test would be 540 – 500, or 40 points.

35. **A** (60)(2) = 120 miles or (60(4) – (60)(2) = 120 miles

36. **D** Similar triangles are in proportion. Note that the height of the large triangle is 20.

$$\frac{\text{side of small triangle}}{\text{side of large triangle}} = \frac{\text{height of small triangle}}{\text{height of large triangle}} = \frac{8}{20} = \frac{x}{15}$$

37. **B** 12 combinations are possible, six with roll and six with rye bread.

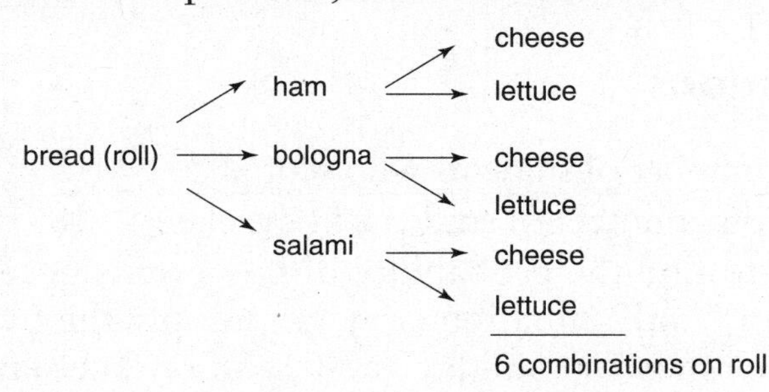

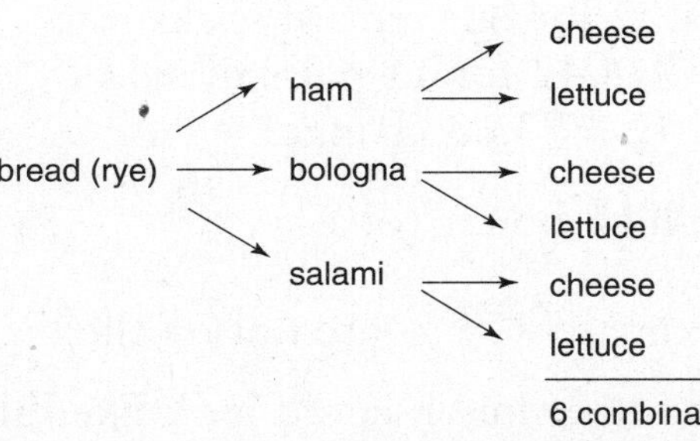

38. $\frac{5}{8}$ inches Set up a proportion: $\frac{0.25 \text{ inches}}{16 \text{ feet}} = \frac{x}{40 \text{ feet}}$

$$16(x) = (.25)(40)$$
$$16x = 10$$
$$x = \frac{10}{16} \text{ or } \frac{5}{8} \text{ inch}$$

Extended Constructed Response Question

39. ■ Volume is 160 cubic inches Volume = (Length)(Width)(Depth) = (4)(5)(8) = 160

■ Surface area is 184 square inches.

Area of top and bottom	2(4)(8) =	64 square inches
Area of front and back	2(4)(5) =	40 square inches
Area of sides	2(5)(8) =	80 square inches
Total surface area		184 square inches

■ Total length of ribbon needed is 42 inches long.

Length of one ribbon	(4 + 5 + 4 + 5) = 18 in. long
Length of second ribbon	(4 + 8 + 4 + 8) = 24 in. long

Total length = 18 + 24 = 42 inches long

PART V

Multiple-Choice Questions

40. C ~54% $\frac{\text{The number of students who scored 85}}{\text{The total number of students in the class}} = \frac{15}{28} = .5357 \sim 54$

41. A When you want to find the probability of two events occurring at the same time, you have an "and" situation, so you multiply the two decimal numbers. The probability of snow on Sunday is .50, and on Tuesday it is .10. The probability of snow on both days is (.50)(.10) = .05 or 5%.

42. B $3.68 $16.88/4 = $4.22 for one pair of socks
($4.22)(11) = $46.42 for 11 pairs of socks
$50.00 – $46.42 = $3.68 change

43. C $\frac{131}{616}$ = 21.2% are on the high honor roll.

100% – 21.2% = approximately 79% **are not** on the high honor roll.

44. B $7^2 = 49$; $8^2 = 64$. On a number line it could look like this: $\sqrt{49}$ $\sqrt{60}$ $\sqrt{64}$

45. B The smallest angle is angle *B* because it is oppositve the shortest side.

46. C $33,743 42 hours × 52 weeks in year × $15.42 per hour = 42 × 52 × 15.42 = 33,742.8

47. B centimeters One inch is approximately $2\frac{1}{2}$ centimeters. A meter is too long, it measures 39 inches (longer than a yard). A gram is something you use to measure how much something weighs; a millimeter is too small for this measurement.

48. C 53.5% 75 (art) + 40 (cooking) = 115; $\frac{115}{215}$ = .535 = 53.5%

49. B 5

Extended Constructed Response Question

50. ■ Stem-and-Leaf plot of the given information:

Stem	Leaves
3	5 6 7 8 8
4	0 0 1 2 2 2 5 5 6 8 9
5	0 0 1 2 2 5 6 8 9 9
6	0

- ■ The median age of the middle school principals is 46 years old. (The number in the middle when all of the ages are listed in numerical order.)
- ■ The mean age of the principals is 46.88, or approximately 47 years old. (All the ages added together = 1,266. 1,266 ÷ 27 = 46.888 or 47.)
- ■ The mode is 42. The mode is the number that appears most often in the data set; the age 42 appears most often.

PART VI

Multiple-Choice Questions

51. D 12 sq. ft. Since the two triangles are congruent, their corresponding sides are the same length. The height of each triangle is 3 feet (since the two smaller triangles are 3–4–5 right triangles). The area of the triangle is $\frac{(base)(height)}{2} = \frac{(8)(3)}{2} = \frac{24}{2}$ = 12 sq. ft.

52. D

53. C 93 To have an 88 average in 6 tests would mean he would have (88)(6) or 528 points in all. He already has 84 + 92 + 90 + 82 + 87 or 435 points. On his sixth test he would need 528 – 435 = 93 points.

54. D $\sqrt{79}$ is between $\sqrt{64}$ and $\sqrt{81}$, which means it is between 8 and 9.

55. D 5:8 There are 5 spaces that are red or yellow and 8 spaces in all.

56. D 676,000 You can generate 676,000 different serial numbers, where the first two characters are letters and the next three characters are single-digit numbers.

$$(26)(26)(10)(10)(10) = 676{,}000$$

First Letter	Second Letter	First Number	Second Number	Third Number
26 choices	26 choices	10 choices	10 choices	10 choices
(A–Z)	(A–Z)	(0–9)	(0–9)	(0–9)

Extended Constructed Response Question

57.
- The diameter of each circle is $\frac{16}{4} = 4$. Therefore the radius of each circle = 2.
- The formula for the area of a circle is πr^2. The area of each circle is $(3.14)(2^2) = (3.14)(4) \approx 12.56$.

  The area of all four circles is about (12.56)(4) = 50.24 sq. units.
- The area of the rectangle is (length)(width) = (16)(4) = 64 sq. units.
- To find the area of the shaded area, subtract the area of the 4 circles from the area of the rectangle. $64 - 50.24 \approx 13.76$ sq. units. The area of the shaded region is about 13.76 sq. units.

Name: ______________________________ Date: ______________

Practice Test 2
Part I (20 minutes)

Use the *NJ ASK 8 Mathematics Reference Sheet* on page 265.
NO CALCULATOR PERMITTED

DIRECTIONS FOR QUESTIONS 1 THROUGH 10: Write your answer neatly on the line provided. No partial credit is given for these questions.

1. In a math classroom of 25 students, 10 of the students are boys. What is the ratio of girls to boys in this class?

Answer: __________

2. On the number line shown, what is the coordinate of point W?

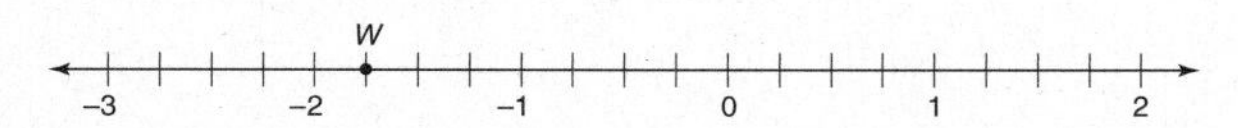

Answer: __________

3. If the area of the rectangle drawn below is 60 sq. inches, what is the perimeter of this rectangle?

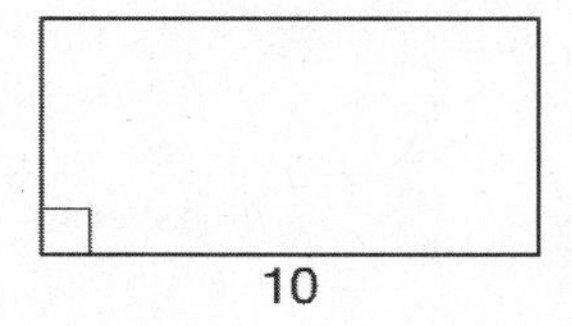

Answer: __________

4. Carly has 6 red marbles, 4 green marbles and 2 yellow marbles in a bag. She picks one marble without looking. Which of the choices below has the highest probability of happening:

A. The probability of choosing a yellow marble
B. The probability of choosing a green or a yellow marble
C. The probability of choosing a green marble
D. The probability of choosing a red or a yellow marble

Ⓐ Ⓑ Ⓒ Ⓓ

5. A new home magazine shows a scale drawing of an apartment. In the drawing, 2 inches represents 10 feet in the actual apartment. If the main bedroom in the apartment is actually 15 feet long, how long should it be on the scale drawing?

Answer: __________

GO ON TO THE NEXT PAGE

6. Which solution is the largest integer?

A. The surface area of the cube shown

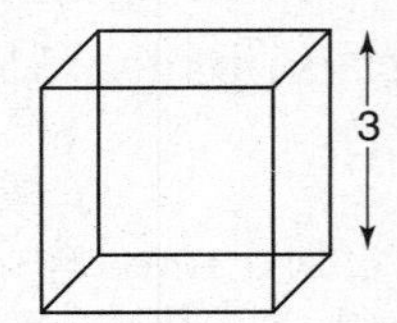

(Figure not drawn to scale.)

B. The volume of the rectangular prism shown.

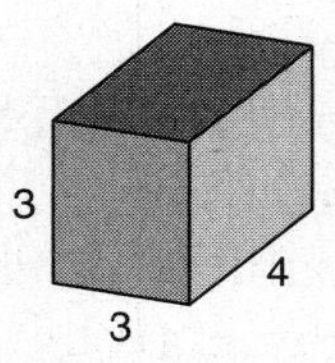

(Figure not drawn to scale.)

C. The volume of the right cylinder shown with radius = 2 and height = 4.

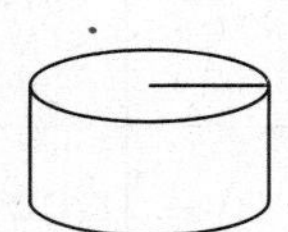

(Figure not drawn to scale.)

D. The perimeter of the isosceles triangle shown.

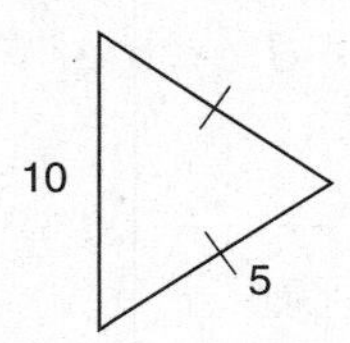

(Figure not drawn to scale.)

Ⓐ Ⓑ Ⓒ Ⓓ

7. Lauren plans to study for 12 days before her final exam. She studied 15 minutes on the first day and will increase this by 15 minutes each day. At this rate, how many minutes will she study on the 12th day?

Answer: __________

8. If the measure of two angles of a triangle are 23° and 56°, what is the measure of the third angle?

Answer: __________

9. If $x = -2$, what is the value of the following expression?

$$3(x - 4) + 10$$

Answer: __________

10. In the figure drawn below, what is the area of the unshaded portion?

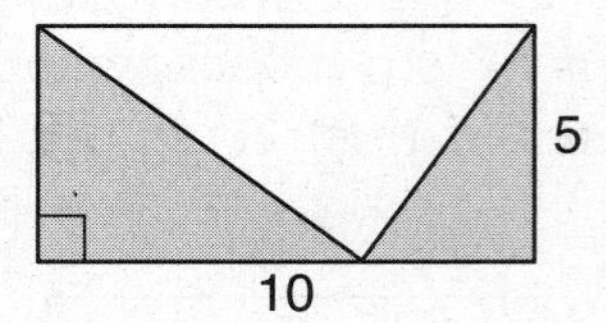

Answer: __________

Part II (22 minutes)

Use the *NJ ASK 8 Mathematics Reference Sheet* on page 265.

NO CALCULATOR PERMITTED

DIRECTIONS FOR QUESTIONS 11 THROUGH 18: Each of the questions or incomplete statements below is followed by four suggested answers. Select the one that is the best in each case, and fill in the corresponding lettered circle. Be sure the circle is filled in completely so you cannot see the letter. Unless you are told to do so in the question, do NOT include sales tax in your answer to questions involving purchases.

11. Which equation represents the perimeter of the rhombus shown?

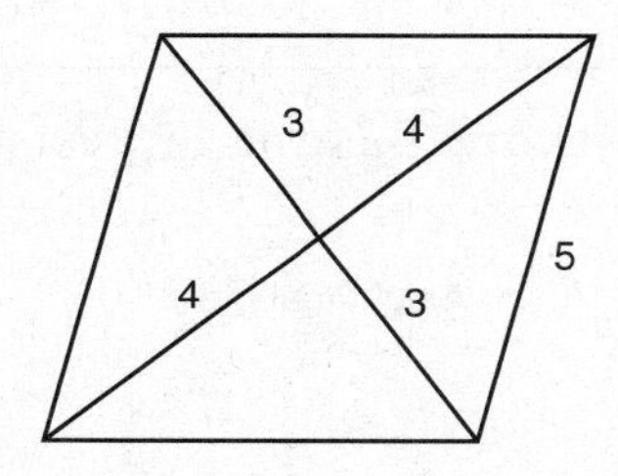

A. $5 \times 4 = 20$
B. $6 + 8 = 14$
C. $(6)(8) = 48$
D. $\frac{(6)(8)}{2} = 24$

---

12. Which expression results in the 8th term in the sequence shown below?

$-8 \quad -3 \quad 2 \quad \_\_ \quad \_\_ \quad \_\_ \quad \_\_ \quad \_\_$

A. $6(2^2)$
B. $3(3^2)$
C. 5^2
D. $(15)(2)$

13. Which diagram below illustrates a and b as vertical angles?

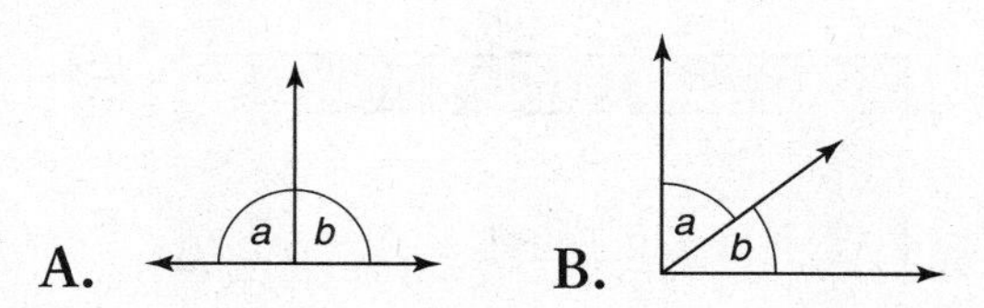

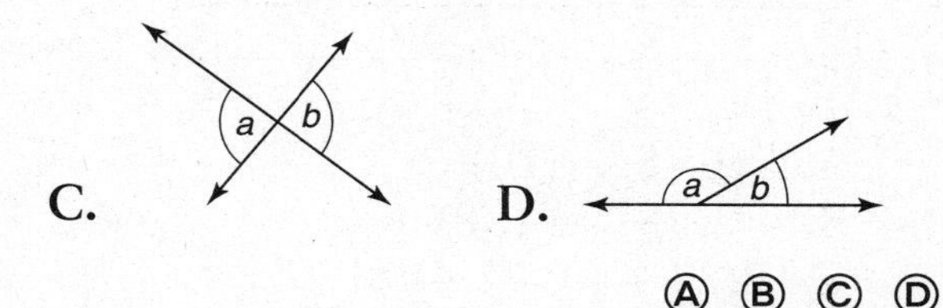

Ⓐ Ⓑ Ⓒ Ⓓ

---

14. What value of x makes the equation true?

$$2(8) + x = 69$$

A. 53
B. 51
C. 85
D. 54

Ⓐ Ⓑ Ⓒ Ⓓ

GO ON TO THE NEXT PAGE ➡

15. There is a two-foot walkway around the central garden area in the diagram shown. Which expression represents the area of the walkway around this garden?

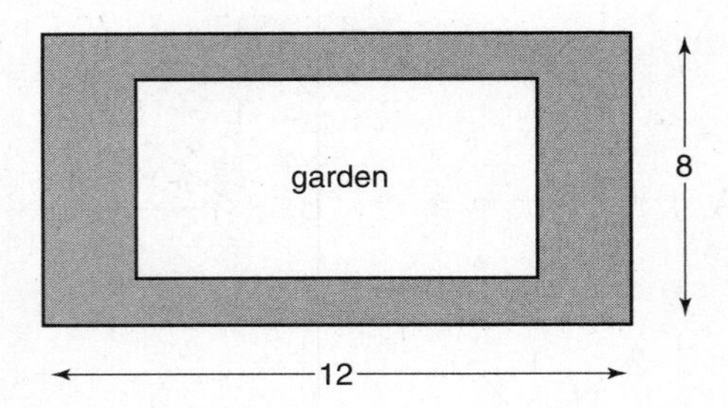

A. (10)(8) – (10)(6)
B. (12)(8) – (4)(8)
C. 2(24 + 8) – 2(4)(8)
D. 10×6

Ⓐ Ⓑ Ⓒ Ⓓ

---

16. Which answer does not match the others?

A. $(\sqrt{100})(2)$
B. $\sqrt{400}$
C. $5(-2)^2$
D. $5^2 - 2^2$

Ⓐ Ⓑ Ⓒ Ⓓ

---

17. The spinner shown will most likely land on which of the following?

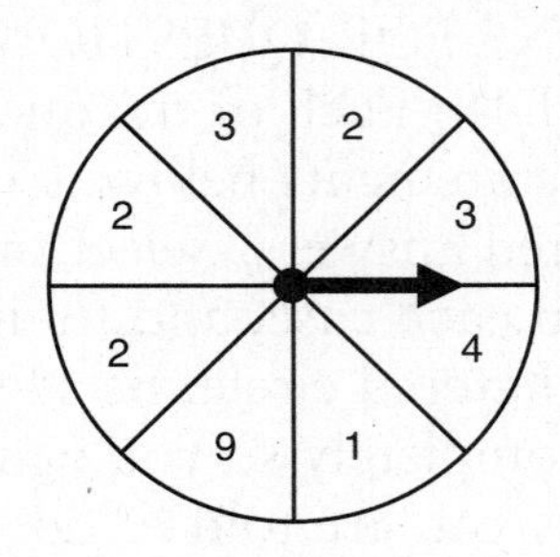

A. A factor of 24
B. An even number
C. A prime number
D. A perfect square number

---

18. Which shape has the largest area?

A.
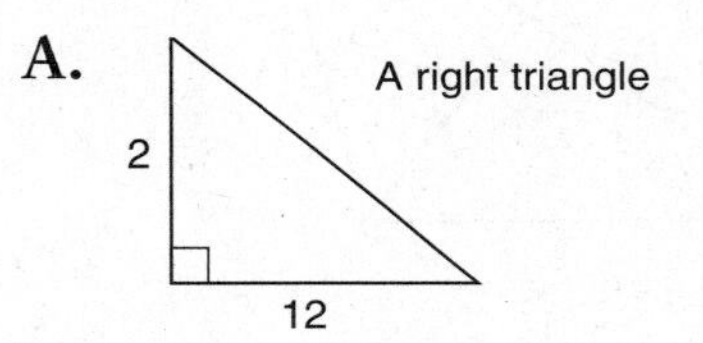

B.
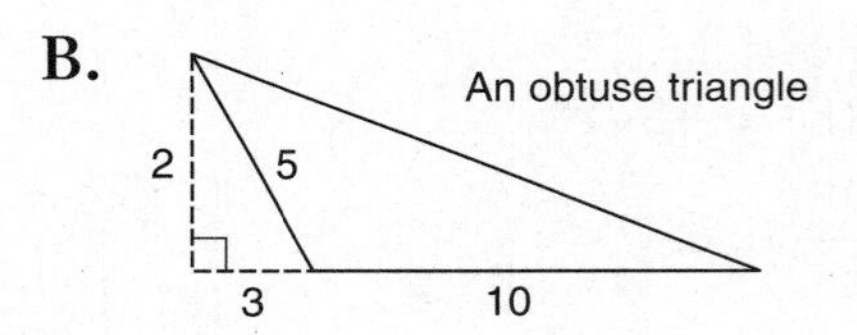

C.
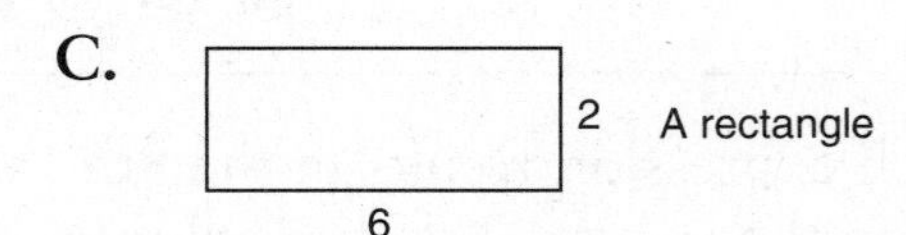

D.
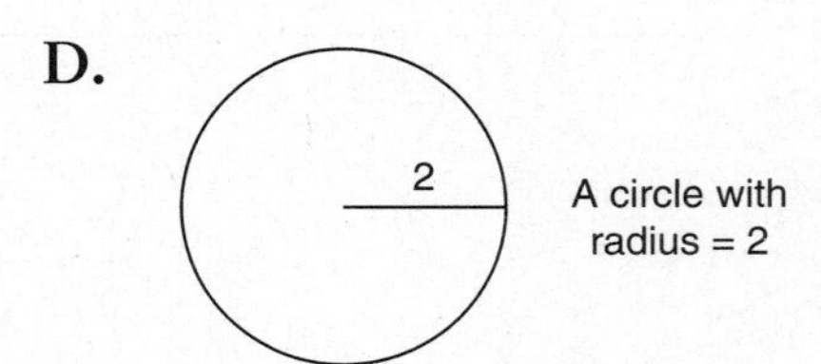

Ⓐ Ⓑ Ⓒ Ⓓ

GO ON TO THE NEXT PAGE ➡

DIRECTIONS FOR QUESTION 19: Respond fully to the extended constructed response question that follows. Show your work and clearly explain your answer. You will be graded on the correctness of your method as well as the accuracy of your answer.

19. Robbie and Anna's new house has a rectangular room that is 14 feet long and 12 feet wide.

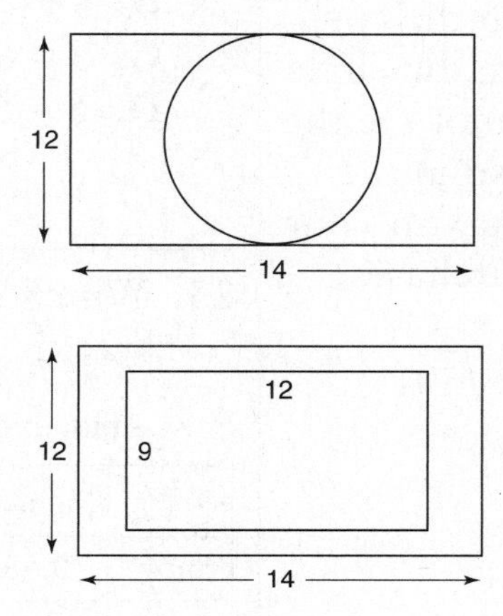

- What is the area of the largest round rug they can fit in this room?
- They are thinking of buying a rectangular area rug instead. This rug measures 9 feet by 12 feet and is priced at $22 a square yard. What is the price for this rug before taxes?
- Since there is a 6% sales tax on home furnishings, what is their actual total cost for the rectangular rug?

Part III (22 minutes)

Use the *NJ ASK 8 Mathematics Reference Sheet* on page 265.

NO CALCULATOR PERMITTED

DIRECTIONS FOR QUESTIONS 20 THROUGH 27: Each of the questions or incomplete statements below is followed by four suggested answers. Select the one that is the best in each case, and fill in the corresponding lettered circle. Be sure the circle is filled in completely so you cannot see the letter. Unless you are told to do so in the question, do NOT include sales tax in your answer to questions involving purchases.

20. Estimate the sum of $\sqrt{24} + \sqrt{50}$.

A. $5 + 7 = 12$
B. $12 + 25 = 37$
C. $\sqrt{69} = 13$
D. $6 + 10 = 16$

Ⓐ Ⓑ Ⓒ Ⓓ

21. Allie and her brother Jeremiah are having a disagreement. Allie says that if she tosses a coin 100 times, she has a better chance of it landing on *heads* than if she tosses it only 50 times. Jeremiah says that is not true. The chances are the same. Who is correct?

A. Allie
B. Jeremiah
C. neither
D. both

Ⓐ Ⓑ Ⓒ Ⓓ

22. *Which* algebraic express says 5 less than the product of a number and 6?

A. $6 + n - 5$
B. $6n - 5$
C. $5 - 6n$
D. $5 - 6 + n$

Ⓐ Ⓑ Ⓒ Ⓓ

23. Which is most likely to happen next week?

Amount of Precipitation This Week (in inches)

	Monday	Tuesday	Wednesday	Thursday	Friday
Snow	0.50	0.60	0.20	0.10	0
Sunny	0.10	0.05	0.50	0.70	1.0

A. It will snow on Monday and Tuesday.
B. It will snow on Monday or Tuesday.
C. It will be sunny on Thursday.
D. It will snow on Friday.

Ⓐ Ⓑ Ⓒ Ⓓ

GO ON TO THE NEXT PAGE

24. Which fraction could be a value for y?

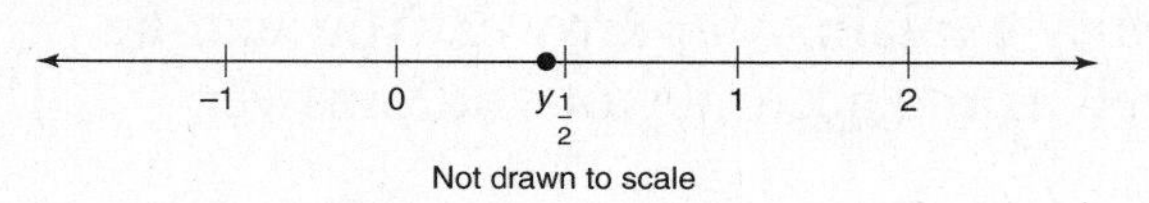

Not drawn to scale

A. $\frac{98}{100}$

B. $\frac{5}{8}$

C. $\frac{3}{4}$

D. $\frac{2}{5}$

Ⓐ Ⓑ Ⓒ Ⓓ

25. This is a rectangular box; which face is NOT perpendicular to the top?

A. top
B. bottom
C. back
D. one of the sides

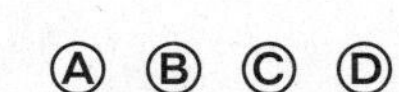

26. Use the frequency table drawn below. How many students scored higher than 29 points?

Scores on Tuesday's Quiz	
Intervals	**Frequency**
1–9	
10–19	3
20–29	5
30–39	8
40–49	10
50–59	2

A. 12
B. 20
C. 25
D. not enough information given

Ⓐ Ⓑ Ⓒ Ⓓ

27. A number was rounded to 22.7. Which of the following could have been the number before it was rounded?

A. 22.07
B. 22.648
C. 22.609
D. 22.651

Ⓐ Ⓑ Ⓒ Ⓓ

GO ON TO THE NEXT PAGE ➡

DIRECTIONS FOR QUESTION 28: Respond fully to the extended constructed response question that follows. Show your work and clearly explain your answer. You will be graded on the correctness of your method as well as the accuracy of your answer.

28. The yearly family-membership to a local swim club in Washington Township is $350 per year for residents. Nonmembers pay $12.00 per visit.

    - What is the least number of times a family of four should go to the pool to make the family membership worthwhile?
    - If grandma was a guest visitor 5 times last summer, what would she have paid for her visits if senior citizens received a 10% discount on the nonmember rate?

Part IV (25 minutes)

Use the *NJ ASK 8 Mathematics Reference Sheet* on page 265.
CALCULATOR ACTIVE

DIRECTIONS FOR QUESTIONS 29 THROUGH 38: Each of the questions or incomplete statements below is followed by four suggested answers. Select the one that is the best in each case, and fill in the corresponding lettered circle. Be sure the circle is filled in completely so you cannot see the letter. Unless you are told to do so in the question, do NOT include sales tax in your answer to questions involving purchases.

29. The school cafeteria sells sandwiches that you can order from the chart below. You want one from each category. How many different combinations of sandwiches are possible?

Meat	Cheese	Other	Bread
Bologna, ham, or turkey	American or Swiss	Lettuce or tomato	Rye, white, wheat, or roll

A. 32
B. 48
C. 11
D. 4

Ⓐ Ⓑ Ⓒ Ⓓ

30. The local lumber yard sells plywood in the following thicknesses. Which is the thickest?

$$\frac{1}{2} \quad \frac{1}{4} \quad \frac{3}{8} \quad \frac{9}{16}$$

A. $\frac{1}{2}$ inch

B. $\frac{1}{4}$ inch

C. $\frac{3}{8}$ inch

D. $\frac{9}{16}$ inch

31. Five hundred students were asked to select their favorite color of the four colors listed. How many students selected purple?

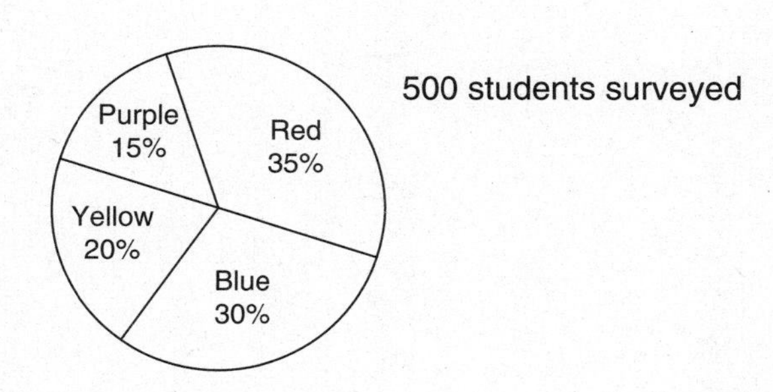

A. 15
B. 30
C. 75
D. 150

Ⓐ Ⓑ Ⓒ Ⓓ

GO ON TO THE NEXT PAGE

32. Jake has $215 to put down as a deposit on a rebuilt computer. The total cost of the computer is $575.00. He plans to make monthly payments of $60. Which equation will tell him how many months he will have to pay off the computer?

A. $575 = x(60) + 215$
B. $575 - 215 = x + 60$
C. $575 = x(215 - 60)$
D. $x = (575 - 60)$

Ⓐ Ⓑ Ⓒ Ⓓ

33. Ethan and Julia took a 3,000-mile trip from New Jersey to Washington state. Ethan drove $\frac{3}{4}$ of the total distance. How many miles did Julia drive?

Answer: ________

34. How does the volume of this cylinder and the volume of this cone compare? (*Hint:* In the diagram, one is measured in inches and the other is in feet.)

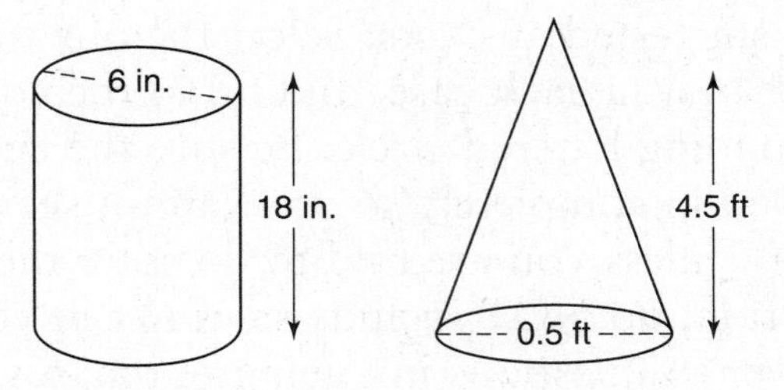

A. They are the same.
B. The volume of the cone is $\frac{1}{3}$ the volume of the cylinder.
C. The volume of the cylinder is 2 times the volume of the cone.
D. The volume of the cone is $\frac{1}{4}$ the volume of the cylinder.

35. Which of the following ratios is NOT in proportion?

A. $\frac{2}{3}$ and $\frac{12}{18}$

B. $\frac{6}{42}$ and $\frac{12}{84}$

C. $\frac{7}{64}$ and $\frac{28}{52}$

D. $\frac{8}{92}$ and $\frac{24}{276}$

GO ON TO THE NEXT PAGE

36. If the perimeter of a square is 32 inches, what is the approximate area of the square?

A. 16 sq. in.
B. 32 sq. in.
C. 36 sq. in.
D. 64 sq. in.

Ⓐ Ⓑ Ⓒ Ⓓ

37. Jill is checking her partner Suzie's math homework. She says that Suzie did the following example incorrectly. What error did Suzie make?

$$\begin{aligned} 3(2x - 9) &= 33 \\ 6x - 27 &= 33 \\ +\ 27 &\quad +\ 27 \\ 6x &= 60 \\ x &= 10 \end{aligned}$$

A. She should have combined the $2x$ and the -9 first.
B. She added 27 to 33 instead of subtracting 27 from 33.
C. She incorrectly distributed the 3.
D. Suzie did not make an error.

38. What is the value of the following expression?

$$3.5 \times 10^3 + |-2| + \sqrt{16} + 3^2$$

A. 3,515
B. 35,015
C. 350,015
D. 3,020

GO ON TO THE NEXT PAGE ➡

DIRECTIONS FOR QUESTION 39: Respond fully to the extended constructed response question that follows. Show your work and clearly explain your answer. You will be graded on the correctness of your method as well as the accuracy of your answer.

39. The new telephone company offers different monthly options for local telephone service.

Option #1: Unlimited local calls for $25 per month.

Option #2: A flat fee of $20 per month plus $0.10 charge for each local call.

- What is the least number of local calls you would have to make in one month so that option #1 would be less expensive than option #2? Show your work and explain your answer.
- If you selected option #2 and made 12 phone calls in the month of April, what is the total of your April bill if there is an additional 8% sales-and-use tax?

Part V (25 minutes)

Use the *NJ ASK 8 Mathematics Reference Sheet* on page 265.
CALCULATOR ACTIVE

DIRECTIONS FOR QUESTIONS 40 THROUGH 49: Each of the questions or incomplete statements below is followed by four suggested answers. Select the one that is the best in each case, and fill in the corresponding lettered circle. Be sure the circle is filled in completely so you cannot see the letter.

40. The transformation of figure *A* to figure *B* used the following translations, rotations, or reflections.

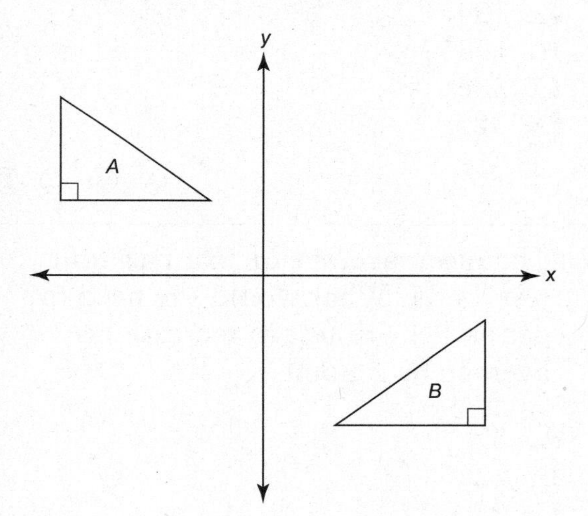

A. *A* was reflected over the *y*-axis, then translated to quadrant IV.
B. *A* was rotated 90° over the *y*-axis, then reflected over the *x*-axis.
C. *A* was reflected over the *x*-axis, then translated over the *y*-axis.
D. *A* was translated over the *x*-axis, then translated over the *y*-axis.

41. Two friends, Mark and Ray, left at 4:00 P.M. and drove west on Route 80 through New Jersey and part of Pennsylvania to attend a special soccer game. Ray drove for 3 hours at 55 miles per hour; Mark drove for $2\frac{1}{2}$ hours at 60 miles per hour.
Which of the following statements is true?

A. Ray drove 15 miles more than Mark.
B. Mark arrived at the game first.
C. They arrived at the game at the same time.
D. They both drove the same distance.

42. What is true for the four polygons shown below?

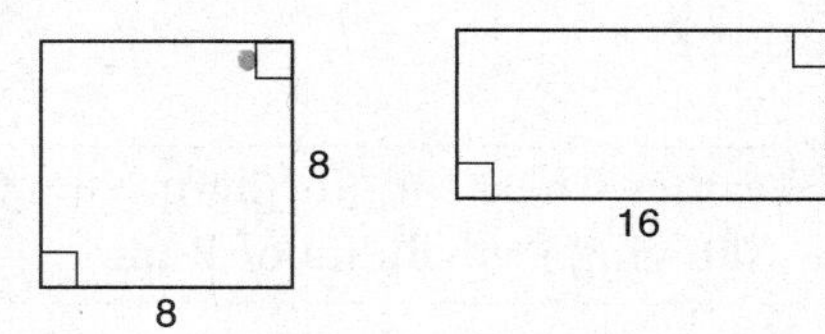

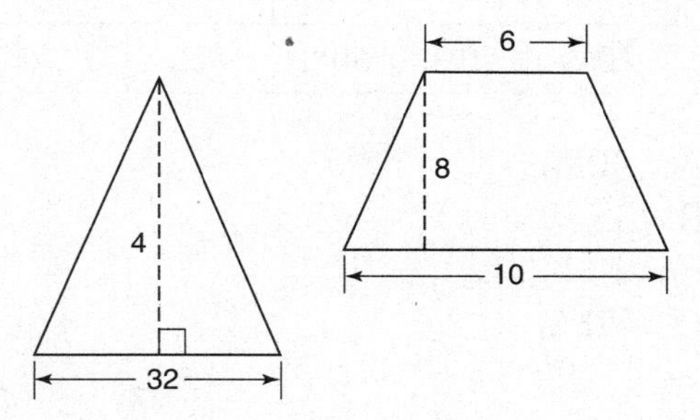

A. They all have the same perimeter.
B. They all have the same area.
C. They are all quadrilaterals.
D. They are similar shapes.

GO ON TO THE NEXT PAGE

43. Put the following in numerical order from lowest to highest.

$\frac{5}{6}, \frac{4}{5}, \frac{7}{8}$, 0.75, and 0.7

A. $\frac{5}{6}, \frac{4}{5}, \frac{7}{8}$, 0.75, 0.7
B. 0.7, 0.75, $\frac{4}{5}, \frac{5}{6}, \frac{7}{8}$
C. $\frac{7}{8}$, 0.7, 0.75, $\frac{4}{5}, \frac{5}{6}$
D. $\frac{7}{8}, \frac{4}{5}, \frac{5}{6}$, 0.7, 0.75

44. Holly is planning her track team's schedule for next week and needs to check the weather carefully. Using the information on the weekly weather chart below, what is the probability that it will rain on Monday or Tuesday?

Weekly Weather Chart for Northern New Jersey Showing Probability of Rain						
Sun.	Mon.	Tues.	Wed.	Thurs.	Fri.	Sat.
5%	20%	50%	60%	5%	0%	10%

A. 30%
B. 35%
C. 40%
D. 70%

Ⓐ Ⓑ Ⓒ Ⓓ

45. Which of the following is not true?

A. $-\sqrt{64} < -12$
B. $|-6| > -3$
C. $-\sqrt{36} < -\sqrt{16}$
D. $|-3 + 5| > -4$

Ⓐ Ⓑ Ⓒ Ⓓ

46. If this pattern continues, what number would be the 8th term?

4 7 13 25

A. 194
B. 193
C. 386
D. 385

47. The mean average of Joanna's four tests is 85. What would she need to get on her 5th test to increase her average by 2 points?

A. 87
B. 92
C. 95
D. 97

Ⓐ Ⓑ Ⓒ Ⓓ

GO ON TO THE NEXT PAGE

48. The school cafeteria offers many choices. If you order a sandwich, you can have it on a roll, rye bread, white bread, or wheat bread. You can also choose to have one meat (ham, turkey, or salami), one cheese (American or Swiss) and have mustard, mayonnaise, or ketchup. How many varieties of sandwiches are possible?

A. 24 varieties
B. 36 varieties
C. 54 varieties
D. 72 varieties

Ⓐ Ⓑ Ⓒ Ⓓ

49. Solve for C in the following equation:

$$\frac{45}{C} + 3 = 18.$$

A. $C = \frac{15}{7}$

B. $C = 3$

C. $C = \frac{8}{3}$

D. $C = \frac{1}{3}$

Ⓐ Ⓑ Ⓒ Ⓓ

GO ON TO THE NEXT PAGE

DIRECTIONS FOR QUESTION 50: Respond fully to the extended constructed response question that follows. Show your work and clearly explain your answer. You will be graded on the correctness of your method as well as the accuracy of your answer.

50. The chart below shows the scores students in my 8th grade class received on a practice NJ ASK 8 *extended constructed response* question. The chart also shows the number of students who received each score.

Score	Number of Students Receiving That Score
0	2
1	6
2	7
3	9

- If one of the students is picked at random, what is the probability that the student's score will be greater than 1?
- What is the *median* score?
- What is the *mean average* of the scores in this class? Round your answer to the nearest whole number.

Part VI (19 minutes)

Use the *NJ ASK 8 Mathematics Reference Sheet* on page 265.
CALCULATOR ACTIVE

DIRECTIONS FOR QUESTIONS 51 THROUGH 56: Each of the questions or incomplete statements below is followed by four suggested answers. Select the one that is the best in each case, and fill in the corresponding lettered circle. Be sure the circle is filled in completely so you cannot see the letter. Unless you are told to do so in the question, do NOT include sales tax in your answer to questions involving purchases.

51. Examine the number below and look for the pattern. What digit will be in the in 43rd position?

0.1717

(*Note*: The first position is "1", the second is "7", etc.)

A. 0
B. 1
C. 7
D. 9

Ⓐ Ⓑ Ⓒ Ⓓ

52. $\sqrt{83}$ is approximately

A. 6
B. 7
C. 8
D. 9

Ⓐ Ⓑ Ⓒ Ⓓ

53. The network below describes Maria's mom's trip to work yesterday. She left home (1), walked to the bus stop (2), got off the bus (3), walked to meet her sister at (4), and then they both walked the rest of the way to work (5). What is this type of network called?

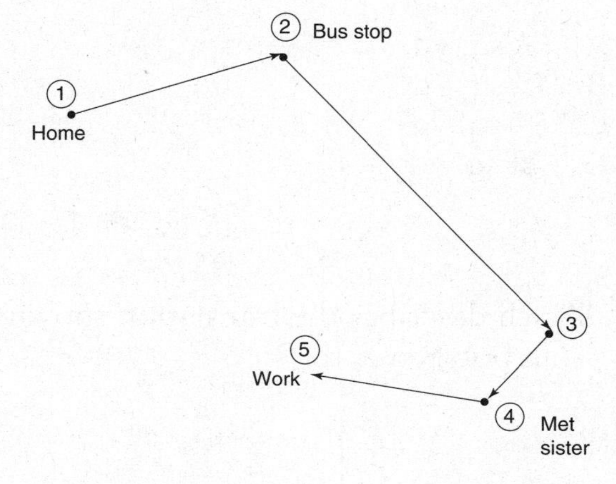

A. a path
B. a circuit
C. a trend line
D. a translation

Ⓐ Ⓑ Ⓒ Ⓓ

GO ON TO THE NEXT PAGE

54. What is the area of this garden if its perimeter is 34 yards? (All angles are right angles.)

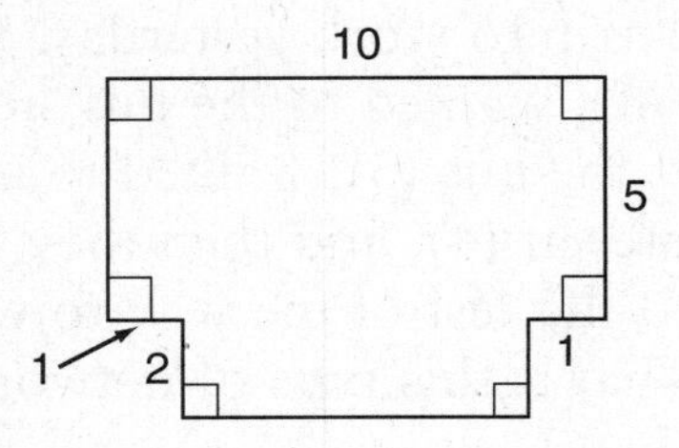

A. 66 sq. yd
B. 59 sq. yd
C. 54 sq. yd
D. 30 sq. yd

55. Which describes the translation shown from polygon *A* to *B*?

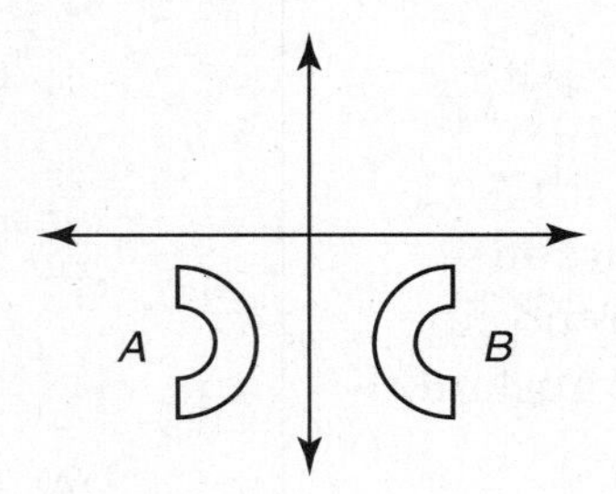

A. a translation over the *y*-axis
B. a rotation over the *y*-axis
C. a reflection over the *y*-axis
D. a reflection over the *x*-axis

56. Each of the four polygons below has the same area. Which one has the largest perimeter?

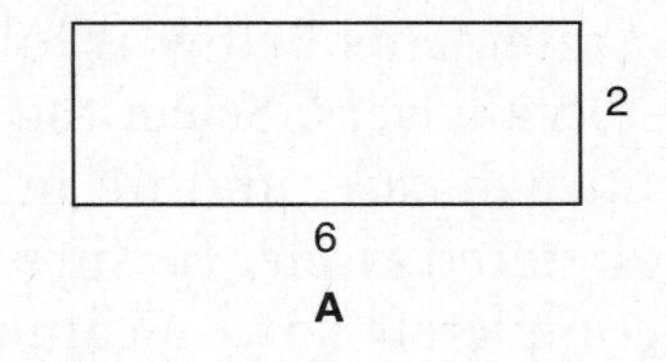

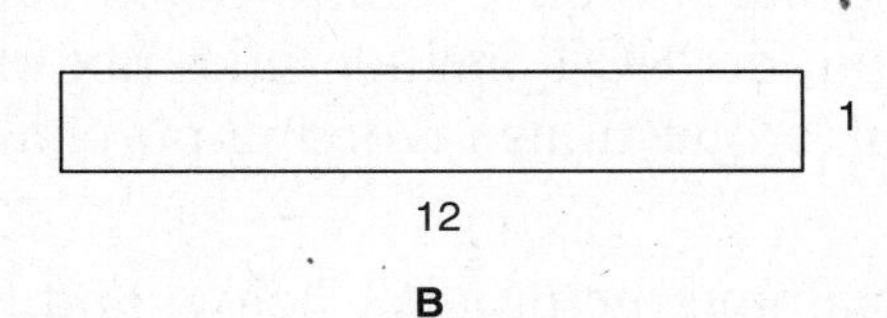

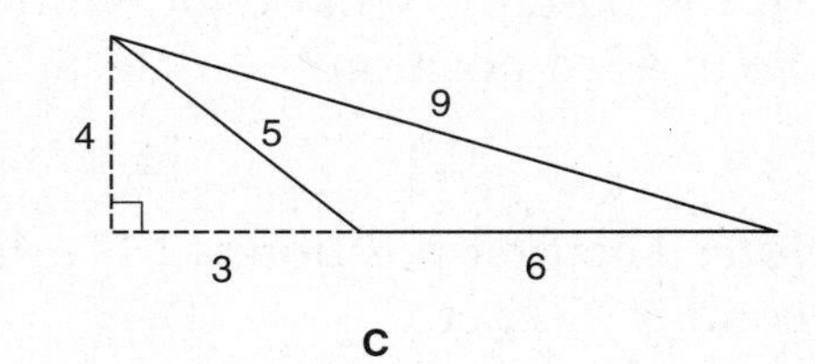

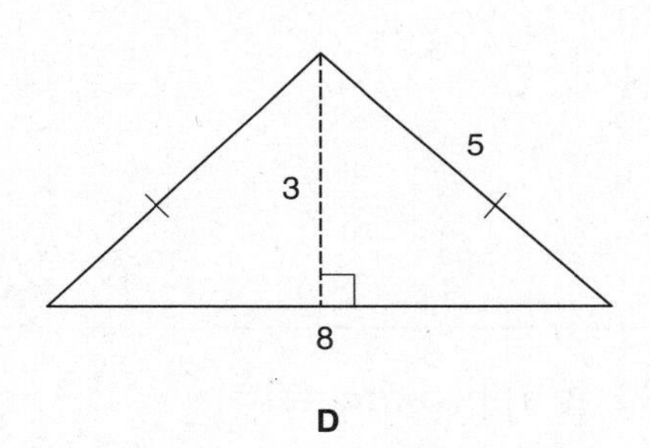

A. Rectangle *A*
B. Rectangle *B*
C. Obtuse triangle *C*
D. Isosceles triangle *D*

GO ON TO THE NEXT PAGE

DIRECTIONS FOR QUESTION 57: Respond fully to the extended constructed response question that follows. Show your work and clearly explain your answer. You will be graded on the correctness of your method as well as the accuracy of your answer.

57. Nancy and Rod are planning a drive between two towns in their home state of New Mexico. Their map shows the towns as 7 inches apart. The scale on the map says $\frac{1}{2}$ inch = 25 miles.

    - How many miles apart are the two towns?
    - If gasoline costs $4.50 per gallon, and they get 35 miles per gallon, how much money should they plan pay for a *round-trip* between the two towns?

SOLUTIONS TO PRACTICE TEST 2

PART I

Short Constructed Response Questions

1. 3:2 or $\frac{3}{2}$ 25 total students – 10 boys = 15 girls $\frac{15\text{ girls}}{10\text{ boys}} = \frac{3}{2}$

2. A $-1\frac{3}{4}$

3. 32 inches If area is 60 and one side is 10, area = (length)(width); 60 = (10)(x), so $x = 6$. The four sides of the rectangle are 6, 6, 10, and 10. The perimeter is the sum of all sides, so 6 + 6 + 10 +10 = 32, or 2(6 + 10) = 32.

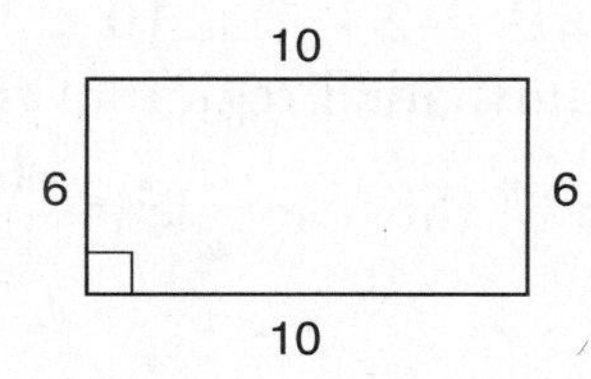

4. D Find the probability of each. Then use 6 as a common denominator to compare each answer.

   A. The probability of choosing a yellow marble:

   $$\frac{2\text{ yellow}}{12\text{ total}} = \frac{1}{6}$$

   B. The probability of choosing a green **or** a yellow marble:

   $$\frac{4\text{ green} + 2\text{ yellow}}{12\text{ total}} = \frac{6}{12} = \frac{1}{2}\ (1/2 = 3/6)$$

   C. The probability of choosing a green marble:

   $$\frac{4\text{ green}}{12\text{ total}} = \frac{4}{12} = \frac{1}{3}\ (1/3 = 2/6)$$

   D. The probability of choosing a red **or** a yellow marble:

   $$\frac{6\text{ red} + 2\text{ yellow}}{12\text{ total}} = \frac{8}{12} = \frac{2}{3}\ (2/3 = 4/6)$$

5. 3 inches You can make a proportion: $\frac{10\text{ feet}}{2\text{ inches}} = \frac{15\text{ feet}}{x\text{ inches}}$; $10x = (2)(15)$; $x = 3''$

6. The largest integer is the surface area of the cube.
   The surface area of the cube is (6 sides)(3)(3) = (6)(9) = 54
   The volume of the rectangular prism is (3)(3)(4) = (9)(4) = 36
   The volume of the right cylinder is about $(3.14)(2^2)(4) = 50.24$
   If you used the π symbol on your calculator, your answer would have been slightly larger; it would have been 50.26548246 or ~50.27.
   The perimeter of the triangle 10 + 5 + 5 = 20
7. 3 hours — Day #1 = 15 min. (or $\frac{3}{4}$ of an hour); on day #4 she would be studying 1 hour; by day #8, 2 hours; and by day #12, 3 hours.
8. 101° — 23° + 56° = 79°. The sum of the three angles in any triangle is 180°; therefore, the missing angle is 180° – 79° or 101°.
9. –8 — $3(x - 4) + 10 = 3.\ (-2 - 4) + 10 = \quad 3(-6) + 10 = \quad -18 + 10 = -8$
10. 25 sq. units is the area of the unshaded region (a triangle).
    Area of a triangle is $\frac{bh}{2}$. Area of this triangle is $\frac{(10)(5)}{2} = 25$

PART II

Multiple-Choice Questions

11. A — $5 \times 4 = 20$ — Perimeter = the sum of the lengths of the sides. Since all sides are the same length in a rhombus you can just multiply 5 × 4 to find the perimeter.
12. B — 27 — Note: The rule is to add 5 to each number. The pattern shows:

    $-8 + 5 = \underline{-3}$ $\quad -3 + 5 = \underline{2}$ $\quad 2 + 5 = \underline{7}$ $\quad 7 + 5 = \underline{12}$
    $12 + 5 = \underline{17}$ $\quad 17 + 5 = \underline{22}$ $\quad 22 + 5 = \underline{27}$

    Choice A: $(6)(2^2) = (6)(4) = 24$
    Choice B: $3(3^2) = 3(9) = 27$
    Choice C: $5^2 = 25$
    Choice D: $(15)(2) = 30$
13. C — Vertical angles are created when two lines intersect as shown in the diagram to the right: $\angle a$ and $\angle b$ are vertical angles.

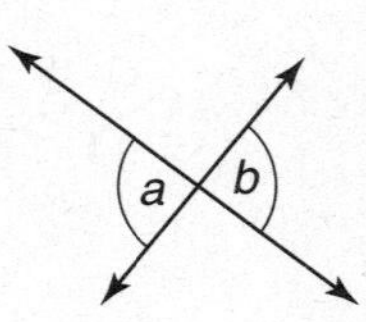

14. A — 53

    $2(8) + n = 69$ — multiply 2 times 8
    $16 + n = 69$ — now subtract 16 from both sides
    $-16 \quad -16$
    $n = 53$
15. B — (12)(8) – (4)(8) — Subtract the area of the small garden, (8 – 2 – 2 = 4) (12 – 2 – 2 = 8) = (4)(8), from the area of the larger rectangle, (12)(8).

16. **D** A. $(\sqrt{100})(2) = (10)(2) = 20$
B. $\sqrt{400} = 20$
C. $5(-2)^2 = 5(-2)(-2) = 5(4) = 20$
D. $5^2 - 2^2 = (5)(5) - (2)(2) = 25 - 4 = 21$

17. **A** There are 7 chances to land on a factor of 24: 2, 2, 3, 2, 3, 4, or 1.
There are only 4 chances to land on an even number: 2, 2, 2, or 4.
There are only 5 chances to land on a prime number: 2, 2, 3, 2, or 3.
There are only 3 chances to land on a perfect square number: 1, 4, or 9.

18. **D** The circle has the largest area.

A. A right triangle: $\frac{(\text{base})(\text{height})}{2}$

$\frac{(12)(2)}{2} = 12$

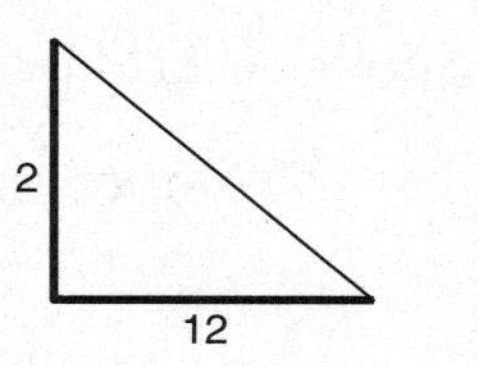

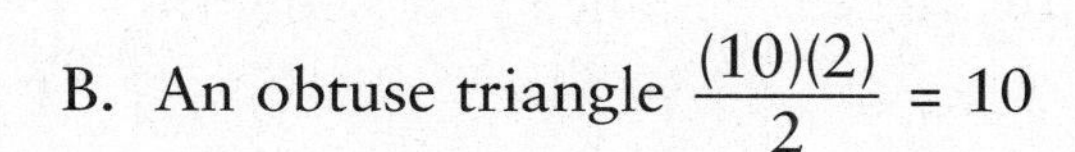
B. An obtuse triangle $\frac{(10)(2)}{2} = 10$

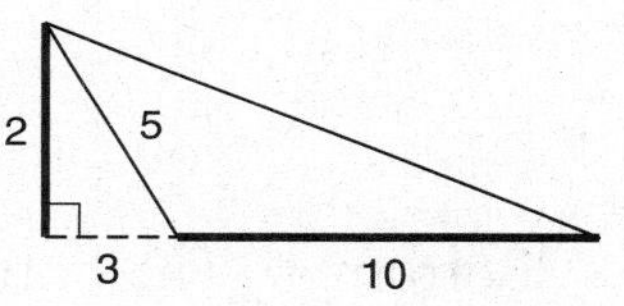

C. A rectangle: (base)(height) = area

(6)(2) = 12

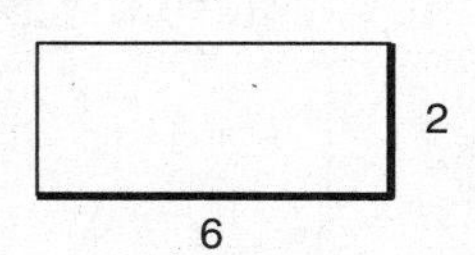

D. a circle (π)(radius squared) = area

$(3.14)(2^2)$ = area
(3.14)(4) is > 12
(3.14)(4) = 12.56

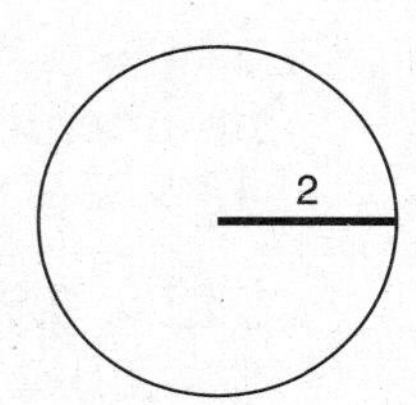

Extended Constructed Response Question

19.
- Area round rug = 113.04 sq. ft; Area circle = $\pi r^2 = (3.14)(6)(6) = 113.04$ sq. yd
- Area rectangular rug = 12 sq. yd; Area = (9 ft)(12 ft) or (3 yd)(4 yd) = 12 sq. yd
- Cost of rectangular rug = (12)($22) = $264
- Cost with tax = ($264) (1.06) = $279.84

PART III

20. A 12 $\sqrt{24} + \sqrt{50} +$ is approximately $\sqrt{25} + \sqrt{49}$ or $5 + 7$

21. B Jeremiah You always have a 50/50 chance of getting *heads* no matter how may times you flip the coin.

22. B $6n - 5$

23. B It is mosy likely that it will snow on Monday **or** Tuesday.
 A. Snow on Monday **and** Tuesday = (0.50)(0.60) = 0.3000 = 30%
 B. Snow on Monday **or** Tuesday = 0.50 + 0.60 = 1.10 = **110%**
 C. Sunny **on** Thursday = 0.70 = 70%
 D. Snow **on** Friday = 0 = 0% (no chance)

24. D $\frac{2}{5}$ Notice that $\frac{2}{5}$ is the only fraction less than $\frac{1}{2}$. It is easier to see this if you change $\frac{2}{5}$ to tenths. $\frac{2}{5} = \frac{4}{10}$. Since you know that $\frac{5}{10} = \frac{1}{2}$, you can see that $\frac{2}{5}$ is less than $\frac{5}{10}$ or $\frac{1}{2}$.

25. B Bottom The bottom face is NOT perpendicular to the top. The bottom is actually parallel to the top; it never touches the top.

26. B 20

27. D 22.651

Extended Constructed Response Question

28. ■ 8 times $24 × 4 = $48 each time family of four goes and pays the nonmember price: $350/48 = 7.29 times; $48 × 8 = $384; 48 × 7 = $336

 ■ $54.00 $12 × 0.90 = $10.80 grandma would pay each time
 $10.80 × 5 = $54.00

PART IV

29. B 48 $3 \times 2 \times 2 \times 4 = 48$

30. D $\frac{9}{16}$ is thickest; $\frac{1}{2} = \frac{8}{16}$ and $\frac{1}{4} = \frac{4}{16}$ and $\frac{3}{8} = \frac{6}{16}$ and $\frac{9}{16} = \frac{9}{16}$

31. C 75 students selected purple 15% of 500 = (0.15) (500) = 75

32. A $575 = x(60) + 215$

33. 750 miles $\frac{3}{4}(3{,}000) = (0.75)(3{,}000) = 2{,}250$ miles, the distance Ethan drove.

$3{,}000 - 2{,}500 = 750$ miles, the distance Julia drove.

Or, use only one step: If Ethan drove $\frac{3}{4}$ of the trip, then Julia drove only $\frac{1}{4}$ of the trip; $\frac{1}{4}(3{,}000) = (0.25)(3{,}000 \text{ miles}) = 750$ miles.

34. A Change all to inches. The volume of this cone is the same as the volume of the cylinder shown.
Volume of cylinder: (Area base)(Height) = (Area circle)(Height) =
$[(3.14)(3)(3)](18) = 508.68$ cubic inches

Volume of cone: $\frac{\text{(Area base)(Height)}}{3} = \frac{[(3.14)(3)(3)](54)}{3} = \frac{1{,}526.04}{3}$

$= 508.68$ cubic inches

35. C $\frac{7}{16}$ and $\frac{28}{52}$ $(7)(52) \neq (28)(16)$; $364 \neq 448$

36. D If perimeter = 32, then one side = $\frac{32}{4} = 8$; Area of square =

(Side)(Side) = $(8)(8) = 64$

37. D Suzie did not make an error.

38. A $(3.5)(10^3) + |-2| + \sqrt{16} + 3^2 = 3{,}500 + 2 + 4 + 9 = 3{,}515$

Extended Constructed Response Question

39. ■ 51 calls minimum $5.00/0.10 = 50 calls would be the same price. You would have to make a minimum of 51 calls to make option #1 less expensive.

■ $20.00 + $1.20 = $21.20 + 8% = $22.896, which is approximately $22.90.

PART V

Multiple-Choice Questions

40. A Reflect *A* over the y-axis, then translate (down) to quadrant IV.

41. A Ray drove 15 miles more than Mark.
Ray: (3 hrs.)(55 mi.) = 165 miles
Mark: (2.5 hrs.)(60 mi.) = 150 miles

42. **B** They all have the same area, 64 square units.

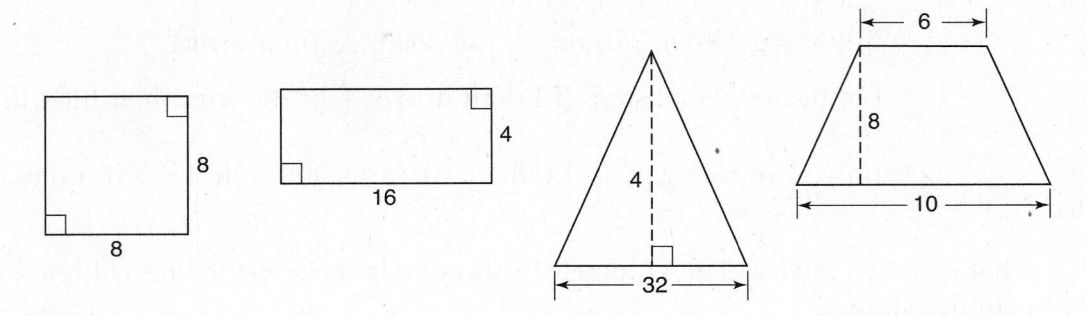

43. **B** This is easiest to do if you change all the numbers to decimal numbers.

$0.7, \frac{3}{4}, \frac{4}{5}, \frac{5}{6}, \frac{7}{8}$

0.7, 0.75, 0.80, 0.833, 0.875

44. **D** When you have an "or" situation, you add the two percents; therefore, 20% + 50% means a 70% probability it will rain on Monday or Tuesday.

45. **A** $-\sqrt{64} < -12$ simplifies to $-8 < -12$ (This is false.)

$|-6| > -3$ simplifies to $6 > -3$ (This is true.)

$-\sqrt{36} < -\sqrt{16}$ simplifies to $-6 < -4$ (This is true.)

$|-3 + 5| > -4$ simplifies to $|2| > -4$ or $2 > -4$ (This is true.)

46. **D** 385 The pattern follows $2x - 1$; (193)(2) = 386; 386 – 1=385.
The series of the first 8 numbers is 4, 7, 13, 25, 49, 97, 193, 385

47. **C** She would need to get a 95 on her 5th test.
To raise four tests from 85 to 87, she would need to get (4)(2) = 8 extra points.
She would need to get 87 + 8 extra points, or 95, on her 5th test.

48. **D** 72 varieties (4 breads, 3 meats, 2 cheeses, 3 garnishes) = (4)(3)(2)(3) = 72

49. **B** C = 3

$$\frac{45}{C} + 3 = 18$$

$$-3 \quad -3$$

$$(C)\frac{45}{C} = 15(C)$$

$$45 = 15C; 3 = C$$

Extended Constructed Response Question

50. ■ The probability that a student will score greater than 1 is about 66.7% or 67%; or you could say there is a $\frac{2}{3}$ chance that a student would score greater than 1 on that quiz.

$$\frac{\text{Number of students scoring greater than 1}}{\text{Total number of students}} = \frac{7+9}{2+6+7+9} = \frac{16}{24} = \frac{2}{3} = 66.667\ \%$$

■ The median score is 2. To determine the median score, first put the test scores in order and see which one is in the middle; if there is no middle number, take the average of the two middle scores.

0 0 1 1 1 1 1 1 2 2 2 2   2 2 2 3 3 3 3 3 3 3 3 3

■ The mean average is 1.958333, which rounded to the nearest whole number = 2. To find the mean average, add all the scores and divide by the total number of scores.

$$\frac{2(0)+6(1)+7(2)+9(3)}{2+6+7+9} = \frac{0+6+14+27}{24} = \frac{47}{24} = 1.958333, \text{ which is } \textit{approximately } 2.0.$$

PART VI

Multiple-Choice Questions

51. B $\frac{17}{99} = 0.17171717$ All the odd-numbered positions have 1 as the digit.

52. D 9 $\sqrt{81} = 9$

53. A path When a network begins at one point and ends at another it is called a path.

54. A 66 sq. yds $10 + 5 + 1 + 2 + 2 + 1 + 5 = 34$; $(10)(5) = 50 + (8)(2) = 16 = 66$; $4 \times 2 = 8$ and $5 \times 10 = 50$; Total area: $8 + 50 = 58$ sq. yds

55. C A reflection (or flip) over the y-axis.

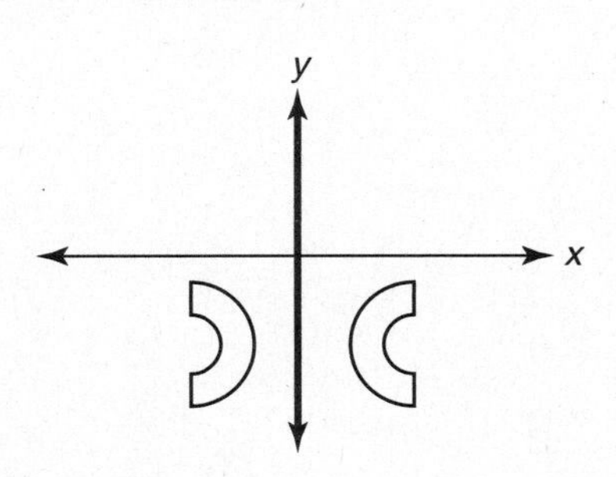

56. B The perimeter of rectangle A is 2(2) + 2(6) = 4 + 12 = 16.
The perimeter of rectangle B is 2(1) + 2(12) = 2 + 24 = 26.
The perimeter of the obtuse triangle C is 5 + 9 + 6 = 20.
The perimeter of the isosceles triangle D is 5 + 5 + 8 = 18.

Extended Constructed Response Question

57. ▪ If $\frac{1}{2}$ inch = 25 miles, then 1 inch = 50 miles.
(7 inches) (50 miles) = (7)(50) = 350 miles. The towns are 350 miles apart.

▪ A round-trip would be (350 miles)(2) = (350)(2) = 700 miles round-trip
To find the number of gallons needed, divide 700 miles by 35 gallons.
$\frac{700}{35}$ = 20 gallons needed.
To find the cost: (20 gallons) ($4.50 each) = (20)(4.50) = $90.00.
Nancy and Rod should expect to pay $90 for gasoline for their round-trip.

Test Specifications

USE OF CALCULATORS

The eighth-grade and eleventh-grade mathematics assessment committees recommended that numerical calculations be embedded in problem-solving situations, and that students be given the opportunity to choose the most appropriate way to perform those operations. Moreover, the committees recommended that the mathematics sections of the eighth-grade and eleventh-grade assessments be constructed to allow for the use of either graphing or scientific calculators.

Students taking any of the New Jersey tests in mathematics should be prepared to use calculators by regularly using those calculators in their instructional programs. Calculators which have been given to students only on the day of the assessment may actually cause them to perform less well than they would have performed without such calculators. Students must be comfortable with and have practice with calculators, or they will be of little benefit during the assessment. Students should be permitted to use their own calculators, rather than the school's, if they choose to do so.

For the eleventh-grade mathematics assessment, use of a graphing calculator is strongly recommended. For both the eighth-grade and the eleventh-grade mathematics assessments, calculators which manipulate fractions without converting them to decimals are permitted. Students taking these tests will be allowed to use graphing or other handheld calculators which have at least these functionalities:

- algebraic logic (i.e., automatically follows the standard order of mathematical operations),
- the ability to do powers and roots of any degree, and
- at least one memory cell.

Note that any device commonly accepted as a handheld calculator and having at least the functionality specified above is acceptable. However, computers (laptops, palmtops, etc.), pocket organizers, and calculators with QWERTY (i.e., typewriter) keyboards are not acceptable under the current guidelines.

As of 2011, students may use a calculator for only Parts IV, V, and VI of the NJ ASK 8 math exam.

HOLISTIC SCORING GUIDE FOR MATHEMATICS EXTENDED CONSTRUCTED RESPONSE (ECR) ITEMS (GENERIC RUBRIC)

3-POINT RESPONSE

The response shows complete understanding of the problem's essential mathematical concepts. The student executes procedures completely and gives relevant responses to all parts of the task. The response contains few minor errors, if any. The response contains a clear, effective explanation detailing how the problem was solved so that the reader does not need to infer how and why decisions were made.

2-POINT RESPONSE

The response shows nearly complete understanding of the problem's essential mathematical concepts. The student executes nearly all procedures and gives relevant responses to most parts of the task. The response may have minor errors. The explanation detailing how the problem was solved may not be clear, causing the reader to make some inferences.

1-POINT RESPONSE

The response shows limited understanding of the problem's essential mathematical concepts. The response and procedures may be incomplete and/or may contain major errors. An incomplete explanation of how the problem was solved may contribute to questions as to how and why decisions were made.

0-POINT RESPONSE

The response shows insufficient understanding of the problem's essential mathematical concepts. The procedures, if any, contain major errors. There may be no explanation of the solution or the reader may not be able to understand the explanation. The reader may not be able to understand how and why decisions were made.

The above generic rubric is used as a guide to develop specific scoring guides or rubrics for each of the Extended Constructed Response (ECR) items which appear on the New Jersey fourth-grade (NJ ASK 4), eighth-grade (NJ ASK 8), and eleventh-grade (HSPA) proficiency assessments in mathematics. The generic rubric helps insure that students are scored in the same way for the same demonstration of knowledge and skills regardless of the test question. More information on Extended Constructed Response items and related scoring is also provided in the *Mathematics Instructional Guide*.

MATHEMATICS MACROS, BY CLUSTER

I. Number Sense, Concepts, and Applications

A. Make appropriate estimations and approximations.

B. Understand numbers, our numeration system, and their applications in real-world situations.

C. Use ratios, proportions, and percents in a variety of situations.

II. Spatial Sense and Geometry

A. Recognize, identify, and represent spatial relationships and geometric properties.

B. Apply the principles of congruence, similarity, symmetry, geometric transformations, and coordinate geometry.

C. Apply the principles of measurement and geometry to solve problems involving direct and indirect measurement.

III. Data Analysis, Probability, Statistics, and Discrete Mathematics

A. Predict, determine, interpret, and use probabilities.

B. Collect, organize, represent, analyze, and evaluate data.

C. Apply the concepts and methods of discrete mathematics to model and explore a variety of practical situations.

D. Use iterative patterns and processes to describe real-world situations and solve problems.

IV. Patterns, Functions, and Algebra

A. Recognize, create, and extend a variety of patterns and use inductive reasoning to understand and represent mathematical and other real-world phenomena.

B. Use algebraic concepts and processes to concisely express, analyze, and model real-world situations.

The above information has been taken directly from the NJDOE website *www.njpep.doe.state.nj.us*

For additional information regarding test specifications or the NJ Math Standards contact:
NJPEP: Virtual Academy, NJ Department of Education
100 Riverview Plaza, Trenton, NJ 08625-0500
Voice: 609.292.9069 Fax: 609.292.7276
NJPEP@doe.state.nj.us

New Jersey Math Core Curriculum Standards and Strands*

Building upon knowledge and skills gained in preceding grades, by the end of Grade 8, students will be familiar with the following.

4.1 Number and Numerical Oeprations

A. *Extend understanding of the number system by constructing meanings for the following*

1. Demonstrate a sense of the relative magnitudes of numbers.
2. Understand and use ratios, proportions, and percents (including percents greater than 100 and less than 1) in a variety of situations.
3. Compare and order numbers of all named types.
4. Use whole numbers, fractions, decimals, and percents to represent equivalent forms of the same number.
5. Recognize that repeating decimals correspond to fractions and determine their fraction equivalents. $5/7 = 0.714285714285 \ldots = 0.714285$
6. Construct meanings for common irrational numbers, such as pi and the square root of 2.

B. *Numerical Operations*

1. Use and explain procedures for performing calculations involving addition, subtraction, multiplication, division, and exponentiation with integers and all number types named above with:

   pencil and paper
   mental math
   calculator

2. Use exponentiation to find whole number powers of numbers.
3. Find square and cube roots of numbers and understand the inverse nature of powers and roots.
4. Solve problems involving proportions and percents.
5. Understand and apply the standard algebraic order of operations, including appropriate use of parentheses.

C. *Estimation*

1. Estimate square and cube roots of numbers.
2. Use equivalent representations of numbers such as fractions, decimals, and percents to facilitate estimation.
3. Recognize the limitations of estimation and assess the amount of error resulting from estimation.

4.2 Geometry and Measurement

A. *Geometric Properties*

1. Understand and apply concepts involving lines, angles, and planes.
   - Complementary and supplementary angles
   - Vertical angles
   - Bisectors and perpendicular bisectors
   - Parallel, perpendicular, and intersecting planes
   - Intersection of plane with cube, cylinder, cone, and sphere
2. Understand and apply the Pythagorean theorem.
3. Understand and apply properties of polygons.
   - Quadrilaterals, including squares, rectangles, parallelograms, trapezoids, rhombi
   - Regular polygons
   - Sum of measures of interior angles of a polygon
   - Which polygons can be used alone to generate a tessellation and why
4. Understand and apply the concept of similarity.
   - Using proportions to find missing measures
   - Scale drawings
   - Models of 3D objects
5. Use logic and reasoning to make and support conjectures about geometric objects.

B. *Transforming Shapes*

1. Understand and apply transformations.
   - Finding the image, given the pre-image, and vice versa
   - Sequence of transformations needed to map one figure onto another
   - Reflections, rotations, and translations result in images congruent to the pre-image
   - Dilations (stretching/shrinking) result in images similar to the pre-image
2. Use iterative procedures to generate geometric patterns.
   - Fractals (e.g., the Koch Snowflake)
   - Self-similarity
   - Construction of initial stages
   - Patterns in successive stages (e.g., number of triangles in each stage of Sierpinski's Triangle)

C. *Coordinate Geometry*

1. Use coordinates in four quadrants to represent geometric concepts.
2. Use a coordinate grid to model and quantify transformations (e.g., translate right 4 units).

D. *Units of Measurement*

1. Solve problems requiring calculations that involve different units of measurement within a measurement system (e.g., 4′3″ plus 7′10″ equals 12′1″).
2. Use approximate equivalents between standard and metric systems to estimate measurements (e.g., 5 kilometers is about 3 miles).
3. Recognize that the degree of precision needed in calculations depends on how the results will be used and the instruments used to generate the measurements.

4. Select and use appropriate units and tools to measure quantities to the degree of precision needed in a particular problem-solving situation.
5. Recognize that all measurements of continuous quantities are approximations.
6. Solve problems that involve compound measurement units, such as speed (miles per hour), air pressure (pounds per square inch), and population density (persons per square mile).

E. *Measuring Geometric Objects*

1. Develop and apply strategies for finding perimeter and area.
   - Geometric figures made by combining triangles, rectangles, and circles or parts of circles
   - Estimation of area using grids of various sizes
   - Impact of a dilation on the perimeter and area of a two-dimensional figure
2. Recognize that the volume of a pyramid or cone is one third of the volume of the prism or cylinder with the same base and height (e.g., use rice to compare volumes of figures with same base and height).
3. Develop and apply strategies and formulas for finding the surface area and volume of a three-dimensional figure.
   - Volume—prism, cone, pyramid
   - Surface area—prism (triangular or rectangular base), pyramid (triangular or rectangular base)
   - Impact of a dilation on the surface area and volume of a three-dimensional figure
4. Use formulas to find the volume and surface area of a sphere.

4.3 Patterns and Algebra

A. *Patterns*

1. Recognize, describe, extend, and create patterns involving whole numbers, rational numbers, and integers.
   - Descriptions using tables, verbal and symbolic rules, graphs, simple equations or expressions
   - Finite and infinite sequences
   - Arithmetic sequences (i.e., sequences generated by repeated addition of a fixed number, positive or negative)
   - Geometric sequences (i.e., sequences generated by repeated multiplication by a fixed positive ratio, greater than 1 or less than 1)
   - Generating sequences by using calculators to repeatedly apply a formula

B. *Functions and Relationships*

1. Graph functions, and understand and describe their general behavior.
   - Equations involving two variables
   - Rates of change (informal notion of slope)
2. Recognize and describe the difference between linear and exponential growth, using tables, graphs, and equations.

C. *Modeling*

1. Analyze functional relationships to explain how a change in one quantity can result in a change in another, using pictures, graphs, charts, and equations.

2. Use patterns, relations, symbolic algebra, and linear functions to model situations.
   - Using concrete materials (manipulatives), tables, graphs, verbal rules, algebraic expressions/equations/inequalities
   - Growth situations, such as population growth and compound interest, using recursive (e.g., NOW-NEXT) formulas (cf. science standard 5.5 and social studies standard 6.6)

D. *Procedures*

1. Use graphing techniques on a number line.
   - Absolute value
   - Arithmetic operations represented by vectors (arrows) (e.g., "–3 + 6" is "left 3, right 6")
2. Solve simple linear equations informally, graphically, and using formal algebraic methods.
   - Multi-step, integer coefficients only (although answers may not be integers)
   - Using paper and pencil, calculators, graphing calculators, spreadsheets, and other technology
3. Solve simple linear inequalities.
4. Create, evaluate, and simplify algebraic expressions involving variables.
   - Order of operations, including appropriate use of parentheses
   - Distributive property
   - Substitution of a number for a variable
   - Translation of a verbal phrase or sentence into an algebraic expression, equation, or inequality, and vice versa
5. Understand and apply the properties of operations, numbers, equations, and inequalities.
   - Additive inverse
   - Multiplicative inverse
   - Addition and multiplication properties of equality
   - Addition and multiplication properties of inequalities

4.4 Analysis, Probability, and Discrete Mathematics

A. *Data Analysis*

1. Select and use appropriate representations for sets of data and measures of central tendency (mean, median, and mode).
   - Type of display most appropriate for given data
   - Box-and-whisker plot, upper quartile, lower quartile
   - Scatter plot
   - Calculators and computer used to record and process information
   - Finding the median and mean (weighted average) using frequency data
   - Effect of additional data on measures of central tendency
2. Make inferences and formulate and evaluate arguments based on displays and analysis of data.
3. Estimate lines-of-best-fit and use them to interpolate within the range of the data.

4. Use surveys and sampling techniques to generate data and draw conclusions about large groups.

B. *Probability*

1. Interpret probabilities as ratios, percents, and decimals.
2. Determine probabilities of compound events.
3. Explore the probabilities of conditional events (e.g., if there are seven marbles in a bag, three red and four green, what is the probability that two marbles picked from the bag, without replacement, are both red).
4. Model situations involving probability with simulations (using spinners, dice, calculators, and computers) and theoretical models.
   - Frequency, relative frequency
5. Estimate probabilities and make predictions based on experimental and theoretical probabilities.
6. Play and analyze probability-based games, and discuss the concepts of fairness and expected value.

C. *Discrete Mathematics*—Systematic Listing and Counting

1. Apply the multiplication principle of counting.
   - Permutations: ordered situations with replacement (e.g., number of possible license plates) vs. ordered situations without replacement (e.g., number of possible slates of 3 class officers from a 23 student class)
   - Factorial notation
   - Concept of combinations (e.g., number of possible delegations of 3 out of 23 students)
2. Explore counting problems involving Venn diagrams with three attributes (e.g., there are 15, 20, and 25 students respectively in the chess club, the debating team, and the engineering society; how many different students belong to the three clubs if there are 6 students in chess and debating, 7 students in chess and engineering, 8 students in debating and engineering, and 2 students in all three?).
3. Apply techniques of systematic listing, counting, and reasoning in a variety of different contexts.

D. *Discrete Mathematics*—Vertex-Edge Graphs and Algorithms

1. Use vertex-edge graphs and algorithmic thinking to represent and find solutions to practical problems.
   - Finding the shortest network connecting specified sites
   - Finding a minimal route that includes every street (e.g., for trash pickup)
   - Finding the shortest route on a map from one site to another
   - Finding the shortest circuit on a map that makes a tour of specified sites
   - Limitations of computers (e.g., the number of routes for a delivery truck visiting n sites is $n!$, so finding the shortest circuit by examining all circuits would overwhelm the capacity of any computer, now or in the future, even if n is less than 100)